UNQUIET PASTS

Heritage, Culture and Identity

Series Editor: Brian Graham,
School of Environmental Sciences, University of Ulster, UK

Other titles in this series

Geographies of Australian Heritages
Loving a Sunburnt Country?
Edited by Roy Jones and Brian J. Shaw
ISBN 978 0 7546 4858 1

Living Ruins, Value Conflicts
Argyro Loukaki
ISBN 978 0 7546 7228 9

Geography and Genealogy
Locating Personal Pasts
Edited by Dallen J. Timothy and Jeanne Kay Guelke
ISBN 978 0 7546 7012 4

Jewish Topographies
Visions of Space, Traditions of Place
Edited by Julia Brauch, Anna Lipphardt and Alexandra Nocke
ISBN 978 0 7546 7118 3

Southeast Asian Culture and Heritage in a Globalising World
Diverging Identities in a Dynamic Region
Edited by Rahil Ismail, Brian Shaw and Ooi Giok Ling
ISBN 978 0 7546 7261 6

Valuing Historic Environments
Edited by Lisanne Gibson and John Pendlebury
ISBN 978 0 7546 7424 5

Sport, Leisure and Culture in the Postmodern City
Edited by Peter Bramham and Stephen Wagg
ISBN 978 0 7546 7274 6

Culture, Heritage and Representation
Perspectives on Visuality and the Past
Edited by Emma Waterton and Steve Watson
ISBN 978 0 7546 7598 3

Unquiet Pasts

Risk Society, Lived Cultural Heritage, Re-designing Reflexivity

Edited by

STEPHANIE KOERNER
University of Manchester, UK

IAN RUSSELL
Brown University, USA

ASHGATE

Published by
Ashgate Publishing Limited
Wey Court East
Union Road
Farnham
Surrey, GU9 7PT
England

Ashgate Publishing Company
Suite 420
101 Cherry Street
Burlington
VT 05401-4405
USA

www.ashgate.com

British Library Cataloguing in Publication Data
Unquiet pasts : risk society, lived cultural heritage,
re-designing reflexivity. -- (Heritage, culture and
identity)
1. Archaeology--Social aspects--Congresses.
I. Series II. Koerner, Stephanie. III. Russell, Ian, 1979-
930.1-dc22

Library of Congress Cataloging-in-Publication Data
Koerner, Stephanie.
Unquiet pasts : risk society, lived cultural heritage, re-designing reflexivity / by Stephanie Koerner and Ian Russell.
p. cm. -- (Heritage, culture and identity)
Includes index.
ISBN 978-0-7546-7548-8 (hardback) 1. Archaeology--Social aspects. 2. Archaeology--Moral and ethical aspects. 3. Archaeology--Philosophy. 4. Archaeologists--Professional ethics. 5. Cultural property--Protection--Moral and ethical aspects. I. Russell, Ian. II. Title.

CC175.K64 2010
930.1--dc22

2009052590

ISBN 9780754675488 (hbk)

Mixed Sources
Product group from well-managed forests and other controlled sources
www.fsc.org Cert no. SGS-COC-2482
© 1996 Forest Stewardship Council
FSC

Printed and bound in Great Britain by
TJ International Ltd, Padstow, Cornwall

Contents

PART 3 RE-DESIGNING REFLEXIVITY, OR CAN THERE BE A CAUTIOUS PROMETHEUS?

List of Figures

List of Tables

Preface
'Playing Statues'

Stephanie Koerner

In a time when 'we can almost no longer assume that anything can be assumed' (Sloderdijk 2005), few themes may be more relevant than the diversity of the past for plurality of future conditions of possibility (Schnapp 1993; Ingold 2000; Thomas 2004). This is the sense in which Stephen Toulmin (1990) envisages the relevance to contemporary problems of appreciating pasts hitherto eclipsed by the 'myth of the clean slate'—the idea that the only way forward under conditions of the 'state of emergency' must involve demolishing everything that went before. Toulmin is, of course not alone in expressing concerns with consequences of such orientations. Walter Benjamin (1940) found in Paul Klee's painting, *Angelus Novus*, an image that expressed for him the cosmological scope of such consequences. Klee's angel is being hurled into the future. This, Benjamin said, was how to depict consequences of policies motivated by beliefs that the promises of techno-science to predict and control progress can be realised only by demolishing everything that went before. While everyday people experience and must broach context-dependent problems as these arise, the furious angel of powerful political ideologies is driven by a trans-historical catastrophic hurling wreckage—the pile of immanent debris growing skyward.

But what sort of image might be appropriate for expressing means to critically and constructively engage such forces. We chose a drawing from the works of Henry Koerner (1946) entitled 'playing statues' for the cover of the volume, at least in part, because it resonates with arguments (with analogues throughout) from one of its inspirations that:

> at the very moment when the scale of what has to be remade has become infinitely larger (no political revolutionary committed to challenging capitalist modes of production has ever considered redesigning the earth's climate), what it means to make something is also being deeply modified. The modification is so deep that things are no longer 'made' or 'fabricated', but rather carefully 'designed', and if I may use the term, precautionary designed. It is as though we had to combine the engineering tradition with the precautionary principle; it is as though we had to imagine Prometheus stealing fire from heaven in a cautious way! (Latour 2008: 4).

Writing in the catalogue of an exhibition of Henry Koerner's works in the Austrian Gallery in Vienna in 1997, entitled *Unheimliche Heimat*, Joseph Leo Koerner remarks:

> Historians are fond of saying that he who does not remember is doomed to repeat...Drawing and painting from life, teaching him to observe the familiar with a stranger's eye, became his tutor in motifs' recollection and forgetfulness...(J. L. Koerner 1997: 25, 67)...Henry Koerner was born in 1915 to Jewish parents in the Leopoldstadt district of Vienna...On March 12, 1938, when Hitler marched into Austria...his family huddled in their apartment while German troops, cheered by crowds, paraded in a procession that ended in Leopoldstadt...The next morning Koerner joined the queues outside the U.S. Embassy to apply for permission to emigrate to America. In early August, and still without a visa, he booked an Air France flight to Venice. At the threshold of their apartment, his parents and brother waved him off. He was never to see them again...(J.L. Koerner 13-15) ...What was a motif for this artist? Which principles of selection determined what he chose to draw, as well as what, from among the drawings, he chose to paint? 'Playing statues' was certainly a motif...Returned to his native city, Koerner remarked, 'Everything in Vienna was changed beyond recognition, but the children playing statues, Just as we did when we were kids' (J.L. Koerner 1997: 25).

In a time when changes in dynamics of pedagogical institutions and public affairs are being transformed by processes by expressions like 'globalisation' and 'risk society' (Beck 1992; Urry 1995; Lash 1999), it is perhaps not surprising that anthropologists and archaeologists are debating the implications for their fields' futures of issues posed by how enmeshed many controversies of questions of for and by whom 'lived heritage' is 'managed' are in deep inequalities regarding exposure to these processes' consequences. One of this volume's aims is to explore the possibility that insights bearing upon these issues might arise—against the grain of the long history of dichotomies of nature:culture, reality:historical contingency, universal:particular, and pure:applied research—out of realms summarised by such expressions as 'plurality of public grounds of truth' (Vico 1744). Writing on the 'needs of a world in which simplicity is a memory of a bygone age' (Funtowicz and Ravetz 1997), Walter Lippman noted:

> Yet it is controversies of this kind, the hardest to disentangle, that the public is called to judge. Where the facts are most obscure, where precedents are lacking, where novelty and confusion pervade everything, the public in all its unfitness is compelled to make its most important decisions. The hardest problems are those which institutions cannot handle. They are the publics' problems (Lippman 1925: 121).

Writing on implications for change in approaches to what archaeologists and anthropologists make of challenges posed by controversies over 'who makes decisions about the future of cultural heritage, what value systems are used and how local, regional and international issues are given appropriate recognition', Tim Darvill (2007: 438) reminds us that, now quite some time ago, Michel Foucault stressed that 'humanity has not inherited domains that are 'given' and have to be filled in; rather it is our task to define and elaborate the intellectual landscape within which we work' (Foucault 1970). For Darvill (2007: 252), as for a number of researchers alert to the extent to which many of today's controversies over heritage are 'quality of life' issues—this is good news; such controversies may help bring greater attention to: 'exemplary situations for the pursuit of research [that] enriches qualities of life by providing and promoting different kinds of knowledge that suit many different needs'.

References

Beck, U. 1992 *Risk Society: Towards a New Modernity*. London: Sage Publications.

Benjamin, W. 1994 [1940] Theses on the philosophy of history, in H. Arendt (ed.), *Illuminations: works of Benjamin*, 245-55. London: Fontana Press.

Ingold, T. 2000 *The Perception of the Environment. Essays in Livelihood, Dwelling and Skill*. London: Routledge.

Koerner, J.L. 1997 Unheimlische Heimat (Uncanny Home), text of catalogue for the exhibition of works of Henry Koerner, *Unheimlische Heimat*, in the Oesterreichischen Gallarie Belvedere. Vienna: Oesterreichischen Gallarie Belvedare.

Lash, S. 1999 *Another Modernity: A Different Rationality*. Oxford: Blackwell.

Latour, B. 2008 A Cautious Prometheus: A few steps toward a philosophy of design (with special attention to Peter Sloterdijk). Keynote lecture for the 3 September 2008 meeting of the Design History Society ('networks of Design'), Cornwall, UK.

Lippman, W. 1993 [1925] *The Phanthom Public*. New Brunswick: Transaction Publishers.

Schnapp, A. 1993 *The Discovery of the Past*, translated by I. Kinnes and G. Varndell. London: British Museum Press.

Sloderdijk, P. 2005 Forward to the Theory of Spheres, in M. Ohanian and J.C. Royoux (eds), *Cosmographs*, 223-41. New York: Lukas and Sternberg.

Thomas, J. 2004 *Archaeology and Modernity*. London: Routledge.

Toulmin, S. 1990 *Cosmopolis. The Hidden Agenda of Modernity*. Chicago: University of Chicago.

Urry, J. 1990 *The Tourist's Gaze*. London: Routledge.

Vico, G. 1948 [1744] *The New Science of the Common Nature of the Nations of Giambattista Vico*, unabridged third edition translated by T.G. Bergin and M. H. Fisch. London: Cornell University Press.

Acknowledgements

We would like to acknowledge the hard work and support of all of the contributors to this volume and to all the participants in the various conference sessions which inspired this development of this project. Without their vision and commitment such debates and discussions would not be possible. We would also like to thank Valerie Rose and Carolyn Court and the rest of the staff of Ashgate Publishing for their support and understanding during the production of the volume.

Introduction
Rethinking 'Crises over Representation'

Stephanie Koerner

> It no longer feasible to treat the human habitat as one of discrete systems and deterministic parts such as material culture and the natural environment, animal and plant species, lithosphere and troposphere. Instead we need to deal with the distributed and underdetermined processes which meld human and machine, nature and culture, challenging all those Cartesian dualisms at the heart of much modernist thinking.
>
> Shanks and Wittmore in this volume

Until quite recently, it is unlikely that many anthropologists or archaeologists are likely to have been receptive to suggestions that some of the most fundamental changes in theoretical and methodological orientations might arise out of areas of so-called 'applied' rather that 'pure' research. It would also have been unlikely that many would have been receptive to suggestions that fresh approaches to controversies over by and for whom cultural heritage is managed might bear directly upon pressing life quality issues in an irreducibly complex world (Funtowicz and Ravetz 1997; Skeates 2000; Layton et al. 2001; Hodder 2001; Carman 2002; Cleere 1984, 2006; Sloderdijk 2005; Darvill 2007).

In retrospect, such negative responses should not be altogether unsurprising. Throughout the histories of today's major academic institutions, the predominant visions of radical change in methodology and theory have been grounded in portrayals of science:

> …as surrounded by a kind of semi-permeable membrane: the results of science flowed outward and influenced ambient society, but there was no traffic in the reverse direction. Science was, on this account, a world unto itself, largely insulated from the society in which it was embedded (Daston 2006: 529).

Today the situation is undergoing remarkable change. There is now widespread appreciation of the importance of 'critical inquiry' into 'the academic cultural production of and representation (of both past and present people) for politically and ethically informed social archaeology' (Marisa Lazzari 2008: 645; see also Stone and Planel (eds) 1999; Buchli and Lucas 2001; Silverman and Ruggles (eds) 2007; David and Thomas 2008). A number of researchers involved in developing cultural heritage based solutions to some of the most pressing social and ecological problems, which have accompanied processes summarised by the expressions 'globalisation' and 'risk society' (Beck 1992; Beck et al. 1996) are challenging the presuppositions about science and the world on which such images of 'pure and applied research' have hinged (Darvill 2007; Koerner 2008). These include presuppositions:

- that the primary task of science is to reducing the plurality of context dependent experiences and phenomena to universally valid concepts, categories and propositions whose explanatory force is independent of any context of application (Ingold 1993; Koerner and Ingold 2009),
- that, despite the variety of physical realms, life forms, cultures and so on, it is possible to assume a 'common world' supposedly divided between the determinism of nature and human capacities for moral freedom and reason (Dupré 1993; Toulmin 1990; Daston 2000; cf. Kant 1784, 1788),
- about obstacles, which are said to impede human knowledge of the timeless principles (or 'foundational laws') of these realms, such as obstacles said to be posed by problems with technological instruments and skills, social interests, cultural biases, and so on.

Such presuppositions have shaped not only dualist characterisations of academic knowledge production, but also some of the most polemically opposed positions on such key methodological and theoretical issues as that of whether humanities and social sciences can or should aspire to be sciences, and associated utopic and dystopic interpreations of modernity (for instance, Durkheim 1895; Boas 1940; Collingwood 1949; Cassirer 1960; Hempel and Oppenheim 1948; Clarke 1968, 1973; Watson et al. 1971; Binford 1982; Binford and Sabloff 1982; Schnapp 1982, 1993; Salmon 1982, 1992; Leone 1987; Shanks and Tilley 1987; Churchland and Churchland 1998; Hodder 1982, 2001; Trigger 1984, 1989; Preucel 1991; Johnson 1999; Thomas 2004; Friedman 2000; Rowlands 2001; Wylie 2002). Today such presuppositions underpin the orientations of many accountability management organisations (Power 1998; Strathern 2000; Felt and Wynne 2008). Reflexivity is said to be an important antidote. But this raises the question of 'what qualifies as an adequate 'constitutive reflexivity' as distinct from a mere self-satisfying introspection?' Marissa Lazzari (2008: 647) suggests that satisfactory answers are likely to hinge upon abandoning 'fear of absence of consensus' and 'harbour[ing] the fantasy that conflictive situations may ever achieve a final moral equilibrium.' Such constructive hopes for the 'indeterminacy of social existence' (Lazzari 2008: 649; cf. De Beauvoir 1948; Deleuze 1990). Such a view goes against the grain of the presuppositions outlined above. For Bruno Latour such a view can be expressed by the idea that:

> A common world is not something we can come to recognize, as though it had always been here…A common world, if there is going to be one, is something we have to build, tooth and nail together (Latour 2004: 455).

And relating to the question Lazzari poses, it opens space for appreciating the relevance for strongly reflective orientations of such insights as that:

- there are no such things as context independent problems. 'We never experience or form judgments about objects and events in isolation, but only in connection with a contextualised whole. The latter is called a situation' (Dewey 1938: 66-67),
- complexity and emergent novelty are the normal state of affairs for reality, and crucial for understanding how we find the world intelligible,
- dichotomies such as those of nature-culture, the global versus the local, the real versus the historically contingent, and so on, prioritise the least (rather than the most) tractable problems, and impede appreciating the importance for sustaining diversity of human life ways of plurality of the past and future aspirations (cf. Benjamin 1940; Whitehead 1978; Rorty 1979; Prigogine and Stengers 1984; Latour 1993; Fardon 1995; Descola and Pálssen 1996; Stengers 1997).

A work by Stephen Toulmin entitled Cosmopolis: the hidden agenda of modernity, which has inspired the present volume indicates that such insights are not at all new. Our use of the expression, 'unquiet pasts' takes its departure from Toulmin's argument that such insights belong to aspects of modernity, which have been hidden by supposed 'settlements' of 'crises over representation' which:

- reduce trust to issues of expert competence,
- hinge upon what he calls the 'myth of the clean slate'—the belief that the 'state of emergency' of contemporary times (if not the human condition, in general) necessitates demolishing all that went before and establishing an altogether new cosmological and political order 'from scratch',
- eclipse the importance for sustaining diversity of future conditions of possibility of plurality of interpretations of the past.

For Toulmin, the unquiet of the past is important because:

> The terms in which we make sense of the past, and the ways in which our view of the past affects our…historical foresight or horizons of expectations. Those horizons mark the limits to the field of action in which we see it possible to change human affairs, and to decide which of our most cherished practical goals can be realized in fact (Toulmin 1990:1).

The present volume builds upon panels organised for meetings of the European Association of Archaeologists (EAA) and the Theoretical Archaeology Group (TAG), as well as discussions and interviews with contributors to the current state of interdisciplinary literatures on the changing roles and aims of academic and other pedagogical institutions, which variously bear upon such questions as:

- What have been the impacts of processes summarised by the expressions 'globalisation' and 'risk society' on directions that archaeological and anthropological research is taking?
- What issues are posed by circumstances under which controversies over by and for whom heritage is being managed are enmeshed in deepening inequalities with regards to exposure to consequences of these processes most contradictory dimensions?
- How important might the ways in which archaeologists and anthropologists address these issues be for fresh approaches to challenges posed by broad changes taking place in the dynamics of academic and other pedagogical institutions, and wider public affairs?

The volume is divided into three parts, entitled, Part 1: Risk Society, Part 2: Lived Cultural Heritage, and Part 3: Re-designing Reflexivity, or Can there be a Cautious Prometheus? Each begins with a summary of materials relating to these themes, and concludes with aspects of discussions with authors of works that have inspired the volume's contributors.

References

Beck, U. 1992 *Risk Society: Towards a New Modernity*. London: Sage Publications.

Beck, U., Giddens, A. and Lash, S. 1994 *Reflexive Modernisation. Politics, Tradition and Aesthetics in the Modern Social Order.* Cambridge: Polity Press.

Benjamin, W. 1994 [1940] Theses on the Philosophy of History, in H. Arendt (ed.), *Illuminations: Works of Benjamin*, 245-255. London: Fontana Press.

Binford, L. 1982 Objectivity—Explanation—Archaeology, in C. Renfrew, M. Rowlands and B.A. Segraves (eds) *Theory and Explanation in Archaeology: The Southampton Conference*, 125-38. New York: Academic Press.

Binford, L. and Sabloff, J. 1982 Paradigms, Systematics, and Archaeology. *Journal of Anthropological Research* 38(2):137-53.

Boas, F. 1940 *Race, Language, and Culture*. New York: Free Press.

Buchli, V. and Lucas, G. (eds) 2001 *Archaeologies of the Contemporary Past*. London: Routledge.

Carman, J. 2002 *Archaeology and Heritage: An Introduction*, Leicester University Press: London.

Cassirer, E. 1960 *The Logic of the Humanities*, translated by C. Smith Howe. London: Yale University Press.

Churchland, P.M. and Churchland, P.S. 1998 Inter-theoretic Reduction: A Neuroscientist's Field Guide, in *On the Contrary Critical Essays, 1987-1997*, 65-79. Cambridge: MIT Press.

Clarke, D. 1968 *Analytic Archaeology*. London: Methuen.

—— 1973 Archaeology: The Loss of Innocence. *Antiquity* 47: 6-18.

Cleere, H. (ed.) 1984 *Approaches to the Archaeological Heritage*. Cambridge: Cambridge University Press.

—— 2006 The World Heritage Convention: Management For and By Whom?, in R. Layton, S. Shennan and P. Stone (eds) *A Future for Archaeology*, 65-74. London: Routledge.

Collingwood, R.G. 1949 *The Idea of History*. Oxford: Oxford University Press.

Darvill, T. 2007 Research Frameworks for World Heritage Sites and the Conceptualization of Archaeological Knowledge. *World Archaeology* Vol. 39, 3: 436-457.

Daston, L. (ed.) 2000 *Biographies of Scientific Objects*. Chicago: University Chicago Press.

—— 2006 The History of Science as European Self Portraiture. *European Review* 14: 523-36.

David, B. and Thomas, J. (eds) 2008 *A Handbook of Landscape Archaeology*. Walnut Creek: New Left Coast Press.

De Beauvoir, S. 1994 [1948] *The Ethics of Ambiguity*. New York: Citadel Press.

Deleuze, G. 1990 *Negotiations: 1972-1990*. New York: Columbia University Press.

Descola, P. and Pálssen, G. (eds) 1996 *Nature and Society. Anthropological Perspectives*. London: Routledge.

Dewey, J. 1927 *The Public and its Problems*. New York: Holt.

—— 1938 *Experience and Education*. New York: Macmillan.

Dupré, L. 1993 *The Passage to Modernity: An Essay in the Hermeneutics of Nature and Culture*. New Haven, CT: Yale University Press.

Durkheim, E. 1938 [1895] *The Rules of the Sociological Method*, translated by S. Solovay and J. Mueller. New York: Free Press.

Fardon, R. 1995 *Counterworks. Managing the Diversity of Knowledge*. London: Routledge.

Felt, U. and Wynne, B. (eds) 2007 *Taking European Knowledge Society Seriously*. Brussels: European Commission of the Directorate-General for Research, Economy and Society.

Friedman, M. 2000 *The Parting of the Ways: Carnap, Cassirer and Heid*egger. Peru, Illinois: Open Court.

Funtowicz, S.O. and Ravetz, J.R. 1997 The Poetry of Thermodynamics. Energy, Entropy/Exergy and Quality. *Futures* 29(9): 791-810.

Hempel, C.G. and Oppenheim, P. 1948 Studies in the Logic of Explanation. *Philosophy of Science* 55: 135-75.

Hodder, I. (ed.) 1982 *Symbolic and Structural Archaeology*. Cambridge: Cambridge University Press.

—— (ed.) 2001 *Archaeological Theory Today*. Cambridge: Polity Press.

Ingold, T. 1993. The Art of Translation in a Continuous World, in G. Pálssen (ed.) *Beyond Boundaries. Understanding, Translation and Anthropological Discourse*, 210-229. Oxford: Berg Publishing.

Kant, I. 1963 [1784] Idea of a Universal History from a Cosmopolitan Point of View, *On History*, essays by Kant edited and translated by L. Beck White. Indiana: Bobbs-Merrill.

—— 1997 (1788) *Critique of Practical Reason*, translated by M. Gregor. Cambridge: Cambridge University Press.

Koerner, S. 2008 Philosophy and Archaeology, in A. Bentley, H. Mascher and C. Chippendale (eds) *Handbook of Archaeological Theories*, 351-374. London: Altamira.

Koerner, S. and Ingold, T. 2009 Emergent Novelty and the Evolutionary Dynamics of Organic and Cultural Life-Forms, a panel in the annual meeting of the Association of Social Anthropology (ASA) meeting, 6-9 April 2009, University of Bristol.

Latour, B. 1993 *We Have Never Been Modern*. Cambridge, MA: Harvard University Press.

—— 2004 Whose Cosmos? Which Cosmopolitics? *Common Knowledge* 10(3): 450-462.

Layton, R., Stone, P. and Thomas, J. (eds) 2001 *Destruction and Conservation of Cultural Property*. London: Routledge.

Lazzari, M. 2008 Topographies of Value: Ethical Issues in Landscape Archaeology, in B. David and J. Thomas (eds) *A Handbook of Landscape Archaeology*, 644-58. Walnut Creek: New Left Coast Press.

Leone, M., Potter, P. and Shackel, P. [1987] 1999 Toward a Critical Archaeology, in J. Thomas (ed.) *Interpretive Archaeology: A Reader*, 458-471. Leicester: Leicester University Press.

Power, M. 1994 *The Audit Society. Rituals of Verification*. Oxford: Oxford University Press.

Prigogine, I. and Stengers, I. 1984 *Order Out of Chaos*. New York: Bantam.

Rorty, R. 1979 *Philosophy and the Mirror of Nature*. Princeton: Princeton University Press.

Rowlands, M. 2004 Anthropology and Archaeology, in J. Bintliff (ed.), *Companion to Archaeology*, 473-489. Oxford: Blackwell.

Salmon, M.H. 1982 *Philosophy and Archaeology*. London: Academic Press.

—— (ed.) 1992 *An Introduction to the History and the Philosophy of Science*. Pittsburgh: University of Pittsburgh Press.

Schnapp, A. 1982 France, in H. Cleere (ed.), *Approaches to the Archaeological Heritage*, 48-53. Cambridge: Cambridge University Press.

—— 1993 *The Discovery of the Past*, translated by I. Kinnes and G. Varndell. London: British Museum Press.

Shanks, M. and Tilley, C. 1987 *Re-constructing Archaeology*. Cambridge: Cambridge University Press.

Silverman, H. and Fairchild Ruggels, D. (eds) 2007 *Cultural Heritage and Human Rights*. London: Springer.

Skeates, R. 2000 *Debating the Archaeological Heritage*. London: Duckworth.

Sloderdijk, P. 2005 Forward to the Theory of Spheres, in M. Ohanian and J.C. Royoux (eds), *Cosmographs*, 223-241. New York: Lukas and Sternberg.

Stengers, I. 1997 *Power and Invention*. Minneapolis: University of Minnesota Press.

Stone, P.G. and Planel, P.G. (eds) 1999 *The Constructed Past. Experimental Archaeology, Education and the Public*. London: Routledge.

Strathern, M. (ed.) 2000 *Audit Cultures: Anthropological Studies in Accountability, Ethics and the Academy*. London: Routledge.

Toulmin, S. 1990 *Cosmopolis. The Hidden Agenda of Modernity*. Chicago: University of Chicago.

Trigger, B. 1984 Alternative Archaeologies: Nationalist, Colonialist, Imperialist. *Man* (NS), 19, 355-370.

—— 1989 *A History of Archaeological Thought*. Cambridge: Cambridge University Press.

Watson, P.J., Leblanc, S.A. and Redman, C.L. (eds) 1971 *Explanation in Archaeology: An Explicitly Scientific Approach*. New York: Columbia University Press.

Whitehead, A.N. 1978 *Process and Reality*. New York: Free Press.

Wylie, A. 2002 *Thinking from Things. Essays in the Philosophy of Archaeology*. Berkeley: University of California Press.

PART 1
Risk Society

Introduction to Part 1

Contextualising Contradictory Trends of Globalisation Risk Society

Stephanie Koerner

In the concluding chapter of his book, *Debating the Archaeological Heritage* (2000), Robin Skeates notes that there are a number of arguments that can be put forward to justify support and appreciation of archaeology, such as that they:

- 'protect and preserve the material remains of the past for the future',
- 'rescue information about the past before it is destroyed',
- 'respond to people's natural curiosity about past populations and their material remains, providing them with reliable answers about the past at the same time as helping to rid them of misconceptions',
- 'provide interpretations of the origins, diversity and long term-term development of human behaviour in the past throughout the world, which can help us to understand and respect for our own societies in the present and future',
- 'contribute to the development of heritage tourism, and hence to local and national economic growth,…offer educators and their pupils a dynamic resource that helps them to unhook their historical imaginations through an integrated variety of active interdisciplinary, problem-orientated approaches to learning' (Skeates 2000: 110).

In this and more recent works, Skeates goes on to ask whether this is enough. He argues that changes taking place in 'public archaeology' can contribute to reappraising the roles of archaeology in the dynamics of the contemporary world.

This first section is made up of chapters variously relating to historical backgrounds and current impacts of processes summarised by the expressions 'globalisation' and 'risk society' on directions that archaeological and anthropological research is taking. The contributions pay attention to the insight, which is developed in Shanks and Witmore's contribution to the third section of this volume, that connecting 'archaeology with an analysis of risk society locates archaeology primarily outside of the academy' and results in 'deep and radical reframing of theory and methodology' under conditions where 'we can almost no longer assume that anything can be assumed' (cf. Sloderdijk 2005).

This first section centres on contradictory trends, which are variously impacting many of the contexts in which archaeological and anthropological research is being pursued:

- conflicting processes and experiences of 'globalisation' and 'risk society' (Harvey 1989; Giddens 1990; Beck 1992; Beck et al. 1994; Hornborg 1998; Friedman 2001),
- the parting of the ways of what might be useful to call 'two cultures': one of 'framework relativism' and/or post-modern theory located largely on academic contexts; the other formed by radical changes taking place in the scales, normative roles and conceptions of sources of the objectivity of science (Galison 1996; Hall 2005; Koerner 2007),
- clashes between concerns to facilitate pluralist dialogical ethics and greater upstream public participation in policy processes, and impacts on relations between pedagogical institutions and public affairs of the roles that expanding expert advisory agencies assign to computer technology based means to adjudicate disagreements.

I will conclude with some observations on the three presuppositions about science and the supposed simplicity of the world mentioned at the onset, with attention to elements that are being called into question in areas of research emerging on interstices not only of physical, life and social sciences and the humanities, but especially fields questioning hitherto predominant conceptions of what divides so-called 'pure' and 'applied research'.

Globalisation and Risk Society

Especially since the collapse of major Cold War institutions, there has been a flood of literatures on globalisation and multi-culturalism. Few images are nowadays more widespread than those of intensification of global interconnectedness—a world full of movement, linkages, persistent cultural interaction and exchange (Inda and Rosaldo 2002). For some we live in a time where no units or scales count for much but the globe. Many images are celebratory and dichotomise: (a) modernity as an era of nation-state thinking—homogeneity, ethnic absolutism, racism, indigenism, essentialism, versus (b) post-modern futures of transnationalism, hybridity, global flows of practically everything, and the univeralisation of liberal democracy through the expansion of techno-science knowledge based political economies (Benhabib (ed.) 1996; Friedman 2001). Discrepant experiences of globalisation are treated as if they were restricted to particular places—especially to 'backward local traditions' (Heatherton 2001). These images proliferate in tandem with the expansion of global tourism industries and the range of regulated, evaluative and highly technology-dependent procedures required to modify and maximise 'global' relationships between places and proximities (Urry 1990, 1995, 2001; Rojek 1993; Crawshaw and Urry 1997; Rojek and Urry 1997; Dennis and Urry 2009).

Not all the images are celebratory. A large inter-disciplinary literature calls into question the ways in which expressions like 'culture' and 'multi-culturalism' render invisible (to the ethical faculties of some) the moral and existential struggles of people—people who are the most exposed to ecological risk, unsustainable economic development projects, exploitation, poverty, dislocation, social exclusion and political conflict (Friedman 2001). For many powerful knowledge-based political economy competitors, 'risk-taking' is indispensable for using science to put the future at the service of the present (Bernstein 1996). In consequence, highly developed government and commercial institutions try to anticipate what cannot be anticipated, forgetting that the first law of risk society goes: catastrophic risk follows the poor and socially vulnerable (Beck 1992; see also Hornborg 1998; Hornborg and Crumley 2006; Nornborg et al. 2007).

The juxtaposition of what David Harvey (1989) called 'time-space compression' and Anthony Giddens (1990) called 'time-space distanciation' with values invested in concrete places and proximities bears underscoring (cf. Lowenthal 1985; Eco 1986; Marres 2005; Werbner 2008). Globalisation traverses national borders, transforms authority and fortifies new boundaries excluding 'others'. The sun never sets on metropolitan centres' shimmering inequality. Super-modern urban castles and subway 'homes' of thousands of beggars have the same geographical coordinates. In many places, heritage conservation clashes with destruction. Several contributions to a recent volume on *Cultural Heritage and Human Rights* (Silverman and Ruggles 2007; see also Cowan et al. 2001) indicate that some of the most sophisticated approaches to both concrete social and environmental problems and to developing alternatives to what have been the most influentially opposed theoretical programmes are coming from places where there are direct and powerful connections between domestic/international conflicts and the state of tourism, cultural heritage preservation and human rights violations (for instance, Wescoat 2008; Silver 2007). These approaches stress the value of concrete, practical (context-dependent) inquiries into widely distributed global problems. (Hastrup 1993; Miller (ed.) 1995; Leach et al. 2005). New light is being thrown on the embeddedness of many controversies over heritage in deepening inequalities with regards to exposure to ecological (climate change and nuclear, chemical and biological industries) hazard, unsustainable development and political conflict—including conflict resulting in grave human rights violations (Meskell and Preucel (eds) 2004; Meskell and Pels (eds) 2005; Silverman and Fairchild Ruggles (eds) 2005). Side-by-side, global tourists and travellers in many 'empty meeting places' or 'non-places' of modernity, such as the airport lounge, the coach station, the railway terminus, the motorway service stations, docks, and so on, are countless global exiles (MacCannell 1992; Fardon 1995; Augé 1995; Urry 2001). These exiles are fleeing from famine, war, torture, persecution, and genocide. Social inequalities and population displacements have magnified and forced mobility upon many (Urry 1995; Friedman 2001). In a conversation between Stuart Hall and Pnina Webner (2008), the latter argues that:

> We must insist on seeing globalisation as a deeply contradictory process. I see the tide of the transnational movement of people—driven by civil war, by ethnic cleansing, famine, poverty and ecological disaster, as well as by the search for economic benefits and a better way of life, as a form of 'globalisation from below'...linked with systems of inequality and power, both historical and contemporary (Hall and Webner 2008: 346).

The publication of Ulrich Beck's work *Risk Society: Towards a New Modernity* (1992) marked a major turning point in approaches to these themes (cf. Beck et al. 1996; Lash et al. 1996; Macnaghten and Urry 1998; Lash 1999; Latour 2004; Webner 2008). For Beck (1992) what is being globalised is 'risk society' and 'reflexive modernisation'. In Beck's account, 'risk society' developed out of ecological and social consequences of instrumental employments of science and technology, which now exceed the available means of control and protection:

> The latency phase of risk threats is coming to an end. The invisible hazards are becoming visible. Damage to and destruction of nature no longer occur outside our personal experience in the sphere of chemical, physical or biological chains of effects; instead they strike more and more clearly our eyes, ears and noses (Beck 1992: 55).

The subtitle of Beck's *Risk Society* (1992) is *Towards a New Modernity*. In Beck's account of 'reflexive modernisation,' the more modern society becomes, the more knowledge it creates about its dynamics and conflicts. He says that this opens new contexts for decisions and action. Individuals are released from structures, and they must redefine their actions under conditions of insecurity in terms of 'reflexive modernity'. This would seem to have much to offer much help with critical reflection upon deepening inequalities with regards to exposure to social and ecological problems, and connections between risk society, expanding tourism industries and controversies over heritage. Such reflection might draw insights from inquiries into close connections between the histories of 'quests to save materials of previous epochs and the natural world' (Macnaghten and Urry 1998). But reflexive modernistion is proceeding in directions that are generating further barriers to such reflectivity. One is that reflexive (unintended) preoccupation with 'bads' rather than 'goods' promotes images of 'living one's own life in a runaway world' (Beck 2001). These engender beliefs in the supposed necessity of the very sorts of practices, which impede critical reflection and further privatise ethics and globalise indifference (cf. Arendt 1958; Bauman 2000; Koerner and Wynne 2008; Koerner and Singleton 2008). Reflexive modernity revolves around paradoxical post-modern interpretations and combinations of some modernity's antithetical elements. These eclipse:

- proliferation of science's normative roles,
- the rationality and logic of every day people's activities and grounds of truth,
- that long before 'reflexive modernity' everyday people have engaged in much critical reflection upon deepening inequalities with regards to social and environmental problems (Irwin and Wynne 1996; Prazniak and Dirlik 2001).

In *Contested Natures* (1998: 24), John Urry and Richard Macnaugten note that 'western consumerism in which nature seems turned into a mere artifact of consumer choice has been extensively critiqued by the environmental movement.' But there are numerous paradoxes, including that:

> …the very development of consumerism has itself helped to generate the current critique of environmental degradation and the cultural focus on nature. This is because environmentalism might be represented as a presupposing a certain kind of consumerism. This is because one element of consumerism is a heightened reflexivity about the places and environments, the goods and services that are consumed, literally through social encounter or visual consumption…[And] such reflexivity in the fields of consumption is clearly connected to recent institutional developments, including what has been characterized as the emergent 'audit society' (Urry and Macnaghten 1998: 25).

Efforts associated with these institutional developments to provide supposedly context independent mathematical and technical models of social and environmental factors often eclipse the complex historical contingencies of 'risk'. Beck is particularly concerned with questions like: What sources of mistakes and errors are systematically built into scientific classification of 'risk', which only become visible in the reference horizon of public controversies? And how does describing these controversies as 'irrational' and/or due to fear of contingency and deficits of scientific understanding impede the constructive potential of critically and constructively reflective efforts to democratise orientations towards sustainability?

Studies of barriers standing in the way of efforts to promote greater upstream public participation in policy processes indicate that these and associated difficulties eclipse remarkable changes taking place in the normative roles of techno-science, the rationality and logic of every day people's activities and grounds of truth. These appreciate that long before 'reflexive modernity', everyday people have not only engaged in much critical reflection upon risk taking and risk managing policies, but also created sophisticated local mean to sustain aspirations and accountabilities (Jasanoff 1995; Irwin and Wynne 1996; Wynne 1992; Hornborg et al. 2007). For Brian Wynne (1996: 59) much discussion of 'risk society' and 'reflexive modernisation' is complicated by the 'juxtaposition of 'propositional' and determinate knowledge of science on the one hand and the 'formulaic', indeterminate knowledge of lay public on the other[:]...the logos of practical and theoretical reason is always already present in the language and truths of lay social actors.' For Uli Felt and Wynne (2007), science is pervaded with diverse indeterminate and formulaic communications and practices, 'ritualisations of rationality' and 'master narratives' that continue to equate instrumental employments of techno-science and 'social progress'.

Risk society's narrowing of modern notions of 'rationality' to a post-modern, predominantly instrumental one is enmeshed in radical changes taking place in forms of control. These include forms of control embedded in and extended by global tourism industries. The work of John Urry (1990, 1995) indicates that images of risk society and 'living your own life in a runaway world' play central roles amongst the tourist industry's symbolic resources and eclipse changes taking place in the forms of control of contemporary 'politics of vision.' Reflexivity is not simply a matter of individuals and their life-possibilities but a series of sets of systematic, regularised and evaluative procedures that enable each 'place' to monitor, modify and maximise their location within the turbulent global order. These procedures 'invent,' produce, market and circulate, especially through global TV and the internet, new or different or repackaged or niche-dependent places and their corresponding visual images. The circulating of such images further performs the very idea of the 'globe' (Urry 2001).

In a discussion that I had with John Urry, I asked that he comment on what he described as the 'omnivorous producing and consuming of places from around the world' (Urry 1995). In numerous works on the emergence of and changes that have taken place in what he calls 'the tourists gaze,' Urry stresses the relevance of locally situated but globally widespread relationships between innumerable forms of occupational specialisation—instruments, objects and places that form today's expanding tourism industries. These industries, Urry says, invent, produce, market and circulate new or different or repackaged niche dependent places, and especially images. These images integrate what are presented as 'unique proximities' within a highly 'mobile sense of vision' in ways that play crucial roles in commodifying, as tourist sites and objects, not only beaches, cultural icons (like the Eiffel Tower), events (a football match between teams from England in Moscow), and activities (like bungee jumping in New Zealand) but also 'intangible heritage' of all sorts—including sites (and peoples) most exposed to risk society's hazards (Urry 2001). For Urry, it is difficult to overstate the implications of globalisation and risk society for inquiries into the proliferation of hitherto unexpected destinations that are now significantly implicated in the patterns of global tourism. These include Alaska, Auschwitz-Birkenau, Antarctica especially in Millennium-year, Changi Jail in Singapore, Nazi Occupation sites in the Channel Islands, Dachau, extinct coal mines, Cuba and especially its 'colonial' and 'American' heritages, Iceland, Mongolia, Mount Everest, northern Ireland, Northern Cyprus under Turkish 'occupation', Pearl Harbour, post communist Russia, Robben Island in South Africa, Sarajevo's 'massacre trail', outer space, Titanic, Vietnam and more (see also Foley and Lennon 2000; Urry 2001).

Beck (1998) has argued for the relevance of inquiries into the deeper historical backgrounds of such developments for promoting strongly reflexive orientation towards contradictory trends of globalisation and risk society. Building upon approaches of Alfred Schutz and Thomas Luckman (1973), Beck suggests that such inquiries might explore conditions under which different forms of unawareness arise, including: processes that contribute to selective reception and transmission of knowledge, intrinsic uncertainty of knowledge, mistakes, inabilities to know, and unwillingness to acknowledge. For Beck (1998), one of Schutz and Luckman's most important insights was that life worlds are apprehended to a considerable extent by awareness that much lies beyond their frames of reference—'we know that we do not know everything,' that there are things we are not aware of, and that these can and do form realms of 'potential knowledge'.

New Versions of 'Two Cultures'

Much of the potential relevance of such insights is recurrently eclipsed by tendencies to assume that we are restricted to vexed options of objectivism and subjectivism and/or relativism and essentialism. Far from disappearing, these options are being opened to new interpretations by new versions of what C.P. Snow (1958) called 'two cultures'. Writing on the problem, Martin Hall (2005) observes that:

> Most anthropological practice is now imbued, in one way or another, with the epistemology of postmodernism and affected by the epistemology of relativism. Modernist concepts such as generalized laws of behaviour and objective truths have been discounted in favour of a larger meta-theory—that there can be no general principles of knowledge. [In these views] ethical codes and concepts of morality are anathema, relics from the grand theories of modernism. At the same time, though, postmodernism seems profoundly ethical, insisting on the equivalence of all social contexts and exposing the 'othering' that is anthropology's burden (Hall 2005: 169).

Hall (2005: 169) notes that, in tandem with these developments:

> as public institutions such as state funded universities and museums have become more commercially oriented, disciplines such as anthropology and archaeology have joined a broad family of consultants that includes psychology cultural resource management, environmental impacts analysis, geology, and the earth and life sciences in general.

He believes that contradictions amongst these developments can be illuminated by Arjun Appadurai's (2001) model of global flows of resources, people and information, Manuel Castells' (1996) 'network society' and Hardt and Negri's (2000) 'empire,' which variously:

> show [how] the ligaments of a new order in which the decentred networks that [once] challenged the ordered world of modernism have been turned to the advantage of...new strategies of rule...[Further,] concepts such as 'transcendental ethics' and truth' may be anathema in some circles, schooled in the tradition of the revolt against modernism. But as Hardt and Negri [2000: 155] point out, forms of resistance may reject post-modernism and valorize absolute principles—in the context of state terror and mystification, clinging too the primacy of the concept of truth can be a powerful and necessary form of resistance (Hall 2005: 169-171).

Hall (2005: 185) is particularly concerned with the implications of post-modern versions of these ethics for archaeology's roles today. Much of his work centres on relationships between researchers and (a) the interests of powerful agencies of today's network society and (b) the communities ('publics') whose histories have become the foci of heritage controversies. Hall (2005) describes an incident in 2000 where two communities, whose histories have been closely related for centuries, made appeals to South Africa's Land Claims Court. The results included that the appeal of the Khoe speaking community of Richterfeld (who made claims on legal property ownership grounds) was rejected, while the Bantu-speaking community of Khomani (who appealed through the National Part Board for a place within a 'natural conservation zone') was accepted because it demonstrated the continuity of the wilderness idea, and a model of sustainable development that fit global elites images Africa's heritage (Hall 2005: 185).

For Hall (2005: 173) at issue is a trope long identified as problematic in anthropology—'spatialised time', in which distancing devices are used to deny coevalness; in Johannes Fabian's (1983: 31) words: the persistent and systematic tendency to place the referents of anthropology in a time other than the present of the producer of anthropological discourse (see also Wolf 1982).

> The rise of network society and the information revolution' offers rehabilitation of anthropology's original sin—the distancing of the exotic through devises that use space and time to draw a boundary between the civilized and primitive worlds. Through the new technical devises—the collapse of time and space in virtual reality—...indigenous communities are increasingly positioned in relation to the ever expanding market in exotic tourism and myriad spin-off industries (Hall 2005: 187).

Today hitherto widespread assumptions about what divides 'pure' from 'applied research' are called into question not least of all by the extraordinary scales on which (to borrow Peter Galison 1996 and Ian Hacking's 1983 terms) 'big science' 'represents and intervenes' in ecological processes and social affairs. There is now a large literature on these scales' implications for policies relating to ecological 'risk issues' (Jasanoff 1995). These may bear upon concerns with the ways in which cultural heritage management (CRM) is being affected by (highly computer technology based) 'risk assessment and management' practices (cf. Holtorf 2001; Carman 2002).

The normative roles of these practices are shaped not only by scales (and associated costs) of technology, but also by changes taking places in conceptions of objectivity. For Lorrain Daston and Peter Galison, much of the tenacity of expert: public and pure:applied research dichotomies has hinged upon the replacement of hitherto predominant conceptions of objectivity by beliefs that computers can achieve a-perspectival objectivity. Daston and Galison's (2007) research suggests that, in such beliefs, computer based 'a-perspectival objectivity' excels in:

- reducing to scientific experiments and universally valid units (often with monetary equivalents) huge scales and complexities of scientific objects and social factors,
- absolute impartiality (views from no where),
- communicating with (and or 'educating') publics.

Challenges 'Ungoverned Reason' Poses for Ethical Reflexivity

Today there is much interest in the relevance for avoiding problems perpetuated by dualist characterisations of pure and 'applied research' of what Julian Thomas (2004) calls a 'pluralist dialogical ethics.'

> Archaeology investigates the past through the medium of material things… Yet it increasingly clear that we do not simply reconstruct the way that things were. Instead we establish a relationship between the past and the present. This relationship can be conceived as a kind of conversation [dialogue], to which we bring a variety of expectations and prejudices, and from which we receive challenges and surprises [Gadamer 1975: 236]. The past never fully reveals itself to us, but through our continued engagement we learn more, both about past worlds and about ourselves…Considered in this way, the perceived distance between the past and the present is not so much a barrier to understanding as a productive space [Gadamer 1975: 264] (Thomas 2004: 1).

For Thomas, key steps towards such an ethics include abandoning approaches that set up a polarity between 'the West and the rest' and then nullify any difference that does not conform to its preconceptions. To this task, he says, it is important to stress that interpretive analogies involve relations between three rather than two contexts:

> …a context within which we archaeologists work; the temporal other of the past; and the ethnographic or spatial other [or the analogue]. With its emphasis on dispassionate observation, processual tended to occlude the first of these contexts, rendering the archaeologists perspective a 'view from nowhere.' From

> a interpretive point of view, the triangulations between contexts becomes quite complex, because much of the point of analysis lies not in demonstrating that the sources and subject contexts are commensurate but that aspects of both are distinct from our everyday modern experience (Thomas 2004: 241).

For Thomas, (2004: 1) this orientation opens reflexive space for a pluralist dialogical appreciation of what Wendy Ashmore calls 'alternative voices.'

> A pluralist dialogical ethics would recognize that different points of view need not be reduced to one another for them to be productive…Without demanding that different regional and national traditions give up their distinctive character, an increased dialogue between…communities would maximise the potential for putting [archaeological] ideas into unfamiliar context, revealing unexpected strengths and weaknesses (Thomas 2004: 242-43).

In tandem with this, it is being argued that many discursive formations concerned with self-reflexivity and the improvement of practice are deeply implicated within a wider culture of auditing and accountability, shaped by transnational managerial domains (Strathern 2000; Lazzari 2008). In 2007, Sheila Jasanoff (Professor of Science and Technology Studies at the John F. Kennedy School of Government, Harvard University, and author of the influential, *The Fifth Sector*) (Jasanoff 1995), presented three lectures to mark her Leverhulme Visiting Professorship at the University of Cambridge, England. Amongst other things, the lectures centred on elements of what Jasanoff refers to as the increasingly 'ungoverned forms of reason' shaping 'politics of public rationality'—including normative aspects of 'evidence,' 'science and citizenship' and 'reason and culture'. Jasanoff explained:

> In democratic societies, rationality is seen as the best safeguard against abuses of power. Decisions that are founded on reason—rather than on passions, emotions, or subjective biases—require no further justification. To say that an action is reasoned is to grant its legitimacy, and effectively to put an end to public debate over it. Claiming reason thus becomes, in effect, a means of taking matters out of the domain of democratic politics. Authorities who bind themselves to the rule of reason need no added political constraints, or so it is thought. In these lectures I question these presumptions about the relationship of power, politics, and public reason in modern democracies. Looking at examples from fields such as the law, environmental policy, and the regulation of biotechnology, I argue that the processes by which we constitute public rationality—or, perhaps more accurately, the semblance of it—are deeply political, as well as culturally specific. Widely prevalent discourses of reason, such as rational choice, risk assessment or bioethics, often conceal underlying political assumptions that were never made explicit or exposed to full deliberation. Moreover, the institutions that should question authoritative claims of rationality are themselves limited in their power to expose rationality's inarticulate and undeliberated foundations.

> Accordingly, reason, while claiming the right to govern, remains itself ungoverned. These lectures are designed to open up the politics of reason to deeper analysis and democratic scrutiny (Jasanoff 2007, Leverhulme Lectures, University of Cambridge, 15-17 May 2007, 'Ungoverned Reason: The Politics of Public Rationality').

What bears noting is that few researchers are likely to find that familiarity with academic debates over influentially opposed relativist and essentialist theoretical positions offers much help under circumstances where some are able to deploy technologies developed in transnational managerial domains amongst means to supposedly settle controversies—for instance, over 'by and for whom heritage is managed' (cf. Cleere 2006).

Paradoxes of Problematical Shared Presuppositions

In his introduction to *The Presented Past* (Stone and Molyneux (eds) 1994), Brian Molyneux's says that the 'notion that a world culture can be created simply through improvements in technology and universal education in the sciences has lost much of its credibility in recent years.' On the other hand, Michio Kaku's best-selling book, *Visions: How Science will Revolutionize the 21st Century* (1997) predicts that, as science progresses from an 'age of discovery to control,' humans will increasingly conduct their affairs through computers; and machines will become 'biological,' develop human traits of recognising objects and meanings, display emotions and form judgements, and play more and more central roles in running society. Such sharply contrasting views may have important implications for some of the tensions summarised by the expression, 'commercial field archeology and academic research' (David and Thomas 2008).

Fortunately, new light is being thrown on highly problematical presuppositions about the tasks of science and the supposed simplicity of the world, which have been shared by many influentially opposed theoretical programmes. A useful way to conclude this section is by exploring three presuppositions, which are variously being called into question not only by new interdisciplinary alliances that are challenging hitherto predominant conceptions of what divides so-called 'pure' and 'applied research'.

Presuppositions about the ultimate tasks of science

Throughout the 19th and 20th century (and to this very day), the most polemical debates over whether humanities and social sciences can or should aspire to the requirements of a science have revolved around shared presuppositions that the ultimate task of science is to establish universally valid concepts, categories and propositions, whose explanatory force is independent of any context of application. The roots of these orientations date back to Plato (427-347 BC) (1999),

Aristotle (384-322 BC) (1984) and their relativist (or skeptical) critics' models of human agency, history and divisions of the world between: (1) absolute unity and permanence, and (2) absolute dis-unity, pure flux (or in modern terms absolute 'randomness'). Aristotle (1984) created a scheme for framing contrasts between opposing paradigms in terms of a question about what sorts of things satisfy the requirements of a science (episteme), namely: If something can be said to be subject to change, what is the essence of that something? (1) The unchanging aspect, (2) the changing aspect, or (3) both, that is, the interaction of changing and unchanging aspects. For foundationalists (like Socrates and Plato), the answer must be (1), and the others have to be reducible to it. Scientific objects must exhibit regularities that are universal and demonstrable by chain of both necessary and sufficient causes. For probabalists (like Aristotle), things that are 'always or for the most part' can satisfy requirements of science if they can be described as examples of essential states or substances (Aristotle, *Metaphysics* 1984, 1027a20-27; cf. Daston 2000). For skeptics—as for 'subjectivists' (Bourdieu 1990) and a number of strong cultural relativists—the only thing that is unchanging is the illusion of unchanging things. All is by chance and/or caused by such things as social interests and values.

Presuppositions about the supposed simplicity of the world

One consequence of these presuppositions is that they reduce ontology (theories about what sorts of things exist and how they came into being), epistemology (theories about knowledge and especially about obstacles to objectivity) and ethics to matters of classification. Examples include the complex systems of dichotomies around which much debate over processualism and post-processualism has been structured (e.g., Augé 1977; Rorty 1979; Godelier 1986; Descola and Pálssen (eds) 1996; Koerner 2008). The presupposition at the heart of this tendency is that the most essential property of the world is simplicity (for instance, Boyle [1686] 1996; Einstein, correspondence 1903-1955, 1972). Ever since the mathematical, mechanical science and natural philosophies of Bacon ([1628] 1884, 1974), Boyle ([1686] 1996), Descartes (1984-1991), Newton ([1687] 1934), Hobbes ([1651] 1962) became something of the 'philosopher king' of the 'modern cosmopolis' (Toulmin 1990). The recurrently predominant ideal vision of the world has been one reducibible to simplicity.

> The accepted model has been one of geometry; and the stated aim was to enable the control of the whole natural world by routine operations, like those of a mechanic. Ever since then, developments in science have been counted as advances if they further articulated that paradigm of simplicity. Its model has been of individuals conceived rather like Hobbes' atomic persons, and it eventually enabled the exclusion of design and purpose from biological explanations. Descartes' 'beau roman de physique' (as his disillusioned follower Huygens called it) finds its contemporary manifestation in that biological science which finds anything bigger than a cell too complicated to be a worthwhile object of study (Funtowicz and Ravetz 1997: 792).

Importantly, reducibility to simplicity is not restricted to Enlightenment rationalism and objectivist philosophy (e.g., Hempel and Oppenheim 1948; Churchland and Churchland 1998). It is likewise a key theme of the beliefs held by many scientists (as well as specialists in the humanities and social sciences) about the nature of the major obstacles to discovering a supposed determinate order 'behind the complexity', including such obstacles as 'inadequate technologies and human 'states of mind'' (Prigogine and Stengers 1984). No less remarkably, it is likewise a presupposition that motivates relativist tendencies to equate the historically contingent with the illusory (cf. Ingold 2000). For Daston, what bears stressing is how, in debates over apparently antithetical paradigms:

> The opposition between nature and culture shadows that between the real and the constructed, nature stands as the eternal, the inexorable, the universal; culture for the variable, the malleable, and the particular. Like the return of the represses, the supra- and sub-lunary spheres of Aristotelian cosmology crop up in a new guise, crystalline nature encircling mutable culture. Both sides of the debate accept the oppositions of the real versus the constructed, the natural versus the constructed. Hence arguments are about which of these two categories notions like 'race' or 'quark' belong—are they real or are they constructed? Discoveries or inventions?—Not about the categories themselves (Daston 2000: 3).

Presuppositions about relationships between contexts and contents of science

All this might not be so relevant for archaeology and anthropology if nature: culture, moderns:pre-moderns, reality:historical contingency dichotomies did not play essential (and highly paradoxical) roles in presuppositions not only about relationships between contents and social contexts of science, but about the contents of the very 'hardest' of sciences (such as physics and some neo-Darwinian paradigms for molecular biology). Here, we explore only two elements of this situation. As noted at the onset, both essentialists and relativists are concerned with relationships between the contents and social contexts of science. The former stresses 'doing boundary work' to separate the two, the latter 'watches boundaries' and argues that the contents of scientific knowledge is reducible to social values and interests (cf. Gieryn 1997). Both eclipse at least two things:

- the importance of dualist characterisations of nature-culture and images of science and modernity to the distinctions 'foundational laws' and 'phenomenological generalisations',
- connections between (a) and problematical images of called 'modern and other world views'.

For several decades, Ilya Prigogine, Isabelle Stengers, Latour and quite a number of others have explored the roles of mainstream experimental and theoretical science in distinctions between 'phenomenological laws' and 'fundamental laws'. Their views state that although the former can describe phenomena mathematically in a rigorous and relevant way, only the latter can claim to unify the diversity of phenomena—that is, to go 'beyond appearances'. Of special relevance for our present concerns, are impacts of these views on mainstream approaches to temporality, irreversible change and the emergence and endurance of relationships and entities in states far from equilibrium (Nicholis and Prigogine 1989). Stengers (1997: 22-23) writes that when she began studying physics, she accepted the mainstream belief that observations on irreversible processes (mixtures that do not un-mix, radical differences between before and after, where it is impossible to overlook the non-equivalence of cause and effect) should be treated as 'merely phenomenological' consequences of our not being 'perfect observers' and inadequate instruments. Working with Prigogine, she began to explore the extent to which the history of paradigms for the tasks of science that deny time (and irreversible change and emergent novelty) has also been the 'history of social and cultural tensions' (Stengers 1997: 42-43).

In the humanities and social sciences (and in debates over influentially opposed positions on experimental archaeology) much disagreement centres on qualitative and quantitative methods. One consequence of these debates is that they eclipse paradoxical consequences of the roles assigned to the equals sign (=), including that of erasing differences between before and after, and cause and effect. Stengers (1997: 22-23) explains that, one the one hand, we are supposed to regard distinctions between 'phenomenological observations' and 'fundamental laws' (laws of changeless change) as crucial. Then we are supposed to use the 'equals sign' to address the supposed cause of these distinctions, namely, supposed 'mere appearances' and instruments. What bears stressing, for Stengers (1997: 22-23), is that, instead of accepting that using the 'equals sign' to erase before and after is theoretically incoherent and empirically refutable, some go on to 'judge the world of phenomena in the name of a normative ideal…more virulently than ever.'

Of course, notions of cause, effect and an 'equals sign' are of extraordinary antiquity. What is new is the idea that one can treat a present state as an effect equal (and reversible) to both a past state and an identical future. Values that have been invested in computer technologies (including values that attribute these technologies with supposed capacities for 'a-perspectival objectivity') (Daston and Galison 2007) have helped facilitate this problematic idea. Another aspect of the history of this idea, which bears stressing, has been applications of problematic interpretations Charles Darwin's notion of the randomness of natural selection to realms studied by physicists and, most recently, by a number of molecular biologists (Prigogine 1997). In several 20th century deterministic interpretations of absolute randomness, all processes end up being treated as though they were time-reversible—that is, as though they can proceed backward as well as forward through time. For Prigogine (1997), determinism is paradoxically, both a way of envisaging trajectory and a fundamental denial of what Arthur Eddington (1958 [1927]) called 'the arrow of time.' One of the consequences of the paradox is that it prohibits common sense experience of change. Determinism prohibits appreciation of an 'event in the present' arising in relation to a 'past that precedes an undetermined future' (Prigogine 1997). It treats time as a paradoxically timeless 'given' and the future as arising—not from a plurality of indeterminate initial conditions—but as if it was as determined as the past.

The implications of dualist characterisations of nature versus culture, the real versus the historically contingent, and 'foundational laws' versus 'phenomenological generalisations' for determinism is difficult to overstate. There may be many ways to illustrate this theme. Our concerns in this chapter make it useful to depart from the notion of 'elimination'. Elimination and reduction are expressions developed by scientists (and analytic philosophers) to talk about objectivist (natural or social) theories about knowledge and explanation (Dennett 1978; Rey 1988; Churchland and Churchland 1998). But recent efforts to 'go beyond' nature–culture and related dichotomies indicate there are subjectivist and cultural relativist versions too (cf. Bourdieu 1970; Latour 1993, 2004; Descola and Pálssen (eds) 1996; Hornborg 1998; Latour and Weibel 2002, 2005). Despite claims to the contrary, in both some of the most influential objectivist and subjectivist views, everyday people's common-sense understandings of the world, the mind, and so on are 'folk psychology' and reducible to supposedly more fundamental substances, like 'the environment,' 'society,' 'collective representations,' and so on. Further—and this cannot be boldly underscored enough—such caricatures of 'folk psychology' (like those of 'pre-modern world views') have played problematical roles in shaping caricatures of 'phenomenological generalisations' as 'mere appearances'.

Things would not be so complicated if the above outlined problems were not so deeply enmeshed in further contradictions. These include contradictions between the above outlined presuppositions about the tasks of science and the supposed simplicity of the world, and about what distinguishes 'scientific' and/or 'modern' from 'other worldviews' and/or 'folk psychology'. Despite claims to the contrary, new versions of the latter are grounded in dualist characterisations of 'reason versus tradition' (Ingold 1993, 2000) and especially the idea that what distinguishes modern reason and science from all 'other' forms of thought and practice is it awareness of the contingency of all human claims to truth (cf. Blumenberg 1983; Funkenstein 1986). In so-called 'standard accounts' of science and modernity (whether interpreted as triumph or tragedy), once this awareness had been established science could be expanded but it could not recover (nor did it wish to return to) the Aristotelian world picture. In such accounts, the most fundamental difference of 'other world views' are the ways in which their being situated in traditions (Ingold 1993) supposedly impedes and/or denies awareness of contingency, the passage of time, even history (Blumenberg 1983; J.L. Koerner 1998).

But how can this be? Would a model of a world view, which supposedly denies contingency, not fit better the deterministic presuppositions about the task of science and simplicity of the world that we have been exploring? Such questions have led Stengers (1997) to argue that complexity is not something that we have to 'discover.' Like awareness of there having always been alternatives to vexed options of objectivism and subjectivism, it is something that we somehow always knew all along. Stengers' approach to the problem departs from observations that:

> We get used to associating the birth of modernity with the 'Copernican revolution,' the substitution of the heliocentric system for the geocentric system. But it was Kepler who made the real difference between the two systems by transforming the significance of the relation between mathematics and astronomy. Kepler divested the circle, and with it, mathematics and geometry of the powers they had been hitherto assigned of 'sufficient reason'—and the status they held as a priori timeless placeless judge of epistemic authority and ascribed political sovereignty ascription (Stengers 1997: 20-22).

What bears stressing is that Kepler's 'breaking the circle of sufficient reason' played essential roles in making explicit the extent to which Bacon's experimental science and Newton's natural philosophy undermined the presuppositions of scholars since antiquity about distinctions between the essential substances of sublunary and celestial realms, and their implications for ontic sources of objectivity. Thus, we should expect the founding figures of early modern experimental science and natural philosophy to have wanted to extend properties hitherto attributed to the contingency of sublunary realms to the ends of the universe. And several argued precisely such aims (for instance, Galileo [1564-1642] 1968). But one of the remarkable things about the presuppositions about the supposed context-independent tasks of science, which we considered above, is that their notions of the supposed simplicity of the world (and of time reversible 'foundational laws') do not resemble historically contingent sublunary realms. To the contrary, they resemble the Aristotelian celestial sphere—'the unchanging and divine world of astronomic trajectories, which was for Aristotle the only world that could be given exact mathematical and geometric description' (Stengers 1997: 34-35). Or put another way, they share features with caricatures of world views of so called 'pre-moderns,' 'others' and 'folk psychology.'

Some Implications

One argument developed in the present volume is that precisely such presuppositions are being challenged by a number of researchers working in contexts where controversies over heritage are enmeshed in deepening inequalities with regards to exposure to hazards of globalisation and risk society. Indeed few circumstances can help bring into sharper relief the roles of the above outlined presuppositions in perpetuating such dichotomies as those of nature:culture, pure:applied research, the real and the historically contingent, contemporary and non-western, and so on. Amongst other things, the contributions to this section variously explore aspects of the backgrounds and current state of such circumstances.

References

Appadurai, A. (ed.) 2001 *Globalization*. Duke University Press.

Arendt, H. [1958] 1989 *The Human Condition*. Chicago: University of Chicago Press.

Aristotle (384-322 BC) 1984 *The Complete Works of Aristotle*. Princeton: Princeton University Press.

Augé, M. 1977 *The Anthropological Circle; Symbol, Function and History*. Cambridge: Cambridge University Press.

Bauman, J. 2000 *Liquid Modernity*. Cambridge: Polity Press.

Beck, U. 1992 *Risk Society: Towards a New Modernity*. London: Sage Publications.

—— 1998 *Democracy without Enemies*, translated by M. Ritter. Oxford: Polity Press.

—— 2001 Living Your Own Life in a Runaway World, in W. Huttons and A. Giddens (eds) *On the Edge*, 164-174. London: Vintage.

Beck, U., Giddens, A. and Lash, S. 1994 *Reflexive Modernisation. Politics, Tradition and Aesthetics in the Modern Social Order.* Cambridge: Polity Press.

Benhabib, S. 1996 *Democracy and Difference: Contesting the Boundaries of the Political*. Princeton: Princeton University Press.

Blumenberg, H. 1983 *The Legitimacy of the Modern Age*, translated by R.M. Wallace. Cambridge, MA: MIT Press.

Bourdieu, P. 1990 *The Logic of Practice*, translated by R. Nice London: Polity Press.

Boyle, R. (1627-1691) [1686] 1996 *A Free Inquiry into the Vulgarly Received Notion of Nature*. Cambridge: Cambridge University Press.

Carman, J. 2002 *Archaeology and Heritage: An Introduction*. Leicester University Press: London.

Castells, M. 1996. *The Rise of Network Society*. Oxford: Blackwell.

Churchland, P.M. and Churchland, P.S. 1998 Inter-theoretic Reduction: A Neuroscientist's Field Guide, in *On the Contrary Critical Essays, 1987-1997*, 65-79. Cambridge: MIT Press.

Cleere, H. 2006 The World Heritage Convention. Management For and By Whom?, in R. Layton, S. Shennan and P. Stone (eds) *A Future for Archaeology*, 65-74. London: Routledge.

Cowan, J.K., Dembour, M.B. and Wilson, R.A. (eds) 2001 *Culture and Rights. Anthropological Perspectives*. Cambridge: Cambridge University Press.

Crawshaw, C. and Urry, J. 1997 Tourism and the Photographic Eye, in C. Rojek and J. Urry (eds), *Touring Cultures. Transformations of Travel and Theory*, 176-195. London: Routledge.

Daston, L. (ed.) 2000 *Biographies of Scientific Objects*. Chicago: University Chicago Press.

Daston, L. and Galison, P. 2007 *Objectivity*. London: Zone Books.

David, B. and Thomas, J. (eds) 2008 *A Handbook of Landscape Archaeology*. Walnut Creek: New Left Coast Press.

Dennett, D. 1978 *The Intentional Stance*. Cambridge: MIT Press.

Dennis, K. and Urry, J. 2009 *After the Car*. Cambridge: Polity Press.

Descartes, R. (1596-1626) 1984-1991 *The Philosophical Writings of Descartes*, translated by J. Cottingham, R. Stoothoff and D. Murdoch. Cambridge: Cambridge University Press.

Descola, P. and Pálssen, G. (eds) 1996 *Nature and Society: Anthropological Perspectives*. London: Routledge.

Eco, U. 1986 *Travels in Hyperreality*. London: Picador.

Eddington, A.S. [1927] 1958 *The Nature of the Physical World*. Ann Arbor: University of Michigan Press.

Einstein, A. and Besso, M. 1972 *Correspondence 1902-1955*, edited by P. Speziali. Paris: Herman.

Fardon, R. 1995 *Counterworks: Managing the Diversity of Knowledge*. London: Routledge.

Foley, M. and Lennon, J. 1997 Dark Tourism—An Ethical Dilemma, in M. Foley and J. Lennon (eds) *Strategic Issues for the Hospitality, Tourism and Leisure Industries*, 153-164. London: Cassell.

Friedman, J. 2001 Indigenous Struggles and the Discreet Charm of the Bourgeosie, in R. Prazniak and A. Dirlik (eds) *Politics and Place in an Age of Globalization*, 53-70. Lanham, MD: Rowman and Littlefield Publishers.

Funkenstein, A. 1986 *Theology and the Scientific Imagination: From the Middle Ages to the Seventeenth Century*. Princeton: Princeton University Press.

Funtowicz, S.O. and Ravetz, J.R. 1997 The Poetry of Thermodynamics. Energy, Entropy/Exergy and Quality. *Futures* 29(9): 791-810.

Galileo, G. (1564-1642) 1968 *Opere*, 20 volumes. Florence: Barbera.

Galison, P. 1996 Computer Simulations and the Trading Zone, in P. Galison and D.J. Stump (eds), *The Disunity of Science: Boundaries, Contexts, and Power*, 119-157. Stanford: Stanford University Press.

Giddens, A. 1990 *The Consequences of Modernity*. Stanford, CA: Stanford University Press.

Gieryn, T.F. [1995] 1997 Boundaries of Science, in A.I. Tauber (ed.) *Science and the Quest for Certainty*, 293-332. London: Mcmillan.

Godelier, M. 1986 *The Mental and the Material*. New York: Verso.

Hall, M. 2005 Situational Ethics and Engaged Practice: The Case of Archaeology in Africa, in L. Meskel and P. Pels (eds), *Embedding Ethics*, 169-194. Oxford: Berg.

Hardt, M. and Negri, A. 2000 *Empire*. Cambridge: Cambridge University Press.

Harvey, D. 1989 *The Condition of Postmodernity*. Oxford: Blackwell Publishers.

Hastrup, K. 1993 Hunger and the Hardness of Facts. *Man* (NS) 28: 727-739.

Heatherton, T. 2001 Ecology, Alterity and Resistance in Sardinia. *Social Anthropology* 9: 289-306.

Hempel, C.G. and Oppenheim, P. 1948 Studies in the Logic of Explanation. *Philosophy of Science* 55: 135-75.

Hobbes, T. (1588-1679) [1651] 1962 *Leviathan*. New York: Collier Books.

Holtorf, C. 2001 Is the Past a Non-renewable Resource?, in J. Bintliff (ed.) *A Companion to Archaeology*, 286-298. Blackwell, Oxford.

Hornborg, A. 1998 Towards an Ecological Theory of Unequal Exchange: Articulating World System Theory and Ecological Economics. *Ecological Economics* 25: 127-136.

Hornborg, A. and Crumley, C. (eds) 2006 *The World System and the Earth System*. Walnut Creek, CA.: Left Coast Press.

Hornborg, A., McNeill, J.R. and Martinez-Alier, J. (eds) 2007 *Rethinking Environmental History: World Systems Histories and Global Environmental Change*. Plymouth: Altamira Press.

Inda, J.X. and Rosaldo, R. (eds) 2002 *The Anthropology of Globalization: A Reader.* Oxford: Blackwell Publishers.

Ingold, T. 1993 The Art of Translation in a Continuous World, in G. Pálssen (ed.) *Beyond Boundaries. Understanding, Translation and Anthropological Discourse*, 210-229. Oxford: Berg Publishing.

—— 2000 *The Perception of the Environment. Essays in Livelihood, Dwelling and Skill*. London: Routledge.

Irwin, A. and Wynne, B. (eds) 1996 *Misunderstanding Science: The Public Reconstruction of Science and Technology*. Cambridge: Cambridge University Press.

Jasanoff, S. 1995 *The Fifth Branch: Science Advisors as Policy Makers*. Cambridge: Harvard University Press.

Kaku, M. 1997 *Visions: How Science will Revolutionize the 21st Century*. New York: Anchor Books.

Koerner, J.L. 1998 Heironymus Bosch's World Picture, in C. Jones and P. Galison (eds), *Picturing Science and Producing Art*, 297-325. London: Routledge University Press.

Koerner, S. 2007 Habitus Unbound. Archaeologies of Plurality of Public Grounds of Truth, in J. Thomas and V. Oliveira Jorge (eds), *Overcoming the Modern Invention of Culture*, 97-120. Porto: ADECAP.

—— 2008 Philosophy and Archaeology, in A. Bentley, H. Mascher and C. Chippendale (eds) *Handbook of Archaeological Theories*, 351-374. London: Altamira.

Koerner, S. and Singleton, L. 2009 *Revisiting Pandora's Hope. Archaeologies of Place and Integrating Difference into Deliberative Democracy*, in D. Georgiou and G. Nash (ed.) *Archaeologies of Place*, final chapter. Budapest: Archaeologia.

Koerner, S. and Wynne, B. 2008 Wittgenstein and Contextualising Reduction and Elimination in the Social Sciences, F. Stadler et al. (eds) *Wittgenstein and Contemporary Debates in the Humanities and Social Sciences*. Vienna: IWS.

Latour, B. 1993 *We Have Never Been Modern*. Cambridge, MA: Harvard University Press.

—— 2004 Whose Cosmos? Which Cosmopolitics? *Common Knowledge* 10(3): 450-462.

Latour, B. and Weibel, P. (eds) 2002 *Iconoclash. Beyond the Image Wars in Science, Religion and Art.* London: MIT Press.

—— (eds) 2005 *Making Things Public. Atmospheres of Democracy*. Cambridge: MIT Press.

Leach, M., Scoones, I. and Wynne, B. (eds) 2005 *Science and Citizens*. London: Zed books.

Lowenthal, D. 1985 *The Past as a Foreign Country*. Cambridge: Cambridge University Press.

MacCannell, D. 1992 *Empty Meeting Grounds*. New York: Routledge.

Macnaghten, R. and Urry, J. 1998 *Contested Natures*. London: Sage Publications.

Marres, N. 2005 *No Issue, No Public: Democratic Deficits after the Displacement of Politics*. Doctoral Dissertation, University of Amsterdam.

Meskell, L. and Pels, P. (eds) 2005 *Embedding Ethics*. Oxford: Berg.

Meskell, L. and Preucel, R. (eds) 2004 *A Companion to Social Archaeology.* Oxford: Blackwell.

Miller, D. (ed.) 1995. *Worlds Apart: Modernity through the Prism of the Local*. London: Routledge.

Newton, I. (1642-1724) [1687] 1934 *Mathematical Principles of Natural Philosophy*, translated by A. Motte and F. Cajori. Berkeley, CA: University of California Press.

Nicholis, G. and Prigogine, I. 1989 *Exploring Complexity: an introduction*. New York: W.H. Freeman and Company.

Plato (427-347 BC) 1999 *Ethics, the Politics, Religion and the Soul*, edited by G. Fine. Oxford: Oxford University Press.

Prazniak, R. and Dirlik, A. (eds) 2001 *Politics and Place in an Age of Globalization*. Lanham, MD: Rowman and Littlefield Publishers.

Prigogine, I. and Stengers, I. 1984 *Order Out of Chaos*. New York: Bantam.

Rojek, C. 1993 *Ways of Seeing—Modern Transformations of Leisure and Travel*. London: Macmillan.

Rojek, C. and Urry, J. (eds) 1997 *Touring Cultures. Transformations of Travel and Theory*. Lomdon: Routledge.

Rorty, R. 1979 *Philosophy and the Mirror of Nature*. Princeton: Princeton University Press.

Schutz, A. and Luckmann, T. 1974 *The Structures of the Life World*. London: Heinemann.

Silver, C. 2007 Tourism, Cultural Heritage and Human Rights in Indonesia: The Challenges of an Emerging Democratic Society, in H. Silverman and D. Fairchild Ruggles (eds) *Cultural Heritage and Human Rights*, 78-91. London: Springer.

Silverman, H. and Fairchild Ruggels, D. (eds) 2007 *Cultural Heritage and Human Rights*. London: Springer.
Skeates, R. 2000 *Debating the Archaeological Heritage*. London: Duckworth.
Sloderdijk, P. 2005 Forward to the Theory of Spheres, in M. Ohanian and J.C. Royoux (eds), *Cosmographs*, 223-241. New York: Lukas and Sternberg.
Snow, C.P. [1959] 1962 *The Two Cultures and the Scientific Revolution*. Cambridge: Cambridge University Press.
Stengers, I. 1997 *Power and Invention*. Minneapolis: University of Minnesota Press.
Stone, P.G. and Molyneaux, B.L. (eds) 1994 *The Presented Past. Heritage, Museums and Education*. London: Routledge.
Strathern, M. (ed.) 2000 *Audit Cultures: Anthropological Studies in Accountability, Ethics and the Academy*. London: Routledge.
Thomas, J. 2004 *Archaeology and Modernity*. London: Routledge.
Toulmin, S. 1990 *Cosmopolis. The Hidden Agenda of Modernity*. Chicago: University of Chicago.
Urry, J. 1990 *The Tourist's Gaze*. London: Routledge.
—— 1995 *Consuming Places*. London: Routledge.
—— 2001 *Globalising the Tourist Gaze*. Lancaster: Lancaster University Papers in Sociology.
Werbner, P. (ed.) 2008 *Anthropology and the New Comopolitanism: Rooted, Feminist and Vernacular Perspectives*. Oxford: Berg.
Wescoat, J.L. 2008 The Indo-Islamic Garden: Conflict, Conservation, and Conciliation in Gujurat, India, in H. Silverman and D. Fairchild Ruggels (eds) *Cultural Heritage and Human Rights*, 53-78. London: Springer.
Wynne, B. 1982 *Rationality and Ritual. The Windscale Inquiry and Nuclear Decisions in Britain*. London: BSHS Monograph 3.
—— 1996 May the Sheep Safely Graze? A Reflexive View of the Expert–Lay Knowledge Divide, in S. Lash, B. Szerzynski and B. Wynne (eds), *Risk, Environment and Modernity. Towards a New Ecology*, 44-83. London: Sage.

Chapter 1

Re-playing the Past in an Age of Mechanical Reproduction

Andrew Cochrane

I delivered this paper at the European Association of Archaeologists conference held at Cork, Ireland, in 2005. Since this time, my memories of the event have become fragmented with gaps often appearing. My notes for the paper were originally typed, and I find that they are covered with handwritten scribbles, the results of last minute adjustments—the overlay and underlay of ideas. As such, and in following this mode, I present this paper as I remember and discover it—as torn thoughts and disconnected connections. Following on from experiments elsewhere (e.g., Cochrane and Russell 2007), and inspired by film media, I opt to present ideas as free floating frame segments. You are invited to read them either in a conventional manner, or to dip in—and if something catches your interest, to maybe cast your eye to the nearby frames, reviewing the ideas there.[1] Such a reading will in fact bring you closer to how I discovered my notes and written thoughts.

Frame 1

What are the relationships between spectator, media presentation and performance within some museum environments? Is it appropriate to separate these media into distinct elements? Why are some museums increasingly attempting to enchant and inform audiences via time-looped film installations? What are cinematic media supposed to do in museums? Such questions are the stimulated responses I have after being exposed to, or immersed within, some film installations for the first time. In attempting to answer some of these queries, emphasis here will focus on the apparatus that deliver images, and on the sites that display them. Engagements with mechanical productions, such as films, have been previously considered in film and gallery space (e.g., Turvey et al. 2003), but rarely with

1 This form of writing is inspired by the novelist J.G. Ballard (2006), and stimulated by necessity; for I am currently writing this paper in a small village in the northern mountains that overlook Kyoto, Japan. As such I have no access to my library—only notes and memory. Referencing here will therefore be less formal with reduced Harvard in-text citation—instead, the references are more an indicator of work that has influenced me.

installations displaying archaeological discourses. Such presentations that focus on the past, attempt to convey data on given topics, by continuously repeating the same programme indefinitely for the viewers. Here, I use two specific film installations as examples—one from the National Museum and Galleries of Wales in Cardiff, and the other from the National Museum of Scotland in Edinburgh, to tease out how we may approach thought or contemplation within the circular logic of repeated film. These films caused me to reflect—as such, my writings here are an expression of these experiences, rather than a verbatim fact. For a sophisticated media aware society, the 'screenage' generation, films are not new; for instance, we can indulge ourselves with them in many conditions, such as at cinemas, in cars, on mobile phones and in the comfort of our own homes. Most people enjoy cinematic images (cf. Sam quoted below)—some people often ask 'what film did you see?'—'have you seen the new so and so?' Some people enjoy in the sharing of these narratives and experiences of looking at moving images and then talking about them—creating fresh connections and performances.

Frame 2

Within museum environments it is possible to generalise that most Western people act with alternative habits to those they may in their home or other public places. Museums produce different spectatorial habits—such as walking politely behind other visitors rather than running or pushing; pausing in front of an image momentarily before moving on; and not touching or sometimes not photographing installations. These habits are interesting as often there are no rules or guidelines present that enforce us to act in these prescribed manners. We often do so of our own free will. This sense of volition is enhanced by the explicit time length of an archaeological film looped installation. The length of time of that the film images play for can make us question amongst other things: how long has this already been on for? Am I at the beginning or near the end? Do I really want to sit or stand near those other people? What is it that I am watching and does it even interest me? If it does, how long will I watch it for?[2] These questions that buzz through ones mind can often make us also think about thinking and where we are within our environment—the actual situation of viewing the installation itself can become central in the thoughts of the spectator—although it is noteworthy that some installations work better than others in stimulating these possible responses. Indeed, if one does stop to gaze, look or glance, it is almost as if one becomes

2 The duration of a looped film can actually stop people from watching it. For instance whilst commenting upon an archaeological film at the *Salisbury and South Wiltshire Museum*, one visitor commented that 'the loop [was] so long and laboured it was dissuasive of watching all the loop had to offer', this viewer did watch the entire film and stated that 'no-one else did whilst I was there. Can understand why, the Orwell 'Winston Smith experience' TV wall didn't work well' (VenerableBottyBurp 2008).

momentarily 'rooted' to the spot or seat, absorbed, and ceases to be the captivated spectator and becomes another captive spectacle in space—albeit temporal (see Figure 1.6). For me, the installation renders the visitors as sculptures in space. Film installations do not favour a museological *flâneur*, although they can through their spectacularity permit the viewer to overlook their status as spectacle.

Frame 3

Some installations assist the viewer in making decisions about whether to stay or go, such as the example in Figure 1.1 that informs when the next show will commence. With others, some spectators are not prepared to wait and take more decisive action by forwarding, rewinding or even stopping the playing disc, as seen in Figure 1.2. In such circumstances, it is the film and the technology that stimulates action and intention. Technology can also influence experience and expectation. For instance, the availability of film for museums and galleries now interrupts the perceived need by some of a constructed 'white-cube' to display contemporary art. The film piece, often contained within it own 'black-cube' generally works in any

Figure 1.1 Installation at the National Museum and Galleries of Wales, 2005, announcing the next film

Source: Digital photo by author.

Figure 1.2 Enmeshed within a film environment, a spectator forwards a film at the National Museum and Galleries of Wales, 2005

Source: Digital photo by author.

given space, be it an old or new building. The film has the ability to absorb viewers and therefore construct its own dimensional awareness.[3] For example, Stoneshifter comments that, 'one test of these installations are how long the images last in one's imagination and that comes down more to the power of the image rather than to the surroundings' (2008). For a museum environment, this can be advantageous, in that the film piece in many respects can be placed anywhere, thereby reducing installation costs and design needs. Indeed, the Llanmaes film, moved around the National Museum and Galleries of Wales (see below, and also Figure 1.2 for example of what be termed *ad hoc* placement).

Frame 4

If one does stop to watch for significant periods of time, one begins to acknowledge the rhythms of the eternal return of the looped film installation. This repetition can force one to attend to the images as presentation in lived time. This aspect of time loop spectatorship might allow the possibility of considering the relations

3 I am aware, however, that this is not always the case. For instance, I have witnessed a film artist declare that their piece 'only' works on high-end widescreen HD flat screens. I found this particularly interesting as other animators and film makers have commented to me that the type of screen is of little concern for them—be it a HD screen or battered old television—they feel that the quality of their work will shine through.

within media through repetition. These looped film installations are not the only exhibitions within museums that can affect our understandings of our contemporary and conceptualised past environments, but are they interesting as they do offer examples of how interpretation can be conveyed via endless media streams.

Frame 5

Within film media there are generally two types of repetition: essential and exact. Essential involves a general reproduction in which one thing replaces a similar thing. An example of this would be television soap operas with similar plots and characters continually being recycled with slightly different sets and actors. Exact repetition differs in that there is no change in the reproduction and it repeats itself within—like the loop. Museum film installations often operate through exact repetition, and as such differ from television in that it does not always fulfil our desires for plot and narrative. Instead, time is used to influence our patterns of thought and movement through our consumption and experience within time (see also Cochrane in press). As a media-literate society, Westerners are often used to breaking down and experiencing time into discrete, enjoyable fragments of entertainment. For instance we can invest half an hour to watch the latest adventures of our favourite television characters. Thus time can be temporally be conceptually broken down into artificially distinct moments—although I acknowledge that all thoughts on time are artificially conceived. Following the writings of Benjamin (1999), one can argue that these distinct moments created by an installation, are received, imagined, disseminated and consumed in a state of distraction. Yet it is distraction that is dynamic and more than mere absent-mindedness, with it also carrying a spatial connotation with dispersive and scattering movements. For Benjamin (1999), the reception of moving images by the spectator is distracted by the shock-effects of the constant and changing images that interrupt the viewers processes of association. How these propositions play through specific examples will now be considered.

Frame 1

The Llanmaes three minute looped installation was originally presented as part of the Welsh National Archaeology Week that ran from 16-24 July 2005, and it attempted to explore the thinking of 10 year old pupils from Year 5 from a local school. The film is based upon their visit to the excavations on a Late Bronze Age and Early Iron Age site with two roundhouses and a mixed layered midden, at Llanmaes in the Vale of Glamorgan, Wales (see Gwilt et al. 2006). Whilst on location the children participate in most aspects of the excavation. They also produced questionnaires and asked the diggers questions such as: why they dig? What they hope to find? What would they do if they found something worth

£1,000? Would you carry it everywhere? Would you pass it on to your kids? The film concludes with the children producing their own material objects and stating what they would like to see preserved for future generations. These items included: mobile phones; Gameboys; a toy dog and a girl's personal diary.

The presentation of the looped film was that of montage and overlay, with the temporal blurring of disparate images, overlaid with the questions and musings of the school pupils. Different shots and scenes were co-joined by the repeated image of an EDM (Electronic Distance Measurement) machine and the surveyor's right ear. It appears that more attention is invested into the aesthetics of the frame, the shot, than the editing than to the context of an overall narrative. One might argue that non-essential frames are added for their mesmerising effects, with the camera sweeping from the EDM to an old rusty tool-box, to peoples feet walking, to some odd bits of pottery laid in a tray and back to the EDM again. This fading and blurring of images, combined with the repetition of the EDM image within a repeating loop created a rhythm punctuated with disruption and continuity as the scenes slowly mutate.

The installation was originally located by the entrance to the museum, and was signposted by two large 'adventure-archaeologists', equipped with all the stereotypes and assemblages that denote an archaeologist in popular culture—here we could see gold, a trowel, a sword, pottery urns, a skull, maps, rugged good looks, and colonial Indiana Jones-esque attire (see Figure 1.3). Adding to what Holtorf terms the 'archaeo-appeal' (2005, 155) of the installation. Attached to his front was general information about the site and the exhibition.

Next to the film installation was presented a more traditional glass display cabinet that featured artefacts from the site and the pupils' questionnaires and interpretative drawings of sections and their experiences. The chairs were arranged in a circular format, and although people were free to alter the arrangement, very few did. The time of day that one visited the installation also affected ones experience, as a certain times the sun shone directly into ones eyes; re-enforcing the notion of changes in time.

Frame 2

The Llanmaes film was displayed on a television that was fixed on wheels. That the installation was mobile enhances Benjamin's notion of exhibition value, where it is not only important that the piece be seen by the public, but also in many places. Creating distractions that incorporate dispersive movements of both information and reception. The location of the installation was also very interesting as it caught the eye of everyone the moment that they entered the museum. This resulted in the spectators that are momentarily fixed to their seats becoming just as interesting to watch (or in my case photograph) as the installation itself (see Figure 1.3).

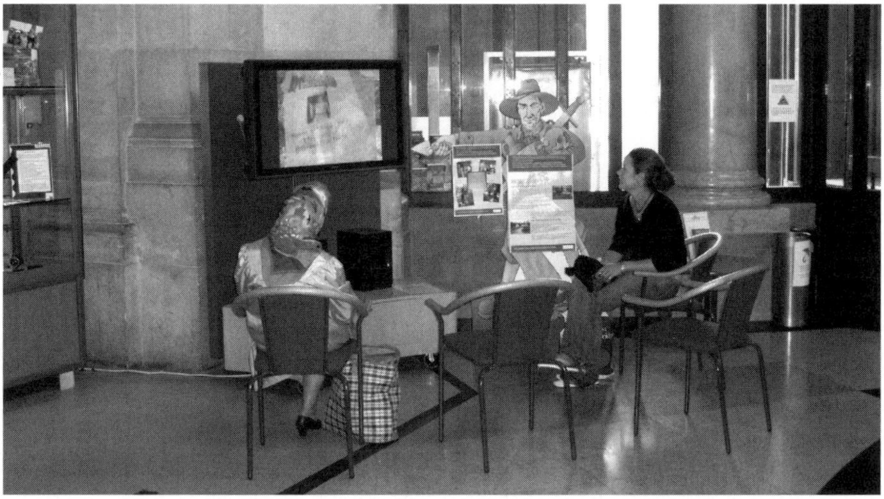

Figure 1.3 Spectators as spectacles in space? Watching the Llanmaes in the National Museum and Galleries of Wales, 2005. Notice the archaeological adventurer in the background

Source: Digital photo by author.

The layering of film looped installation, textual information notices, display cabinets, questionnaires and drawings might be argued to be similar, at some level, to Pearson's and Shanks's Deep Mapping project. As such the combination installation augments and percolates with temporalities and senses of place. The exhibition adds to and comments upon what it presents—evoking rather than describing the excavations at Llanmaes. The installation was later moved to what was then the Prehistoric section of the museum, minus the seats,[4] cut-outs and the school pupils display cabinet. The movement of the installation was interesting as it dramatically altered the dynamics of its deliverance and reception. Whereas before, everyone noticed the installation—now it could only be observed by visitors to the archaeology section.

So what do I think of the installation? Well, from an archaeological background I am interested in the ways in which the film through juxtaposition, attempted to penetration both the past and the contemporary, and how it mixed material evidence with oral accounts of excavators and school pupils. Such mechanisms generated senses of place—in another place.

4 The replacement of seats with wooden church-like benches made the experience more austere, with the linear arrangement restricting ones seating choices. Whilst there, I noticed that fewer people were actual inclined to stop and watch.

Frame 3

The ephemeral nature of the images and of the installation within the museum, combined with the evanescence of the transient museum spectator, for me resonated with the percolation of time and events within life (see also Witmore 2006). The incessant revolutions of the looped image offer the audience the opportunity to witness ambiguities in different instances layered upon layer. You see repetitions before your eyes and are forced to deal with them (see Cochrane 2009)—to use Shanks's (2004) term, it 'augments reality', or more to the point it augments interpretation of a reality, by building upon sensations and thought, as we experience images within a continually shifting present (see also Cochrane 2006, 2008). There can be an intensification of self-reflective viewing, in which one is intent on understanding the images.

Frame 4

The film loop can create frustrations for the spectator, disrupting our comfort, the rapid movement of the changing images never allows one to verify the identity of a particular voice or person. The recycling of images within the loop forces understandings out of a particular context and combined with the repetition of the EDM and the montage blurring and merging of seemingly random images, helps create new juxtapositions, ruptures, distractions, connections, thoughts and dialogues about time and our interactions with it. Ultimately of not being there but seeing more of there.

Frame 5

So what do other people think of it? Well unfortunately most of the general public that I asked disliked it. People seemed to be confused by the voice-overs of the children—for some it was not loud enough and created a white-noise, and for others it just simply did not relate to the images. The lack of linear narrative structure, combined with the repetitive looping of the film also made is difficult for some to understand. One viewer, stated that it 'looked like kids on a picnic', rather than an archaeological dig.

Most people asked wondered 'who' the installation was aimed at—for adults, for children or both? Since few felt that they could relate to it. The lack of archaeological context left many questioning what they looking at, where was it from. People seemed to want hard scientific facts (realism) rather than ephemeral images (expressionism). The lack of images of archaeologists actually digging in the ground, left some feeling that the film was too ambiguous, relying too much on its context in the museum, to portray any information. In essence it could be argued that people were dissatisfied with the montage effect as it failed to analyse and expose particular elements—presenting parts freed from the associations and complexities of their existence outside the film installation. All those asked were displeased with the EDM repetition. Interestingly, the museum security guards reported that very few people actually watched the entire three minute installation from start to finish.

Frame 1

My next example is from the National Museum in Scotland and is entitled 'A Generous Land: About the House'. This looped film installation is presented on three televisual screens as part of a permanent exhibition. The screens are set back from the audience and raised on a plinth and enclosed by walls. The installation is introduced by an information sign, and visitor can listen to a short introductory voice-over if they wish. The wall directly behind it is covered in river clay (see Figure 1.4).

Figure 1.4 'A Generous Land: About the House' film installation at the National Museum in Scotland, 2005

Source: Digital photo by author.

Frame 2

The wall behind the film installation was created Andy Goldsworthy in 1998. The orange clay wall presents the temporalities of change and decay—even though the wall can appear static. That the wall can change was an integral part of its purpose, so that the piece becomes stronger and more complete as it falls apart, fragments and disappears. Goldsworthy stated that this wall was about the anticipation of the moment, rather than the control of the future—for the piece does not crack evenly and one does not know if it will indeed stick. As we can see, in this instance it did.

Whereas the Cardiff installation was influenced by the changing sunlight within the foyer, this installation is permanently illuminated by fixed spot-lights—creating a kind of stasis that is juxtaposed to the continually changing images. The film looped installation is primarily about some people's engagements with a place and how they created different types of dwellings and fixtures. The installation gives insights into the timescales and temporalities involved in the formations of what can be termed the Scottish environment and its people, as well as hints to the various processes leading to the current archaeological record—such as, transience in the Mesolithic period with temporary structures and the limited survival of remains to the more enduring dwellings and structures of the Iron Age that appear to decay more slowly (see Figure 1.5).

Figure 1.5 Scene from 'A Generous Land: About the House' film installation at the National Museum in Scotland, 2005

Source: Digital photo by author.

These transformations are also punctuated with text to further promote understandings of the moving images, but that also split the viewer's attention, dividing it between text and image in a linkage of information. In fact during the entire process, emphasis seems to shift from the three screens, as you are invited to consider the piece as part of a single integrated installation.

Frame 3

The looped film is presented in a linear-chronological manner—starting in early prehistory with the introduction of people into the landscape through to the Norse invasions or migrations—the images convey a sense of the evolution and dynamism of a landscape. As the succeeding frames are very similar to the previous, there is less ambiguity for the viewer to experience than in the Llanmaes example. The evolution of histories through the establishment of relations between the frames assists in the interpretation of the piece for the audience. The translation of the past is augmented, in that the particular frames or fragments of time combine to create a larger image or composition. In many ways it is like the time root or creeper that swells as it grows, moves and gnaws at the future, that Ingold describes (2007). Combined with film immersion, new senses can be imagined; as Stoneshifter proposes, 'we've been well trained to suspend reality, in a cinema, and can enter and leave that state quite easily' (2008). The continual scene movement therefore creates momentary journeys from disorientation to orientation, in a dislocated place to create better understandings of Scotland as a place. It is a luminous window to a now pixelated past.

Frame 4

As this film runs for slightly longer than the Llanmaes one (being over seven minutes long), there is an enhanced sensation that the loop is everlasting—it probably does stop when the museum closes, but theoretically it could run all night long. This continuity can create for the spectator the feeling that one can come and go as one pleases, with the film always playing the same rhythmic loop. At this installation the spectators are able to stand or sit—with most people preferring to stand (see Figure 1.6).

Frame 5

So what did I think about this film? Well, I enjoyed the linear narrative and the ways in which you can watch time unfold within a flow of sustained repetition. What did other people think of it? Well, most people I spoke to loved it and the experiences it generated (especially the people in Figure 1.6). Some commented

Figure 1.6 Momentarily frozen in space at the National Museum in Scotland, 2005

Source: Digital photo by author.

that they found it 'enlightening'; others enjoyed the freedom to photograph it; most enjoyed the linear performance of the piece and felt it contrasted well with the organisation of the rest of the gallery; several stated that they liked being told what happened in the past—definite statements—'it could only be better if Tony Robinson was here' someone exclaimed. Some disliked the black and white images, and one visitor said that the installation generated the question in her head, 'what am I responding to?' So I asked what she thought she was, she replied that he was responding more to the people around her than the exhibition itself. Not everyone is so receptive to films in museums though. For instance, Sam states that her 'main problem with the film (and in fact all the ones I've seen) are that they are an attempt to make museums all modern and exciting, whereas what I actually want is to see more 'stuff'' (2008). For Sam (2008), film installations are a noisy distraction, 'just a load of pictures' that do not add to knowledge or positive experience. The commentator Sam does, however, conclude that she does not really like films in general, but that as her mum likes them in some museums, they cannot all be bad (2008).

Frame 1

> …I, the machine, show you the world the way only I can see it. I free myself for today and forever from human immobility…Thus, I explain in a new way the world unknown to you… (Dziga Vertov).[5]

This discussion with particular case studies has demonstrated that the creation of film looped installations is often a collective and sometimes contested affair. The primary difference between these types of archaeological installation and the more tradition 'static' displays, is that when you watch and listen to a looped film, you have the opportunity to enter into the elsewhere of the moving looped image. This might be termed 'technoetic'[6] immersion (Ascott 2000; 2004) or *techné* (see also Koerner; Russell 2006). This elsewhere could be an imagined past or it could just as easily be daydreams of anticipated ventures—often this can happen whilst one is momentarily still in a particular space. With more traditional installations, you are generally stimulated by the exhibitions to keep walking for longer moments in time, measuring what you see with this movement—creating different types of performance.

Within most film installations, there is continual oscillation between images of easily understood embedded features—such as a Roundhouse or a surveyor's EDM, to meaningless images that just mesmerise, and then back again. Most images in life can affect the same oscillation—but witnessing such changes in museum film installations, I suggest presents the opportunity to think through the processes differently.

Both the film examples created temporalities that brought the ephemeral to centre stage—indicating the change that is always coming. At times, both examples focused on prehistoric Roundhouses; but the Llanmaes one forced people to think in new ways about the past. This is maybe one of the reasons it was less popular than the Edinburgh film, which generally re-enforced or re-played what most people already think they remember about the past. Ultimately I feel that the navigation of these presented film spaces moves visitors beyond representation itself and towards technologies or mechanical arts, that trigger differences and previously unrealisable expressions within museums—ultimately for dynamic interaction within knowledge.

5 Taken from Berger, J. 1972 *Ways of Seeing*. London: Penguin, 17.

6 'Technoetics' is a term used to describe practice that seeks to explore how people engage with different spaces and environments, be it digital, chemical, cybernetic, genetically engineered, musical, mental, theoretical, textual, oral, spiritual and so on—it attempts to dissolve or at least delve into the gaps between such distinctions.

Acknowledgements

Thank you to the anonymous visitors who provided their thoughts and views—thank you also to the members of the Modern Antiquarian Forum for engaging in discussion on the topic of film installation. Special thanks to James Marshall and Kate Waddington for assisting in data collecting, and providing friendship, love and support. Alison Sheridan was kind enough to discuss the Edinburgh film at great length over many occasions. Thank you, as always, to Ian Russell for lively discussions! Cole and Peta Henley were kind enough to put me up for several weeks, over the course of two research trips to Edinburgh. The members of The Research Institute for Humanity and Nature (Chikyûken) deserve many thanks for looking after me and making me feel welcome in Japan. Thank you to Marcus Brittain and Stephanie Koerner for inviting me to participate in the Cork EAA session. A big thank you to Ian Russell and Stephanie Koerner for making this volume happen—respect!

References

Ascott, R. 2000 Edge-life: technoetic structures and moist media. In R. Ascott (ed.), *Art, Technology, Consciousness: mind@large*, 2-6. Bristol: Intellect Books.

—— 2004 Planetary technoetics: art, technology and consciousness. *Leonardo* 37(2), 111-16.

Ballard, J.G. 2006 *The Atrocity Exhibition*. London: Harper Perennial.

Benjamin, W. 1999 *The Arcades Project*. London: Harvard Belknap.

Cochrane, A. 2006 The Simulacra and simulations of Irish Neolithic passage tombs. In I. Russell (ed.), *Images, Representations and Heritage: moving beyond a modern approach to archaeology*, 251-82. New York: Springer-Kluwer.

—— 2008 Some stimulating solutions. In C. Knappett and L. Malafouris (eds), *Material Agency: towards a non-anthropocentric approach*, 157-86. New York: Springer-Kluwer.

—— 2009 Additive subtraction: addressing pick-dressing in Irish passage tombs. In J. Thomas and V. Oliveira Jorge (eds), *Archaeology and the Politics of Vision in a Post-modern Context*, 163-85. Cambridge: Cambridge Scholars Publishing.

—— In press What i think about when i think about time. In J. Savage (ed.), *Depending upon Time*. Cardiff: Safle.

Cochrane, A. and Russell, I. 2007 Visualizing archaeologies: a manifesto. *Cambridge Archaeological Journal* 17(1), 3-19.

Cooley, R.H. 2004 Its all about the *fit*: the hand, the mobile screenic device and tactile vision. *Journal of Visual Culture* 3(2), 133-55.

Cubitt, S. 2002 Visual and audiovisual: from image to moving image. *Journal of Visual Culture* 1(3), 359-68.

Dolinsky, M. Transformative navigation: energizing imagery for perceptual shifts. *Technoetic Arts: Journal of Speculative Research* 7.1, 49-64.

Everett, A. 2004 Trading private and public spaces @ HGTV and TLC: on new genre formations in transformation TV. *Journal of Visual Culture* 3(2), 157-81.

Gwilt, A., Lodwick, M. and Deacon, J. 2006 Excavation at Llanmaes, Vale of Glamorgan, *Archaeology in Wales* 46.

Hamilton, J. 2004 The way we loop 'now': eddying in the flows of media. *Invisible Culture* 8. Available at: <http://www.rochester.edu/in_visible_culture/Issue_8/hamilton.html>. Accessed 4 January 2005.

Hastie, A. 2007 Detritus and the moving image: ephemera, materiality, history. *Journal of Visual Culture* 6(2), 171-4.

Holtorf, C. 2005 *From Stonehenge to Las Vegas: archaeology as popular culture*, AltaMira Press, Oxford.

Ingold, T. 2007 *Lines: a brief history*. London: Routledge.

Martin, S. 2006 *Video Art*. Köln: Taschen.

Pearson, M. and Shanks, M. 2001 *Theatre/Archaeology*. London: Routledge.

Rodowick, D.N. 2000 Unthinkable sex: conceptual personae and the time-image. *Invisible Culture* 3. Available at: <http://www.rochester.edu/in_visible_culture/issue3/rodowick.htm>. Accessed 5 January 2005.

Russell, I. 2006 Imagining the past: moving beyond modern approaches to archaeology, in I. Russell (ed.) *Images, Representations and Heritage: moving beyond modern approaches to archaeology*, Springer, 361-66.

Rutsky, R.L. 2002 Pop-up theory: distraction and consumption in the age of meta-information. *Journal of Visual Culture* 1(3), 279-94.

Sam 2008 Re: Museum Film Installations. Available at: <http://www.themodernantiquarian.com/forum/?thread=50258&message=625717>. Accessed 4 November 2008.

Shanks, M. 2006 Sense of place—matters of resolution and augmented reality. Available at: <http://metamedia.stanford.edu/~mshanks/weblog/?p=98>. Accessed 3 May 2005.

Stoneshifter. 2008 Re: Museum Film Installations. Available at: <http://www.themodernantiquarian.com/forum/?thread=50258&message=625701>. Accessed 4 November 2008.

Turvey, M., Foster, H., Iles, C., Baker, G. and Buckingham, M. 2003 Round table: the projected image in contemporary art, *October* 104, 71-96.

VenerableBottyBurp. 2008 Re: Museum Film Installations. Available at: <http://www.themodernantiquarian.com/forum/?thread=50258&message=626214>. Accessed 6 November 2008.

Witmore, C. 2006 Vision, media, noise and the percolation of time. *Journal of Material Culture* 11(3), 267-92.

Chapter 2
Nation/First Nations: Conflicts in Identity and the Role of Archaeology

Roberta Robin Dods

What of a country composed from the national identities of people gathered through the emigration/immigration policies of governments and where 'multi-culturalism' has been the theme and politically devised policy? Such is Canada—the home of people defined by ethnic origins—French Canadians, English Canadians, Italian Canadians, Japanese Canadians, German Canadians…The archives of their origins, their commencements rest elsewhere—they live in truncated sequentia. Within this country in search of nationhood, however, there are those whose origins lie within its boundaries. These are the people who came to be called the First Nations, these are the original peoples, the indigenous peoples. They stand apart from, but within, Canada and define themselves as Nations on the basis of shared culture, language and territory—their ethnicity. They do not define themselves as Cree-Canadians, Mohawk-Canadians, or through any such hyphenated hybrid term. However, until recently the place and the taking place of their lives has been ordered by those dominant in political, economic, and academic institutions composed by and for the immigrant peoples. Here archaeology has been part of the process of dispatched truth in support of a historical and political dogma. Emerging from this is a 'new' view on the role of archaeology driven by the worldviews of the First Nation Peoples themselves. This movement by indigenous peoples to control the interpretation of their pasts is echoed elsewhere in the world and presents to the archaeologists of dominant groups a serious ethical challenge.

Clifford Geertz's book *Available Light* (2000) on first reading looked interesting. Later it became more challenging with a critical examination of the section where he discussed Canada in the context of the interpretation of the concepts of country and nation. Canada is put in the company of Sri Lanka and Yugoslavia as being (or having been) one of three examples of countries in, what he terms, nation phased tension (ibid.: 237). He deciphers each of our collective agencies and notes for Canada the interplay between the terrain of politics and their complexion…the arrangement and disarrangement of the rifts and solidarities that language, decent, race, religion, and so on generate and the spaces and edges within those rifts and solidarities are so arranged and disarranged…in what he calls, in the Canadian instance, a very large country, unevenly occupied (ibid.: 237-238).

And while Geertz continues to superficially negotiate the language, descent, race, religion spaces and edges (ironically using the words of Conrad Black, the English Quebecer who has become a British citizen to sit in the House of Lords who now sits in a prison in Florida—incidentally an excellent example of the situation of identity elsewhere that I will discuss in some detail shortly) Geertz misses the important ground on which the spaces and edges are actually negotiated and defined—the great divide of us, the people from away, and them, the Others, the Indigenous Peoples—First Nation (FN) Peoples of Canada!

As I have noted elsewhere (Dods 2000), the emerging Eurocanadian[1] worldview marked, however, a borderland as well. This was and remains the edge, not to an expanding frontier as found in the United Stated in the past, but to Eurocanadian defined wilderness[2] where Aboriginal Peoples were and remain shadows in a land of liminality.[3] This liminal place is composed from the political realities of Geertz's nation phase tension where the politically dominant population of Geertz's uneven occupation live in a string of settlements, large and small, along the American border—as south as they can get! It is a wilderness posed to us by Northrup Frye as the '…unvisualized land…' (1971: 201), that causes for Canadians the '…dissatisfaction one feels about a country which has not been lived in: the tension between the mind and a surrounding not integrated with it' (200). Here we have an enduring Eurocanadian political construct of terra nullius in the context of Canadian concepts of a primordial wilderness. This wilderness is the dangerous place of fact and fancy that sustained the political decisions on the 'disposition' of Indians as described in Barron and Waldram (1986) and the establishment and justification of repressive measures against Native Peoples described by Carter (1997). This wilderness is seen in written descriptions of explorers and discussed as a cardinal truth in our literary tradition. McLaren notes the genesis of this approach when discussing Samuel Black's journals on his 1824 Finlay River canoe trip. Here, in the writing of landscape description:

1 I use the term Eurocanadian for the dominant population in the political economy that is Canada. This does not mean that people from places other than Europe have not contributed to this construct—one termed multicultural in its configuration. Rather, this reflects the fact that the political system is constructed on the British parliamentary model, the society is dominated by Euro-centric jurisprudence (mostly British and French) and English and French are the official languages.

2 I contend that the term *wilderness*, first coming from the European context of land outside central control, reached its deepest meaning of domination in the period of colonization. It created the *terra nullius* places, and was an effective description as a political weapon.

3 I challenge you to follow the vector of the H_1N_1 (Swine Flu) virus in the Canadian context if you want to see an uneven landscape of health care and a land of liminality! Consider this in the context of the 1918-1919 influenza pandemic at Norway House (Herring 1994).

> ...the Canadian landscape bursts the bounds...(since)...almost a complete abandonment of grammar [English grammar of course] is needed for the narrative depiction to capture a mood of titanic and indomitable natural power (McLaren 1985: 575).

Margaret Atwood remarked that the 'central symbol...is undoubtedly Survival, la Survivance...' (1972: 32) and this is the survival of 'the land, the climate... (a) spiritual survival...' (33) '...with the human psyche attempting to preserve its integrity in the face of *an alien*, encompassing nature' (McGregor 1985: 5) (emphasis added). It is my contention that ownership but not actual occupation of the place beyond has been supported by traditional anthropological and archaeological investigations. Further, with global warming, the opening of the arctic waters to possible year round navigation and the potential for undersea oil exploration, the extreme north has become the immediate and latest *terra nullius* under contention by nation states. Here is the recent ground on which political agendas of domination are played out.

To understand what I mean in questioning Geertz's ground missed we need to consider two important questions:

- What of a country whose national identity is composed from the national identities of people initially gathered from Great Britain and France, and subsequently from various countries from around the world?
- And is this more so when these identities are superimposed on or used to negate the realities of aboriginal life?

In actuality Canada in its modern form is a nation of people from elsewhere, since the aboriginal peoples are one of those edges not negotiated successfully. Canada as a nation state is composed of people now gathered from all corners of the earth through the emigration/immigration policies of governments. It is a country where multi-culturalism has become the theme and a politically devised programme used as both a pacifying and unifying metaphor. Indeed, multi-culturalism has been touted as the distinguishing feature of Canadian immigration policy and newcomer integration. It is that which made us most distinctly not American. The United States is as diverse in its foreign origins but it has self-defined as functioning within the metaphor of the melting pot, however valid this metaphor remains as we see the great divides of race in the African American and Latino aspects of American life (noteworthy here is the lack of Indians included in such discussions of American socio-political and economic identity and fissures). In Canada any analysis that hyped the ability to keep identities of origin somewhat soothed the actual identity stressed psyche of Canadians who remained in angst over the specific uniqueness that distinguished Canadians from Americans.

Canada was and still is being composed of people from a multitude of cultures from elsewhere. These people are not tied to specific North American land over deep time and it is not rare for them to bring with them their ancient animosities over the land of their ancestors—that land over there, back home. Many old and newcomers know through written records to the day, month, year when first they or their forefathers and mothers landed. In such a context I am an English-Canadian, for like most Canadians from elsewhere, even if ancestry in Canada is through multiple-generations, I hyphenate my national identity to state my ethnic place of origin either through language or actual location. I could also indicate my Irish ancestry but unlike the Scots component of this ancestry this stands aside, until recently, in the North American context first on grounds of religious differences in the English speaking community but mostly, I think, on the basis of British colonial constructs and the marking of the Irish as other,[4] a practiced process repeated with the aboriginals of Canada. The reality of the Irish troubles has played out through the years for those attached to the ancient sod, through the support of the IRA in the 20th century and earlier, in the mid-19th century Fenian[5] raids. These incursions into Canada from the United States were in no small way influential in the pro-Confederation movement that came to the forefront in 1866. I suppose for anyone familiar with Canadian politics this could be called the McGee problem. In more recent history there has been the involvement of the Canadian Sikh community in the movement in India for an independent Sikh state. The most familiar upshot of this for Canadians was the Air India bombing of 22 June 1985 off the coast of Ireland. For most of a quarter of a century we have had investigations, court cases, and memorials none actually bringing any closure.

To continue to a modern context where the English/French debate has played out in the realms of referendums, I along with other English-Quebecers have been described by Jacques Parizeau, the head of the Quebec government in 1995 as les maudits ethnics—'the cursed [euphemism for @#&*] ethnics' after the loss of the last referendum for separation of Quebec from the confederation that is Canada. Here again the play of where we are from not where we are 'at' based on old animosities from lands of origin.

Regardless how any of us non-Native from away people describe ourselves, the archives of our origins as distinct peoples will always rest elsewhere. Our commencements, that we cling to, are in the archaeological records encased in the soils from which the bodies of our ancestors were composed and to which their bones returned. We from away people live in a condition of truncated sequentia in a land we know in our guts, blood and bones, belonged to and indeed really belongs to those we have defined as Other—a distinction made here in the singular rather than the plural. As from away people we live in a plural world and wear our ethnicities as cloaks of identity, emblems of inclusion somewhere. These cloaks are woven from the fabric of history, myth and fancy of elsewhere places. This

4 Consider Jonathan Swift's *Modest Proposal.*

5 Fenians: A secret society founded in Ireland in the 1850s fighting for Irish independence. It expanded into the United States by recruiting Civil War veterans particularly drawing on Irish immigrant populations for membership and financial support.

fabric defines identities, and functions as emblems of community thus constructing metaphors of existence now nurtured and encompassed in the bounded club of Canadian multi-culturalism.

Much of this ties us to and into the Western Tradition that has centred our concerns on an almost uniquely Western intellectual obsession with our past and the past of others that plays out in the field of archaeology. Our metaphors and thus our identities from elsewhere create a sense of entitlement that plays out in the political sphere and impacts directly the terms of existence of a Native populations rooted in a sequences that link through the soil and through indigenous metaphors to the creator(s) and places not situated in our linear continuum of past—present—future but rather in traditional interpretations that link to the seamlessness 'ingredient of being' (Bielawski 1989: 229-30).

The conditions of place for First Nation Peoples are not truncated in the same fashion as those from elsewhere. Rather FN conditions of place have been truncated by another form of displacement caused by our very occupancy and redefinition of their ancestral settings. In the socially constructed landscapes of the colonized, native people were forcibly sent to residential schools where their circles of life were broken, their life affirming metaphors of unity and wholeness were usurped and Eurocanadian 'circle games' were substituted by the 'empty nonexistence of zero', the 'vacuity of circular definitions or circular arguments' (Chrisjohn, et al. 1997: 115) as cultural identity was mangled in the factories of cultural genocide designed by the Government of Canada to eliminate the 'Indian Problem' one way or another, once and for all (Chrisjohn, et al. 1997: 42). Or as Morris noted in 1877 '...let us have a wise and paternal Government...doing its utmost to help and elevate the Indian population who have been cast upon our care (doing)...our duty by the red man' (1877: 296-97) and the utmost was to bring civilization and civilization cannot be pluralized: it does not come in the rainbow flavours of culture. Indeed, here succinctly stated is the 'white man's burden'! Of course First Nation peoples have noted that the actual agenda was the good Indian and of course they sardonically observe the only good Indian is a dead Indian, 'Not dead physically, but dead spiritually, mentally, economically, and socially' (Waubageshig 1970: vi) (Dods 2000).

How is entitlement reinforced in a context of truncated sequentia? Here we can turn to the interpretation for public consumption of various archaeo-historical sites such as they are, and indeed they are marked in the discipline as colonial archaeology or historical archaeology to set them apart from the prehistoric archaeology conducted to investigate the ancestry of the indigenous populations. This distinction made academically through the demarcation of prehistory, proto-history, and history from each other has, essentially, marked those spaces and edges missed by Geertz. Here is the edge that marks truncated sequentia from rooted sequentia thus marking us from them, defining them as Other. In North America it also marks the boundary between the people of the literate tradition and the oral tradition. The latter became devalued and the record in writing became the documents of possession and the canon that interprets conditions of existence.

I want to give you three examples of such 'sites' that are used to bemuse us into the idea that 300-400 years counts more than forever:

- Louisburg is an eastern Canadian site (Nova Scotia) dating from the time of the colonial expansion of France into North America. Destroyed by the English and associated with the expulsion of the Acadians, it has since been excavated and reconstructed to display the fort-settlement and has been the 'set' of an historical drama broadcast on both French and English Canadian television networks. The reconstruction of the site was funded by the federal government and presented as part of a national-historical continuum but not as a rational for an independent Quebec.
- Queenston Heights near Niagara-on-the-Lake (Ontario) has the monument to General Brock (English) who died in the battle that occurred here. He is buried on the site. This battlefield is presented as proof of national identity wrested from the American threat in the War of 1812-1814—a war that both Americans and Canadians claim to have won although it actually was between the Americans and the British. Mythic constructs are formed around this event as evidenced in the Laura Secord and Cow story, a tele of her hike through enemy American lines to deliver valuable information to the Brock troops. This Secord story has been an illustrative short fill in piece used on the state television network, the CBC, and is part of the official construction of national identity. Indeed, the potential return of ownership of the modern Laura Secord Company (manufacturers of chocolate) to Canadian hands has been a matter of comment numerous times in Canadian business reporting!
- Upper Canada Village (eastern Ontario) is a compilation of buildings from various villages and towns now submerged under the waters of the St. Lawrence Seaway. Here is the quaint white clapboard presentation of United Empire Loyalists Upper Canada (Ontario as distinct from Lower Canada Quebec). This is the now composed home of the British patriot refugees from the emerging United States of 1776. It is also the town used by American movie producers as the set for the quintessential 19th century American 'hometown' of Tom Sawyer and Huckleberry Finn of the Mark Twain stories of the same name.

These and diverse other locations such as the Log Farm Site outside the nation's capital of Ottawa, on which I worked, are presented as marks of an emerging identity and landscape/townscape of place in the sense of belonging to the land when history of deep time emphatically insists that this is not the case. So thusly is patched together a country in nation-state identity crisis and the more so as First Nation peoples effectively challenge the authority of the nation-seekers.

There is no nationally known and substantially funded Louisburg equivalent of First Nation interpretation—and, mostly lacking the substantial archaeological ruins such as found in the Anasazi—Hohokam—Mogollon Southwest of the

United States, there are no Canadian equivalents of American National Monument Sites as testament to the pre-Columbian past. In 1979 I was involved in the pilot project Ontario Prehistory Thematic Review and Site Assessment (Latta, et al. 1979) funded by PARKS CANADA. It was designed to identify and define specific prehistoric archaeological sites for marking as historic monument sites. In Canada such a monument consists of a stone cairn with a bronze plaque on which is written a brief description of the reason for the cairn. The report outlined criteria of selection based, usually, on the accident of being the first site of a specific cultural manifestation, essentially the type site, this itself most likely the accident of scant excavation in relation to that very large country, unevenly occupied syndrome defined by Geertz. Ontario was used as the spatial test case and NW Ontario was my area of survey. The research was completed and the report was submitted with well defined and illustrated recommendations. It never resulted in action. I think that the reasons for this are basically two fold:

- First is the fact that so many projects fit our definition of political and academic action but are essentially constructed busy work in a system of advancement based on the reporting of such work—they are make work projects for civil servants and the inmates of academic departments, like myself. In this case archaeological interpretation is for us and does not reach the public generally and more specifically the people whose ancestors we are discussing.
- Second, there is a historical context in which the interpretation of the pre-Columbian past has come to fit. And this interpretation fits our historical constructs and political agendas. Thus such sites are not seen as part of the construction of a national identity.

First Nation Peoples are defined by us and for us as the disappeared past in contrast to us as real and here, although initially from elsewhere. They are indigenous peoples, while I am struck by the fact that the English (but not the Scots and Welsh as they too have been colonized by the English) are not defined as indigenous with respect to their occupation of that sceptred isle off the coast of Europe—you know—just west of France, although England was historically and remains presently very much a multi-ethnic composed nation with significant immigrant populations in much the way of the more recent composition of Canada. First Nation Peoples have been categorized by their linguistic affiliations[6] like Algonquian, Athabaskan, Iroquois, Salish, etc...They are mapped as regionally distinct on the basis of environment and traditional economy and more recently on the territories delineated in treaties in much of Canada (British Columbia being the noted exception). Certainly archaeological work has focused on the lining of these cultural sequences through

6 This is not different from the European experience of language classification as seen in such descriptive terms as Germanic Languages, Romance Languages, or even the various uses of Irish, Manx, and Scottish Gaelic of the Celtic language family group.

space and time—tied thus to specific geographical locations. However, this usually has not translated into positive results in land claims cases in our federal and provincial courts. Examples of territory delineation from artifact types can be found in Eastern Canada where pottery (Iroquois and/or Algonquian) has been used for both diachronic and synchronic distinctions interpreted as cultural sequences and variations in what could be termed ethnicity. Such archaeological categories, Flinders Petrie aside, have a history in the discipline of archaeology in the United States, for example Kidder, Haury and the Galdwins in the Southwest and with later schemes such as McKern's Midwest Taxonomic System. These theoretical and methodological constructs were imported directly into Canada with the cadre of American nationals hired to run our national and regional museums and teach in our major universities in the middle decades of the 20th century. Such archaeology takes its lead from earlier developments in Europe, as Bruce Trigger pointed out to us on numerous occasions but most specifically in his 1980 American Antiquity article. Thus it was (and somewhat remains) tied to traditional institutional forms seen in European museums. The most influential in the Canadian context was the British Museum although this was superseded somewhat by American input and the example of the Smithsonian in Washington, itself not too dissimilar to a compilation of the British Museum, the Natural History Museum, and the Victoria and Albert of London. Here displays showed the exotic of time and/or place for the entertainment and somewhat incidentally, the education of the public. In Canada the public almost invariably was composed of those from away—the elsewhere people and for many years these were white and of European ancestry. Dominant in the system was the National Museum of Man now seemingly politically correct in its new name of Museum of Civilization, until one recognizes the historical-political content of the word civilization and the echoes of Lewis Henry Morgan's ethnical stages (1877) and the genesis of American Anthropology in the 19th century. Here goodies were collected in various ways beyond official excavations, themselves sanctioned by agencies of the away people. A prime example is the confiscation of the masks and coppers used in ceremonies such as the North West Coast Potlatch under what was termed the Potlatch Law of 1884 where violators of Section 3 of the Indian Act were subject to a prison term of two to six months for dancing, singing and the wearing of masks in Potlatch ceremonies. Some of these sacred material culture items disappeared into private collections. Not infrequently confiscated masks came to be displayed in traditional European museum formats. The interpretation of these materials initially was in what could be called the Franz Boas model. Boas, characterized as the grandfather of American Anthropology, thus North American archaeology, true to his European renaissance man roots, accepted the prevailing ethos of 19th-century North American academics that Amerindian chances at a future were nonexistent. In this view North America was a world for us from away people—we were and remain the inheritors of this earth who hold various versions of the myth that we moved onto a continent devoid of human interaction in any meaningful way—a world where remnant Indian populations were considered to be markers of a distant past that we had already superceded, and who as representatives of this past would soon disappear into the tomb of prehistory (Hinsely 1981) as they were savages that had all past and no future (Pearce 1988: 135). It was our obligation, in the Boasian view, to collect as much as we could of these people before they were gone! So totem poles, coppers, and masks for example, could be acquired for display. Masks are particularly cogent in the context of this discussion since it was the collection of West Coast masks held in a German museum that finally captured Boas's imagination to what he termed the great work of gathering ALL on and about the Indian for posterity. The mask—a component of a specific configuration of cultural/social/oral tradition record keeping constructs founded both in the mundane and the metaphysical/metaphorical became displayed essentially as an art piece. Usually a brief description was provided of place and time of collection with one or two short sentences of the supposed meaning of the symbols used. I want to linger briefly on the mask as a metaphor and exemplar of the differences between culture as a performance art in the enacted lives of the people of the culture and the presentation of their past as defined by us for our consumption. This is quite a different process than having the story told, if it is appropriate to do so for strangers, by those whose only truncation from their past is effected by the interpretation of them as Other by people transported from elsewhere. For a mask—such as those of international fame from the North West Coast of North America—is a physical representation of a multiplicity of cultural meanings found in a number of questions not exhausted in the list I am now to provide:

- Who made the mask?
- From what materials and are they significant to the process?
- What is the mode of manufacture and its significance?
- How was it decorated (colour, shape conventions)?
- What are the symbols used and what is their meaning?
- How do these meanings encode specific information of both the transcendental and the mundane?
- Who wears the mask?
- In what context is it worn?
- What do the ceremonies of use mean?
- What is the performance content?
- Does it include song, dance, rhetoric that are part of a wider meaning?
- How is the mask marked as belonging and how is it inherited?
- Does the mask have a 'lifetime'?
- Does it go into mythic time in some fashion?
- How is it remembered or is it to be remembered?
- Indeed, are these questions that we should ask at all?

And then when we are done we still do not really know that much, for the mask has a life in the wearing by a specific individual in the performance art of a specific ceremony in a specific context within his or her culture. To this we are most likely denied *entrée*. So there is a configuration that causes the mask only to have meaning in specific contexts that it is impossible to display in the essentially static constructs of the museum. Making such as the mask into an exotic for our consumption marks it as outside the boundaries of the national identity being composed by those whose identity is still based on categories formed elsewhere, categories that support the institutionalization of things maintaining the development of centralized concepts of national self.

Now here is the rub, if we follow through with the Geertz analysis. He sees culture as symbolic and interpretive—culture as a story or, as I have termed it, performance art. This fits with the post-modern analysis. So culture becomes a story that is never ending in that each interpretation is both:

- Partial—a concept that is very immediate for the archaeologist; and
- Subjective—another concept that is immediate but whose acceptance is problematic for those who believe that science is neither cultural nor essentially less objective than is traditionally espouse in Western Tradition.

Culture being a story that is partial and subjective means that it is contingent not on the truth but on the situation of the teller in a social/cultural construct since it becomes a story based on individual or at most restricted group identity and perspective. Now this is an idea well understood by peoples called Indians—a name itself an accident of European history and the subjective responses of these Europeans to the accident of ending up elsewhere that they expected (Mason 1990). What is understood by each First Nation leader, each Grand Chief, is that culture is the performance art I have named it and that the show is produced and directed for public consumption by and for the dominant group in any nation state construct such as Canada. To the Grand Chief of the day we from away folk are not a disparate set of people from elsewhere but a monolith regardless our hyphenated self-descriptions, a unity whose political identity itself attempts to define the existence and indeed the survival of First Nation Peoples. The irony is that the First Nations are themselves defined as a monolith by the dominant political group. What First Nation Peoples want is the production and direction of their stories, their identities—control of their own show. The institutions that have worked against this have been controlled directly and indirectly by the political sector in Canada through such tools as the Indian Act, the Department of Indian Affairs (DIA), National and Provincial Museums, the Archaeological Survey of Canada, various funding engines of universities and research, in other words, the assorted agencies of the cultural and political hegemony. At provincial levels, where increasingly the federal government has been dumping things for funding purposes, this plays out in the licensing processes and extensive contract work, as seen in British Columbia, designed to interpret cultural landscapes (e.g., culturally modified old trees) while pursuing land claim settlements in the courts and treaty processes all the while attempting to pacify the environmentalists and the forest industry sector of the economy.

Regardless, there arises, in such contexts questions about the moral—indeed ethical constructs of the discipline of archaeology. We have professional codes of conduct put forth by the Canadian Archaeological Association and by the Research Council of Canada. These affirm the dignity of the peoples we study. But they really do not address in a direct way what our role is when we are confronted with the ethos that insists that the place for the past is where it rests—that what we term archaeological materials are truly cultural and as such should remain in the realm of the ancestors—a place of mystery and metaphor not to be disturbed except through the telling/re-telling and perhaps glimpsed through the portal of dreams. Would we not be out of jobs if we followed the not infrequent view of First Nations that dignity of ancestors is maintained by leaving them and their material culture in eternal rest?

This is the edge of space that becomes the true terrain of difference in political representation between a dominant group, regardless of their self-definition as a amalgam of multiple ethnicities, in a politically constructed multi-cultural nation-state, with the oppressed original peoples defined as Other who are kept beyond the edge of the discussion of ethnicity as part of the emerging national identity. It is true that our work can be used to support various land claims being brought to court by specific First Nation groups and this is now not infrequently being

Figure 2.1 Sun Peaks protest, 2004

Source: Photograph by Janice Brilly. Courtesy the Secwepemc.

done by invitation. But as one media defined FN militant, demonstrating against the development of Sun Peaks near Kamloops. B.C., said: if we want to discuss ownership of the land let us return to our points of origin and negotiate nation to nation (see Figure 2.1).

Archaeologists can be of service by invitation but this means we must move from being charmed by the artifact to a true enchantment with the process of support for those who have been defined, even by us, as distinctly Other. This means that we must demure with grace.

Our acceptance that the First Nation peoples have the right to define their own past, on their terms, in their way is a political act on our part as much as it is one on their part. With this comes the ceding of our pose as the occupiers and definers of time and space. Think of the realms of possibility for Aboriginal self-interpretation and the positioning of indigenous metaphors within the constructs of any inclusive nation-identity. Thus FN metaphors become components of political action and solution. This removes us from the position of being cultural wardens—the tools,[7] even indirectly, of the agencies of government and academia. That we have been co-opted into the position of cultural gate-keepers is the pity for archaeology so associated with anthropology in North America although this position has allowed for serious discussion of issues of appropriation perhaps not afforded in European archaeology. Sadly beyond Geertz's analysis of the relationship of the researcher to the field of research and indeed beyond Trigger's historiography, fieldwork operated, and in some instances still operates, as an extension of the colonial/neo-colonial world and the more so when aboriginal peoples live in Forth World conditions in lands of wealth. What I want to emphasize is that, sadly, archaeology has the potential to function in an economy of trinkets and beads to use Geertz's analogy in an apt context, but more so when we offer bits and pieces of ourselves and our projects to various interests of the dominant culture without making substantive contributions to the economic, social, political, metaphorical well being of the peoples whose archaeological resources we consume. For us from elsewhere there is no link between the metaphorical and the real cognated world when we long for place as strangers in a strange land.

References

Atwood, M. 1972 *Survival: A Thematic Guide To Canadian Literature*. Toronto: The House of Anansi Press Limited.

Barron, L. and Waldram, J.B. (eds) 1986 *1885 and After: Native Society in Transition*. Regina: University of Regina Press.

7 Actually I like the feminist Mary Daly's term for any person who is the tool of the patriarchy: *snool* (1973). Of course I would use it here in terms of a specific hegemony. It is so very succinct and economically descriptive.

Bielawski, E. 1989 Dual Perceptions of the Past: archaeology and Inuit culture. In *Conflict in the Archaeology of Living Traditions*, edited by R. Layton. London: Unwin Hyman, pp. 228-36.

Carter, S. 1997 *Capturing Women: The Manipulation of Cultural Imagery in Canada's Prairie West*. Montreal and Kingston: McGill-Queens University Press.

Chrisjohn, R., Young, S. and Maraun, M. 1997 *The Circle Game: Shadows and Substance in the Indian Residential School Experience in Canada*. Penticton: Theytus Books Limited.

Daly, M. 1973 *Beyond God the Father: Toward a Philosophy of Women's Liberation*. Beacon Press.

Dods, R. 2000 *Visions Of Wilderness And The Savage Other*. MS of Guest Lectures at Department of Archaeology, University of Sheffield and Department of Archaeology University of Durham, England winter of 2000-2001.

Frye, N. 1971 *The Bush Garden: Essays on The Canadian Imagination*. Toronto: House of Anansi Press Limited.

Geertz, C. 2000 *Available Light*. Cambridge, MA: Princeton University Press.

Herring, D.A. 1994 There Were Young People and Old People and Babies Dying Every Week: The 1918-1919 Influenza Pandemic at Norway House. *Ethnohistory* 41(1): 73-105.

Hinsley Jr., C.M. 1981 *Savages and Scientists: The Smithsonian Institution and the Development of American Anthropology 1846-1910*. Washington: Smithsonian Institution Press.

Latta, M., Dods, R. and Haley, S. 1979 *Ontario Prehistory Thematic Review and Site Assessment*. Ottawa: Parks Canada Manuscript.

Layton, R. (ed.) 1994 *Who Needs The Past? Indigenous Values and Archaeology*. London: Routledge.

Maclaren, I.S. 1985 Literary Landscapes in the Writings of Fur Traders. In *Le Castor Fait Tout: Selected Papers of the Fifth North American Fur Trade Conference*, B.G. Trigger, T. Morantz and D. Dechêne (eds) Montreal: The Lake St. Louis Historical Society. Pp 566-86.

Mason, P. 1990 *Deconstructing America: Representations of the Other*. New York: Routledge.

McGregor, G. 1985 *The Wacousta Syndrome*. Toronto: University of Toronto Press.

Morgan, L.H. 1877 *Ancient Society: Researches in the Lines of Human Progress from Savagery through Barbarism to Civilization*. New York: Holt.

Morris, The Hon. A. 1877 *The Treaties of Canada with The Indians of Manitoba and The North-West Territories, including The Negotiations on Which They Were Based, and Other Information relating Thereto*. Toronto: Belfords, Clarke and Company Publishers.

Pearce, R.H. 1988 S*avagism and Civilization: A Study of the Indian and the American Mind*. Berkeley: University of California Press.

Trigger, B.G. 1980 Archaeology and the Image of the American Indian. *American Antiquity* Vol. 45: 662-76.

Waubageshig (ed.) 1970 *The Only Good Indian*. Toronto: New Press.

Chapter 3

Nation, Identity and Ideology: *Romanità* and *Portugalidade* under the Fascist Dictatorships

Sergio Gomes

Introduction

The countries of Italy and Portugal underwent quite different processes as national constructs, despite presenting several common aspects during the Fascist and Salazarist dictatorships. In this chapter I will analyse how the regimes established several concepts that became the basis of nationalistic thought, and how these conceptions allowed for the development of a particular type of National Identity. In the Italian case, there was an emphasis on *romanità*, an effort to construct the new National Identity on the ancient values of the Roman Empire. In Portugal, Salazar had promoted the *reaportuguesamento* of the nation, i.e., arguing that the Portuguese needed to learn again how to be Portuguese. In both cases, the political agenda used several historical, anthropological and philosophical concepts in order to create a discourse on the Nation. Although brief, this study aims at connecting some aspects of the history of ideas with political history, in order to understand how those concepts were used by the State policies in to create the Nation they needed, and legitimate their interests within a supposed 'true' idea of the national character.

Nationalism and National Identity

Nationalism and National Identity are often discussed due to their importance in the contemporary world. Nationalism, as an ideology, may provide an explanation for several aspects of human life, and set values that lead people into how to act. As any ideology, it enforces the gathering of several aspects of a community, linking territories, habits, language, and a narrative where those links are explained. This narrative provides people with an image of themselves as an element of a community and allows an organization of interests and projects. Therefore, Nationalism can produce an Identity's speech that would be used by several political projects and which would support the unity that they want to promote. Throughout this process local Identities are summoned in a larger and political Identity: the National Identity.

There are many theories about the rise of nationalistic movements and the way National Identity originates. Anthony Smith argues that it arises from the necessity that communities have for a ground plan which they can develop to defend themselves as a unity. Among the possibilities considered by Smith for that ground plan, he pointed at the ethnic/symbolic aspects as a source of continuity between generations (2001). Ernest Gellner (1983, 1998), Benedict Anderson (2005[1983]) and Eric Hobsbawn (1994[1990]) argued that Nationalism promoted the 'creation', 'imagination' or 'invention' of culture, community or traditions that were summoned in those National Identities. They explained that 'Nationality' is a phenomenon connected with the rise of political movements and new institutions that would promote collective identity as the greatest concept in the Nation. Therefore, National Identity seems to be an essentially modern construct. In this line, John Hutchinson (1992) emphasizes National Identity as a 'moral regeneration' that, in spite of being constructed by politicians and legislators, would be the work of historical and artistic scholars or 'moral innovators,' in Hutchinson's terminology. From a different viewpoint, James Edensor (2002) argues that the construction of National Identity cannot be seen only as a question of will or strategy. He states that, although a process of identifying, it happens in different scales and presents several performative processes.

National identity arises from a conflict between a sense of duty—which offers order, stability and legitimacy—and an unstable and arbitrary chaotic reality. National identity can therefore be understood as a constant task to integrate elements that it seeks to associate with a political unit (Bauman 2005: 26-29). In this article, I will address the way Italian Fascism and Portuguese Salazarism undertook this task. In fact, the nationalist movements that preceded these political projects had already collected raw material beforehand. Moreover, there were always dynamics of production of regional, or even supra-regional, identities whose symbols, monuments and commemorations constituted an integral part of the collective memory. Therefore, their work consisted in adjusting those elements to the specificity of the political scheme they wanted to develop. In the case of historic monuments, for example, politicians prefered to define their historical meaning from their own version of the past, rather than investigating its historicity. As explained by Marc Guillaume 'a monument [or the group of elements that form the National Identity] cannot lie, since it escapes language, of which the lie is a constituent dimension (…). But, correlationally, although having no univocal meaning, a monument can support different fictions of the past [or any kind of Nationalism]' (2003 [1980]: 143). In the ordering of historical, ethnographic or artistic elements, Fascism and Salazarism would promote a discourse based on their authenticity. Their approach was based on an essentialist concept of culture, seen as an isolated and predestined development of an essence. In this line, the authenticity of the available elements was connected with the way that an element expressed that essence. And so process for the identification/selection of elements was dependent on the political aims of its promoter. As a matter of fact, the emphasis on the *romanità* or the *reaportuguesamento* were policies that held defined political projects, whose legitimization rested in a self-promoted National Identity. Thus, their versions of national identities were merely the reflection of their ideology.

National Identity under Fascism and Salazarism's Nationalism

In 1922 in Italy, and 1926 in Portugal, dictatorial regimes came to power in the aftermath of distinct scenarios of instability and popular discontent: Fascism and the *Estado Novo* (New State). These political events were reactions of some social and political movements that used nationalistic values to contest democratic, parliamentary and liberal principles. They witnessed the rise to power of political groups that advocated a line of historically-driven Nationalism developed during the 19th century in connection with Romanticism and opposing the democratic Nationalism associated with the French Revolution (Rémond 1994: 239).

Between the Congress of Vienna in 1814-1815 and the eve of World War I, national sentiment was invoked in the discourses of political projects from both the revolutionary left and the traditionalistic right wing. However, the political and social transformations that took place in Europe during this period would result in the establishment of a tighter relationship between Nationalism and the conservative right. This relationship was due partially to the evolution of leftist movements. In fact, the birth of a working class conscience and the diffusion of socialism (internationalist by nature) would result in a perspectivation of the nationalistic spirit as an element of the capitalistic system and bourgeois domination (ibid.: 241-245). However, one must emphasize that it is in the assimilation and adequateness of national sentiment by disparate political movements that one witnesses its complexity. As it is argued by Guibernau:

> While the Nationalism of the French Revolution focused on a political dimension, accentuating equality among men (women were not yet included at the time) and popular sovereignty as the only way to legitimize the power held by their rulers, the ideas of the German romanticism added a new strength and a new character to Nationalism, pointing out the common language, blood and territory as elements that constitute the *Volk* (Guibernau 1996: 66).

At the end of the 19th century, international relations between European countries become tense as a consequence of the multiple conflicts triggered by territorial aspirations within and outside European space. It is in this context that we find the development of several political doctrines that were based on an authoritarian model and legitimized in the defence of a nationalistic ideal. The French case exemplifies this evolution: as a consequence of their defeat in 1871 and the territory loosed to the German state, the universalistic Nationalism of 1848 was replaced by Boulangism, Barres' or Maurras' thought.[1] These discourses of denouncing the nation's enemies as decadent, together with the suggestion of isolationist and xenophobic measures, pre-shaped the fascist movements of the 20th century (Rémond, ibid.: 244). After World War I, national spirit stills the engine of a process of mediation between governors and constituents. In fact, when in 1922 Mussolini proclaimed that:

> We see in Rome the preparation of the future. Rome is our myth. We dream of a Roman Italy, that is, a wise, strong, disciplined and imperial Rome. A great part of the immortal Rome spirit is reborn with fascism: Roman is the fascio, Roman is our organization of combat (quoted by Díaz-Andreu 2003: 35).

He uses the national spirit as a tool to legitimize his political project. By invoking continuity in time and differentiation from others, Mussolini produces a discourse based on sovereignty, territory, history, culture, and social *imaginarium*, in order to connect its meaning to the elements that constitute the Nation-State. In the same sense, in 1936, ten years after the 1926 revolution, Salazar declares that:

> Without fear we place Portuguese Nationalism at the indestructible base of the Estado Novo; first, because it is the most clear imperative of our History; secondly, because it is a most valuable factor of progress and social elevation; and thirdly, because we are the living example of how the sentiment of country, through the action we exerted in all continents, has served the interest of Mankind (quoted by Melo 2001: 46).

This extract from one of Salazar's discourses offers four fundamental aspects regarding the conception that dictators had of Nationalism. Firstly, Nationalism was seen as an imperative of the history of the nation. This connection would lead to a particular interpretation of history understood as a cyclic continuum, where a nation rises and falls down, and is inspired by Vico's Cyclic theory of history. Secondly, Salazar also affirmed that Nationalism is a factor of progress, but not only from material perspective, since the growing of Portuguese national culture was understood as an essence. The progress and social elevation were reflexes of

1 Charles Maurras (1868-1952) was a French politician who defended the Nation as a primary value. He claimed that France should have a strong, authoritarian and centralized power, be a Christian Roman Catholic community and practice a discrimination policy based on racial, religious and xenophobic principles. Maurras' ideas would spread throughout Portugal during the first decades of the 20th century and lead to a social-political movement called *Integralismo Lusitano* which presented the same ideas (being less radical in terms of anti-Semitism). They just defended that Semitism or Reform movements were a threat to the Church and traditions (Marques 1981: 274-77). Among the group of persons that defended those values, Oliveira Salazar should be highlighted as he would maintain many aspects of Maurras' thought once he reached the power (Pinto 2004:44).

the development of that essence, thus using Herder's conception of culture. Thirdly, the sentiment of the country is also one of the main aspects to retain from this kind of Nationalism, for it is the expression of the Fichte's organic concept of nation and the nation's right to self-determination. Finally, he mentions that Portuguese Nationalism served the interest of Mankind, leading us to another central aspect of the Salazarist or fascist paradigm, as in both cases the growing of the Nation is a mythical civilized mission that conveys the salvation of mankind. All these aspects are relevant in order to understand the kind of National Identity that each dictatorship produced. In the following sections these aspects will be developed in an attempt to explain the way in which they combine in the construction of National Identity.

Cyclic Theory of History

This is the new consciousness; our race's new masculine and warlike pride which, thanks to fascism, has brought back the Roman tradition. Here are the fascist battalions marching past: and here, in perfect marching order, we can see the most beautiful, noble and generous aspects of our people. ('Superba dimostrazione a Milano', Il *Popolo d'Italia* (4 October 1921) quoted by Gentile 1990: 244).

This extract is the description of a fascist march in 1921 presented in the Italian fascist journal *Il Popolo d'Italia*. Gentile presented this description concerning what he considered one of the main aspects of fascist rhetoric: 'the myth of the new fascist state, in terms of it being a 'moral community' founded on a common faith, which united both diverse classes and generations in the cult of the nation' (ibid.). In this way, it should be pointed that *romanità* was a key element in the establishment of a consensus that the fascist movement needed to promote an internal cohesion. The cult of *romanità*, which was deep-rooted before the rise of fascist movement, favoured the 'coalition' between the right-wing (Catholic) intellectuals and fascism (Visser 1992: 10). This coalition, that led to 'years of consensus' during the 1930s, was based on the belief that fascism would revive the spirit of the Roman period to modern Italy, and would bring about a new *Golden Age*. In this sense, invoking the Roman warlike tradition, and making a comparison between the ancient legions and the new battalions, fascists presented themselves as the new cycle in which the Roman spirit could rise.

This cyclic view of history can be found in Vico's work, as he proposed that history could be divided into periods of flourishing and decadence. This and other aspects of his work would be studied and promoted in the early 20th century in Italy by Benedetto Croce (Berlin 2006a [1969]: 81), and would witness the erroneous use of some of his ideas by Mussolini's regime and, consequently, of Vico's thought. Note that, as Visser refers, 'corrupting Croce's maxim that all history is contemporary history, they [the fascists] freely presented aspects of classical history as metaphors in order to prove the historical necessity and inevitability of Mussolini's 'Roman Empire', often conceived as the final stage of the *Risorgimento*' (1992: 12). Vico's *corsi* and *ricorsi* can be put into perspective as the dynamics that underlie the historical narrative publicized by Italian fascism. In fact, as the extract above exemplifies, fascist Italy arises as a promise of a new *Golden Age*, as the movement grants Italy's salvation from the mediaeval political atomization or from communism, which threatens the nation's unity (ibid.).

It is in this same line that in Portugal, on 1 December 1909—before rising to power—while celebrating the *Restauração da Independência* (the Reinstallment of the Independence) of 1640, Salazar declared:

> No! Portugal must not die! Portugal must live for the worlds it has discovered, for the nations that it has marveled with the scintillations of its grandeur and heroism. There are no new worlds to discover, nor strange nationalities to make wars with: but there is a grandiose work of peace to make, there is the formation of citizens that are as good twentieth century Portuguese as those of the seventeenth century (quoted by Henriques, Mello 2007: 39).

In this speech the dictator invoked two *Golden Ages* of Portugal's past—the discoveries and the war of independence with Spain in 1640—just to oppose the greatness of the past against the social, economic and political instability that characterized the country at that time. So, he stated that, if people wanted to live again in a flourishing period, they must be like the citizens that fought against Spain in the 17th century, but this time as part of a project of peace and development. Thus, Salazar refers to a new cycle as opposed to the instability that Portugal was suffering. By mentioning those flourishing cycles of the Past, the dictator was able to depict the situation which preceded his regime as a period of decadence. Therefore, memory arises as a fundamental instrument in the confrontation between him and the degeneration that he had to face. Flourishing cycles are used as evidence that his promises can be fulfilled and, consequently, the Past becomes an argument which legitimizes political practice (Cunha 1994: 116). As an example, by recalling the discoveries, Salazar legitimized his colonial policy, just as Mussolini had recollected the Roman Empire to support his Mediterranean policy.

The Nation's Spirit and the Essentialist Concept of Culture

In the cyclic view of the past, the nation's spirit, that is, a coherent set of values and attitudes repeated throughout the Nation's *Golden Ages*, is exalted. The past is seen as a sequence of episodes of flourishing in which perennial elements are expressed as those elements that characterize the people and confer resistance to the nation as a political, territorial and cultural unit. This *expression of the nation's spirit* constitutes one of the key points in the thought of the 18th-century German philosopher Johann Gottfried Herder. Isaiah Berlin considered him the father of cultural Nationalism (Berlin 2006b [1998]: 28-29) stressing in his thought two fundamental ideas of Nationalism: the 'expressionism' and the 'sense of belonging' (Berlin 1999: 58-62). Herder understood that one of the fundamental needs of a human being is to express his nature and to communicate with others, and in this activity reference points are used—namely territory and its relationships to other elements of the community—that define his culture. From this perspective, each object, although made by a particular human being, carries within itself elements of its entire community—or its entire nation—allowing that community to be reflected in that object and other communities to identify it in its differences and singularity. 'The whole notion of being at home, or being cut off from one's natural roots, the whole idea of roots, the whole idea of belonging to a group, a sect, a movement, was invented by Herder' (ibid.:60).[2] This essentialist view of culture [as something emerging from within a community conferring its uniqueness and enhancing its singular character] is one of the key characteristics of National Identity promoted by Mussolini and Salazar. To this effect, on the preface to volume I of his discourses, Salazar defends the *reaportuguesamento* which is an appeal that he made to Portuguese people to become authentic Portuguese:

> I do not even know in what manner the work of *reaportuguesamento* of our social and political institutions, and the cult of good, sane, fertile national traditions, so suited to confer us with originality and character, will surpass great difficulties and not be preferred to the submissive copy of all that is thought and done abroad, the maximum inspirational source of our activity for a long time. Moreover, this effort is a homage to the creator spirit of the Lusitana race, to its initiative power, which shall bear fruits if the persistent work of 'inner' discovery does not give in to the lazy imitation of strange creations (quoted by Henriques, Mello 2007: 42).

In this extract we find two features that have often been remarked on regarding the way these political regimes articulated the past with National Identity: the cyclic vision of the past and Herder's essentialist perspective of national culture. The word *reaportuguesamento* considers that Portuguese people were far from being authentic Portuguese in the way they needed to understand and act according to their cultural essence. In the same way, the crisis which Portugal was living in—the decadent part of the cycle—was a consequence of the forgetfulness relative to the '*creative spirit of the Lusitana race*'. Therefore, Salazar argues that if they were to achieve a better situation, Portuguese people would have to be 'real Portuguese' again. It should be noted that Salazar also defended the notion that it is not by 'the lazy imitation of strange creations' that this could be done. What people should do is look towards the essence of being Portuguese—the essence of the nation—and expand that essence in order to reach another flourishing period.

---

2 To this effect one must note that, as pointed out by Isaiah Berlin, these aspects were already part of Vico's thought (Berlin 1999: 60).

This essentialist view of national culture would have some consequences in the way monuments would be seen under those dictatorships. The will to materialize that essence was reflected in the destruction of all elements which might be considered as dangerous towards that essence. Monuments were treated in terms of their historical significance (a the significance that the regime defined) rather than through their historicity. In fact, 'in restoring monuments [or restoring some aspects of a monument], a dictator seeks to link his regime to what is perceived to be a glorious past, one in which he seeks a connection and from which, in forging a connection, he gains legitimacy' (Galaty and Watkinson 2004: 3). This explains Mussolini's urban reforms in Rome during the 1920's as he proposed that:

> All that arose during the times of decadence must be put aside and the millenary monuments of our history must reappear in their splendid isolation (quoted by Guillaume ibid.: 142).

Concerning fascist urbanistic interventions, it should be pointed out that the regime had not a consistent project to the city, instead there was an opportunistic and political use of several ideas coming from the *Instituto di Studi Romani* (Roman Study Institute) which defended a Roman character to Rome (Visser ibid.: 11). These projects would also provide several archaeological excavations, that within the regime's discourse would be promoted as 'symbolic archaeology, inspired by the mythical attraction towards a 'sacred centre' and a desire to come into contact with its magical power' (Gentile ibid.: 245). Mussolini himself had participated in the inauguration of the Forum's excavations delivering the first blow of the pick-axe (Guidi 1996: 114). This focus on a Classical Roman environment can be seen as a reflex of a political use of a scholarship's essentialist concept of culture. As a matter of fact, this was also one of the strategies that fascists used in their consolidation in the colonies. Thus, as pointed out by Visser, 'this 'Roman' type of colonialism was closely linked with an ideological concept of Italian culture as the vanguard in the defence of European culture threatened by 'oriental' decadence (the Ottoman Empire) and 'Gothic' materialism (Protestantism and socialism/communism)' (ibid.: 7).

Organic Concept of Nation

As Salazar argued in the last quotation, the nation's salvation—its regeneration—would have to involve the *reaportuguesamento* of Portugal, meaning the '*inner discovery*' of *portugalidade* or, as he previously mentioned, '*the formation of citizens that are as good twentieth century Portuguese as those of the seventeenth century*'. Salazar's reference to this date, and the emphasis placed on the celebration of Portugal's independence during the regime, is a symptom of another aspect of the nationalistic thought that underlies Fascism and *Estado Novo*—the right to self-determination.[3] This aspect is rooted in the thought of the German nationalist Johann Gottlieb Fichte. In fact, Fichte, in line with Herder, would interpret the German language and spirit as unifying aspects for human groups, which would thus have legitimacy to claim political independence (Díaz-Andreu 2000: 159).

The nation was understood as a living organism, as well through relationships between the *collective being* and the *individual*. As is explained by Isaiah Berlin, Fichte argued that everything about an individual is an extention of the community in which he lives; thus, he forms with it an organic unity—the nation. Like with other organic beings, the nation has its own purpose, has its own will, namely, the will to be free from other nations (ibid.: 91). This 'will of the nation' allows the development of Nationalism as a process of collective identification. In this process, culture and history are the work and the instrument of a right to self-determination. Therefore, although cultural unity allows a political argument for demanding independence, the nation contemplates 'an internal solidarity that stems from its difference in relation to other systems of the same type. This internal solidarity is not essentially cultural (as it is for the word *pátria*), but is above all political' (Gil 1989: 285).

In political discourses, this dimension is legitimized in identity, cultural and historical terms. Figure 3.1 is symptomatic of this situation, before the undeniable cultural differences that characterize the Empire, Salazar overlays the union of the different regions as an agglutinating element, as is suggested by the national flag behind the three multi-continental children. Simultaneously, while alleging to the universal and civilizing spirit of Portuguese people, and expanding the nation in territory, the political project of the nation is legitimized into an anthropological and historical project. It is in this sense that Salazar would defend:

> Portugal constitutes a whole with its colonies, due to a political thought that has become political reality over time (quoted by Lira 2000: 3).

Fascism also used this nation's organic conception. Figure 3.2 shows a propagandistic flyer in which Mussolini's facial profile is shaped within the crowd, as to express the communion between leader and people. This message is reinforced at the bottom of the flyer by the slogan: One Heart, One Will, One Decision. In the image, one can also read the organic vision of the society promoted by the regime: the leader overlays an atomized group of individuals, giving it

---

3 Independence is celebrated on 1 December. This commemoration goes back to the re-establishment of Portugal as an independent country in 1640 after 60 years of Spanish rule. Note that on this date it is precisely this right to self-determinantion that is celebrated, that is, in a more juditial assertion of the term, Nation is celebrated. In fact, on 10 June the Nation is celebrated as well, however, as the designation of this bank holiday implies—'Day of the Race' during the *Estado Novo* and 'Day of Camões and the Portuguese Communities' after the deposition of the fascist regime—the motif for the celebration has more to do with Portuguese culture and communities.

Figure 3.1 *Estado Novo*'s school figure of a 1930s primary school reading book

Source: Reproduced courtesy of www.oliveirasalazar.org.

cohesion as a unity, that is, fighting liberalistic atomization in defence of the nation as an organic unit. Concerning this conception of the nation, Visser emphasizes the importance of Alfredo Rocco's[4] organic state theory published in *La dottrina del fascismo e il suo posto nella storia de pensiero* in 1925 (ibid.: 11). In this proposal, the corporation's fascist politic was seen as the final stage in a tradition of thought that begins with Plato and Aristotle, and was present in Thomas Aquinas' philosophy and in the work of Dante, Machiavelli and Vico. To Rocco all these authors defended a model of the state where the will of the state or the spirit of

4 Rocco was one of the most important Fascist intellectuals. He started his political career in the Nationalist Party and then joined the Italian Fascist Party. In Mussolini's government he was the main responsible for the legislative programme, particularly its repressive legal code.

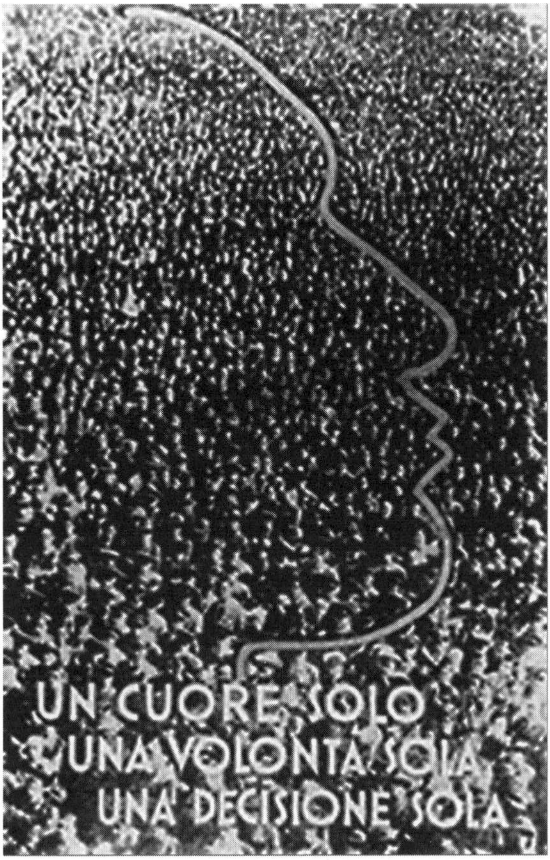

Figure 3.2 Political flyer used by the Fascist Party during the 1920s
Source: Reproduced courtesy of Fernanda Viana (2003: 42, Fig. 17).

the nation took the place of individualism. According to this perspective, Fascism brought to modern Italy the *romanità* values,where politics based on collective interests also expressed an organic state conception (ibid.: 12).

Concerning Rocco's perspective on the history of the spirit of *romanità* there are some aspects that should be emphasized because they also express the importance of the essentialist concept of culture used by a scholarship that supported fascism. As Visser pointed, Rocco made an opposition between the organic state model of the Latin authors and what he called the mechanical/atomical state conception. According to him, these last ones were typical from 'Germanic Culture', which would be seen as a major threat to *romanità* (ibid.: 12). Considering this, it is interesting to note that I used Fichte, a Germanic romanticist philosopher, to explain this organic concept of nation. In the same way, I selected Herder's essentialist conception of culture intending to systematize how politicians had conceived their culture in their speeches. In fact, when Rocco had associated those atomical models of state to 'Germanic Culture' he was thinking about those countries that were corrupted by Reformation, 'natural law', the Enlightment, The French Revolution, liberalism, socialism and parliamentary democracy (ibid.:11). Probably Rocco had known Herder and Fichte's political thought but his concern with the uniqueness of *romanità* had led him to ignore them and, in spite of exploring its similarity with some aspects of Germanic romanticist culture, he had preferred to made connections with several Latin philosophers to show that all the values of *romanità* had specific Latin values. Thus, he had created two separated cultures, in essence connected with two geographical unities and had proposed two different histories to each other. Although the arguments expressed in this paragraph should be taken as a mere hypothesis that calls for more study on Rocco's work, considering its validity it should be pointed out that Rocco, an Italian Fascist scholar that had worked in Mussolini's government, was using in his research a form of conceptualizing nation, culture and history that fascist propaganda had adopted.

The Mythical Vision of the Nation's History

The combination of an essence concept of culture and an organic concept of the nation, provides political groups with the arguments they need to validate their territorial expansion (in the Italian case) or maintenance (in the Portuguese example) since every nation has this kind of doom that is inscribed in its essence. In fact, the identification of that essence in several regions (in the Mediterranean world in the Fascist colonial policy) was used as an argument for territorial expansion. In the same sense, the idea of a nation's fate provided by the Divinity, as found in this discourse, was also used to support the territorial purposes of these regimes (this argument was adopted by Salazar who had presented the colonial policy as the last moment of Portuguese's civilizing challenge mainly related to the spread of Christian Catholic faith). In this manner, the nation's destiny is a sacred project in which all citizens must participate because their salvation depends on that of the nation. In fact, when a nation becomes the Divinity's project, its singularity becomes a universal value and the interests of the nation are no longer the interests of a particular unity but Divinity's will.

This universalism of the nation in these regimes is illustrative of the development of Nationalism through the years. Universalism is a characteristic of the revolutionary Nationalism of the 18th century in opposition to the traditionalistic movements of the 19th century. However, the synthesis these regimes had operated about the Nationalism, juxtaposes the poles of this opposition associating the particularism of the Nation with its universal destiny. In this kind of argumentation, the cyclic theory of history is used to show that this universal project—the salvation of Nation—is possible and that the past must be an example for the future and, because of that, the citizens must accept the authority of the dictator because he is the heir of the spirit of those past time and knows the path to the Nation's safekeeping. Consequently, these dictatorships provide the Nation with a mythical sphere that would be the source of its legitimating strategies. As José Gil explains:

> this logic of the discourse creates the necessary temporality to consolidate the identification of individuals and groups within the Nation and to justify the nationalistic action. The «moments» indicated lead to a final stage that reiterates the mythical time 'before the fall', before all is threatened; thus, a temporality is forged, uniting real time and mythical time, which acquires a mythical configuration: the present (real) must warn of the future (real) that shall realize the potential inscribed in the original myth. The last stage, in which reality joins myth, represents also the 'moment' in which the end joins the beginning. Real history should, thus, be interpreted, perspectivated according to the logic of the nationalist discourse, so that the sacred history of the Nation takes shape and so that the historical time may identify with the mythic time. Then, the Nation will be saved, because it will be completely fulfilled (ibid.: 299).

Regarding the mythical approach to a Nation's history, it should be noted that emphasis was placed on Fascism as a new kind of religion, in the way that it provided a political experience based on rituals and commemorations as a sort of liturgy which acted as a sort of initiation and political integration. Giovanni Gentile (ibid.) designated this phenomenon as a process of 'sacralisation of politics' which consisted in the form of politicians using the nationalistic ideology and turning it into a Nation's cult. In this process they created a type of citizenship based on the cult of the nation's past, martyrs and heroes that had acted as a way of promoting social adherence to political projects. The fascist slogan 'believe, obey, fight' summarizes this strategy of making people come together in rituals and celebrations and join them under the government's proposal. In the words of Giovanni Gentile:

> fascism was the first totalitarian nationalistic movement of this century which used the power of a modern state in an attempt at bring up millions of men and women to the cult of the Nation and the state as being supreme and absolute values (ibid.: 248).

Conclusion

The Fascist movement created a type of citizenship that, instead of having a mere juridical form of defining people, sought to provide people with a way of living. In fact, if the State is a juridical institution, and if Nation is the key element of an ideology, the Nation-State is a juxtaposition whereby these aspects are joined to create a juridical-ideological community that supports each other. When ideology is spread through all the community's activities it becomes a representation of all human dimensions. This step, as pointed by Hannah Arendt, is the moment when the idea becomes a premise allowing reality to become the viewpoint which is provided by ideology (2004 [1948]: 622). In this scheme, ideology acts as a total explanation—'the total explanation of the Past, the total knowledge of the Present and the safety prevision for the Future' (ibid.: 623). In this sense, the Nation, with the key-element of Nationalism as an ideology, becomes the source of explanation. Even Divinity, which by definition would be the highest idea to support all kinds of explanation, was shown as a supporting element of Nation's purpose. This 'total explanation' trend would allow the development of a National Identity that its not only a juridical, historical or cultural form of being but, as it was a consequence of an ideology that seeks to interpret everything, would be a Total Identity.

To conclude, it should be emphasized that such identities aimed to establish a quiet form of living, a reference imposed on the plurality of subjectivities, and that aimed to erase the unquietness of life in the community. Moreover, it is through the concerns of community life, that scholars and politicians develop a model of perfection, and which they legitimate through a fusion of time scales: a continued spread of an essence, yet composed of things from different times, that builds the real way to be Portuguese or Italian. This is the manifestation of a desire for unity, for harmony, a desire for an absolute where everything makes sense. The fallacy of a belief that each action is surely the correct way of being in the world, an act that exceeds the contingency in which we are placed, and where we feel assured of its perfectibility by connecting it to other scales of time (e.g., it has been done in this way through the ages). At the core to this way of giving meaning to past materials, is a regime of the hereditarity—or an ideology based on inheritance—that allows for the idea of the possibility to cross time in order to show the ancestral origins of any Present, hiding the disjunctions in which people live their lives. It is at this point, that we should ask *how* the study of the past is able to unpick the processes that underpin those practices, so that it becomes a resistance to ever new modes of tyranny.

Acknowledgements

I started my research on this article during my stay in the Department of Archaeology, Durham University, during the months of May and June 2007; this stay was funded by Fundação para a Ciência e a Tecnologia concerning my PhD programme. I would like to thank Dr Margarita Díaz-Andreu for her guidance during that period and for the encouragement both then and afterwards to write this article and for her many suggestions made to drafts of this article. I gratefully acknowledge Susana Oliveira Jorge, my PhD supervisor, for her continual support. I'm also grateful to Beccy, Andy, Ana, Irene, Francisco and Susana for their support while I was in Durham. My deepest gratitude to Lesley McFadyen for her accurate comments on this chapter. Finally, I'm thankful to Stephanie Koerner and Ian Russell for giving me the opportunity to participate in this volume.

References

Anderson, B. 2005 [1983] *Comunidades Imaginadas—Reflexões sobre a origem e a expansão do nacionalismo*, Lisboa: Edições 70.

Altekamp, S. 2004 Italian Colonial Archaology in Lybia 1912-1924 in M.L. Galaty and C. Watkinson (eds) *Archaeology under Dictatorship*, 55-71, New York: Academic/Plenun Publishers.

Arendt, H. 2006 [1948] *As Origens do Totalitarismo*, Lisboa: Dom Quixote.

Baczko, B. 1985 Imaginação Social in Fernando Gil (coor.) *Enciclopédia Einaudi* Vol. 5 *Anthropos-Homem*, 296-332, Lisboa: Imprensa Nacional—Casa da Moeda.

Bauman, Z. 2005 *Identidade*, Rio de Janeiro: Jorge Zahar Editor.

Begg, D. 1999 *The Roots of Romanticism*, London: Pimlico.

—— 2004 Fascism in Desert: a microcosmic view of archaeological politics in M.L. Galaty and C. Watkinson (eds) *Archaeology under Dictatorship*, 19-31, New York: Academic/Plenun Publishers.

—— 2006a Um dos mais audaciosos inovadores da História do Pensamento Humano in I. Berlin *O Poder das Ideias*, 81-98, Lisboa: Relógio d'Água.

—— 2006b O meu Itinerário Intelectual in I. Berlin *O Poder das Ideias*, 19-36, Lisboa: Relógio d'Água.

Cunha, L. 1994 *A Nação nas Malhas da sua Identidade: O Estado Novo e a construção da Identidade Nacional*, Braga: Universidade do Minho.

Díaz-Andreu, M. 1997 Conflict and Inovation—the development of archaeological traditions in Iberia in M. Díaz-Andreu and S. Keay (eds) *The Archaeology of Iberia—The Dynamics of change*, 6-33, London/New York: Routledge.

—— 2000 Cultura y Nácion: una mirada historiográfica in M. Díaz-Andreu *Historia de la Arqueología.* Estudios, Madrid: Ediciones Clásicas.

—— 2003 Arqueología y Dictaduras: Italia, Alemania y España in F. Wulff Alons and M.A. Martí-Aguilar (eds) *Antigüedad y Franquismo (1936-1975)*, 33-73, Málaga: Centro de Ediciones de la Diputación de Málaga.

Díaz-Andreu, M. and Champion, T. 1996 Nationalism and archaeology in Europe: an introduction in M. Díaz-Andreu and T. Champion (eds) *Archaeology and Nationalism in Europe*, 1-23, London: UCL Press.

Edensor, T. 2002 *National Identity, Popular Culture and Everyday Life*, Oxford/New York: Berg.

Fabião, C. 1989 Para a História da Arqueologia em Portugal, *Penélope. Fazer e desfazer história* No 2, 10-26.

—— 1996 Archaeology and Nationalism: the Portuguese case in M. Díaz-Andreu and T. Champion (eds) *Archaeology and Nationalism in Europe*, 169-178, London: UCL Press.

—— 1999 Um século de Arqueologia em Portugal I, *Almadan* IIª Série No 8, 104-132.

Fabião, C. and Guerra, A. 1992 Viriato: Genealogia de um Mito, *Penélope—Fazer e desfazer a História* No 8, 9-23.

Galaty, M. and Watkinson, C. (2004) The practice of Archaeology under Dictatorship in M.L. Galaty and C. Watkinson (eds) *Archaeology under Dictatorship*, 1-17, New York: Academic/Plenun Publishers.

Gellner, E. 1983 *Nations and Nationalism*, Oxford: Blackwell.

—— 1998 *Nationalism*, London: Phoenix.

Gentile, E. 1990 Fascism as Political Religion, *Journal of Contemporary History*, Vol. 25, No. 2/3, 229-51, London: SAGE.

Ghirardo, D.Y. 1996 Citta Fascista: Surveillance and Spectacle, *Journal of Contemporary History*, Vol. 31, No. 2, Special Issue: *The Aesthetics of Fascism*, 347-72.

Gil, J. 1989 Nação in F. Gil (coor.) *Enciclopédia Einaudi* Vol. 14 *Estado-Guerra*, 276-305, Lisboa: Imprensa Nacional—Casa da Moeda.

Gilkes, O. 2004 The Trojans in Epirus: Archaeology, Myth and Identity in Inter-War Albania in M.L. Galaty and C. Watkinson (eds) *Archaeology under Dictatorship*, 33-54, New York: Academic/Plenun Publishers.

Guibernau, M. 1996 *Nacionalismo—O Estado Nacional e o Nacionalismo no século XX*, Rio de Janeiro: Jorge Zahar Editor.

Guidi, A. 1996 Nationalism without a nation: the Italian case in M. Díaz-Andreu and T. Champion (eds) *Archaeology and Nationalism in Europe*, 108-118, London: UCL Press.

Guillaume, M. 2003 *A Política do Património*, Porto: Campo das Letras.

Henriques, M. and Melle, G. 2007 *Salazar: Pensamento e Doutrina política—Textos antológicos*, Lisboa/São Paulo: Verbo.

Hobsbawm, E.J. 1994 [1990] *Nations and Nationalism since 1780*, Cambridge: Cambridge University Press.

Hutchinson, J. 1992 Moral innovators and the politics of regeneration: the distinctive role of cultural Nationalism in nation-building, *International Journal of Comparative Sociology*, XXX111 (1-2): 101-117.

Jorge, V.O. 2006 Quatro décadas depois. Alguns percursos, encruzilhadas, perspectivas e contributos no âmbito da arqueologia portuguesa—breve exercício de autor reflexão retrospectiva in V.O. Jorge *Fragmentos, Memórias, Incisões—novos contributos para pensar a arqueologia como domínio da cultura*, 77-105, Lisboa: edições Colibri.

Jorge, V.O. and Jorge, S.O. 1996 Arqueologia Portuguesa no Século XX: alguns tópicos para um balanço, *Revista de Antropologia e Etnologia* Vol. 36, 143-58, Porto: SPAE.

Kohl, P.L. 1998 Nationalism and Archaeology. On the constructions of nations and the reconstructions of the Remote Past, *Annual Review of Anthropology*, vol. 12, 223-46.

Kohl, P.L. and Fawcett, C. 1995 Archaeology in the service of the state: theoretical considerations in P.L. Kohl and C. Fawcett (eds) *Nationalism, Politics and the Practice of Archaeology*, 3-18, Cambridge: Cambridge University Press.

Lilios, K.T. 1995 Nationalism and Copper Age research in Portugal during Salazar regime (1932-1974) in P.L. Kohl and C. Fawcett (eds) *Nationalism, Politics and the practice of Archaeology*, 57-69, Cambridge: Cambridge University Press.

Lira, S. 2000 *Identidade territorial portuguesa sob o nacionalismo do Estado Novo—mensagens ideológicas nos museus e exposições temporárias*, http://www2.ufp.pt/~slira/artigos/culturasaloia.pdf.

Matos, S. 2002 História e identidade nacional—A formação de Portugal na historiografia contemporânea, *Lusotopie* 2002, 123-39.

Mattoso, J. 1992 As três faces de Afonso Henriques, *Penélope—Fazer e desfazer a História* No 8, 25-49.

Melo, D. 2001 *Salazarismo e Cultura popular (1933-1958)*, Lisboa: Imprensa de Ciências Sociais.

Minor, H. 1999 Mapping Mussolini: Ritual and Cartography in Public Art during the Second Roman Empire, *Imago Mundi*, Vol. 51, 147-62.

Monteiro, N. and Pinto, A. 2004 A Identidade Nacional portuguesa in A.C. Pinto (coord.) *Portugal Contemporâneo*, 51-66, Lisboa: Publicações Dom Quixote.

Martin, M.L. 1992 ?A Dona de Casa e a Caravela Transatlântica. Leitura socio-antropológica do imaginário salazarista', *Cadernos do Noroeste*, Vol. 5 (1-2), 191-204.

Munzi, M. 2004 Italian Archaeology in Lybia: from colonial Romanità to decolonization of the Past in M.L. Galaty and C. Watkinson (eds) *Archaeology under Dictatorship*, 73-107, New York: Academic/Plenun Publishers.

Neto, M.J.B. 1999 A Direcção-Geral dos Edifícios e Monumentos nacionais e a intervenção no património arquitectónico em Portugal, 1929-1999 in Margarida Alçada e Maria Inácia Teles Grilo *Caminhos do Património*, 23-43, Lisboa. DGEMN/Livros Horizonte.

Oliveira Martins, G. 2007 *Portugal—Identidade e Diferença*, Lisboa: Gradiva.

Payne, S. 1995 *History of Fascism, 1914-1945*, Madison, Wisc.: University of Wisconsin Press.

Pinto, A.C. 2005 Portugal Contemporâneo: uma introdução in António Costa Pinto (coord.) *Portugal Contemporâneo*, 11-65, Lisboa: Publicações Dom Quixote

Rémond, R. 1994 *Introdução à História do nosso tempo—Do Antigo Regime aos nossos dias*, Lisboa: Gradiva.

Rodrigues, J. 1999 A Direcção Geral dos Edifícios e Monumentos Nacionais e o restauro dos monumentos medievais durante o Estado Novo in Margarida Alçada e Maria Inácia Teles Grilo (coord.) *Caminhos do Património*, 69-82 Lisboa: Livros Horizonte/DGEMN.

Schnapp, J.T. 1996 Fascinating Fascism, *Journal of Contemporary Hist*ory, Vol. 31, No. 2, Special Issue: *The Aesthetics of Fascism* 235-44.

Smith, A. 2001 *Nacionalismo*, Lisboa: Teorema.

Sobral, J. 2003 A formação das nações e o nacionalismo: os paradigmas explicativos e o caso português, *Análise Social* Vol. 165, 1093-1126.

—— 2006 *Memória e Identidade Nacional: considerações de carácter geral e ocaso português*, <http://www.ics.ul.pt/publicacoes/workingpapers/wp2006/wp2006_4.pdf>.

Thomas, J. 2004 *Archaeology and Modernity*, London: Routledge.

Trigger, B.G. 1995 Romanticism, Nationalism, and archaeology in P.L. Kohl and C. Fawcett (eds) *Nationalism, Politics and the practice of Archaeology*, 263-79, Cambridge: Cambridge University press.

Viana, F. 2003 *O cartaz e o outdoor ao serviço da comunicação política*. Available at: <http://www.bocc.ubi.pt/pag/viana-fernanda-cartaz-outdoor.pdf>. Accessed 9 October 2009.

Visser, R. 1992 Fascist Doctrine and the Cult of the Romanita, *Journal of Contemporary History*, Vol. 27, 1, 5-22.

Chapter 4

The Social and Political Significance of Prehistoric Archaeology in Modern and Post-modern Societies

Ulf Ickerodt

The social and political influence of archaeology within nationalist states is widely recognized, but we have to take it a step further than this. Archaeology can be said to be one of the factors that helped to shape modern Western society. In this connection we should recognize the fact that the growth of scientific archaeology is very closely related to the emergence of the economic thinking and social structures characterising modernity. Not only in a nationalist context are pictures of prehistoric man and the image of the archaeologist used as a point of social reference. Moreover, it is crucial to understand that images of the archaeologist and prehistoric man, in view of their many-facetted utilization by Western society, in a sense reflect the values and norms of Western society. In this connection, acceptance of the results of archaeological research affects a very specific portion of European space and time perception, and helps—on a meta-level—to channel human behaviour, which is primarily based on society's ideas of progress and evolution, i.e., things are they way they are because they have succeeded in competition. As a social inspiration (*Leitbild*), the socially integrative potential of archaeological research is concerned with the provision of an identity at all levels and the structuring of social behaviour. Thus, archaeology has a considerable responsibility to face up to.

Introduction

> 'It's the economy, stupid!' Bill Clinton

As early as the 1930s, the British archaeologist Stuart Piggott (1937), in his article on 'Prehistory and the Romantic Movement', described the birth of archaeology as a natural outcome of the social and industrial development of the 19th century. The relatively young European states, which had come into existence after the Napoleonic Wars, needed new kinds of social inspiration and modes of self-legitimation for the simple reason that old traditions had been discarded (Ickerodt 2005a; 2006a).

The mediaeval idea of continuity developed into the European 'continuity paradigm',[1] functioning as a method of social reassurance based on scientific discoveries and producing a new cultural self-confidence (Burmeister 2000: 583). The continuity paradigm, at the social level, supported European expansionist efforts and national self-assurance (Ickerodt 2006a). Against this background, Ernst Wahle (1953: 503), in his commentary on the history of research, described 'archaeology' as 'a weapon in the arsenal of nationalist states'. This idea can also be extended to Euro-American imperialism, which was developing in the late 19th century (Ickerodt 2004a).

This is only half of the picture, and this reduces complex historic process that led to the development of modern archaeology to the social and political factors mentioned above. A further factor was the rapidly changing economic structure and its influence on the social framework, which in turn is based on it. This development away from the traditional agricultural society towards industrial, service, and information societies has, since the beginning of the 19th century, caused an increasingly strong separation of much of European and North American society from its natural roots and from the feedback effect, which in the past controlled man's relationship with his environment. This decoupling process affected the different social levels in completely different ways, and since the 1980s we have observed another decoupling effect. The increasingly rapid change to a service and information society accompanying the advance of globalization coincides with a process of decoupling of the economy from its social and geographical environment.

The issue dealt with below concerns the influence of archaeology on modern and post-modern societies. The aspect of the distortion of archaeological research for nationalist purposes is deliberately omitted here as it has already been studied extensively.

Initially, before we discuss the social relevance of archaeology in the 19th and 20th centuries, it is necessary to agree on three premises:

1. When we talk about archaeology, we are considering the social system 'archaeology' with all its conventions, hierarchies and power structures with respect to the actors. It is in this environment that the rules and methods of empirical research are forged.

1 In Germany, Gustav Kossinna (1858-1931) developed a method of ethnic evaluation (*ethnische Methode*) as a tool to search for the roots of the Germanic race. In contrast, the 'direct historical approach' (1916), which was developed at the same time in the United States, aimed to examine the origin of the aboriginal population of North America. In both cases written documents are studied systematically in order to identify ethnic units and to document their evolution using archaeological methods. 'Methodically, the direct historical approach involves the elementary logic of working from the known to the unknown' (Steward 1942: 337).

2. Archaeology is, in a sense, a social system which is embedded in an overall social system, which more or less gives science its framework. These two systems, science and society, are interdependent and interact in different ways and on different levels.
3. The scientific discipline of 'archaeology' is—like all living systems—a historic and therefore an evolving discipline and reflects the adaptive process with respect to point (1) and (2).

The main starting point of this chapter is the thesis that archaeology (restyled in its social function of producing narratives or myths describing our origins) symbolizes Western scientific positivism.[2] As science appears to have replaced the religious myth of the origin of man, science is seen here as reflecting the progress made by society as a whole, whose prime achievement has been the overcoming of superstition (Ickerodt 2004a: 122 ff, 158ff). This naturally applies to archaeology itself with respect to its credibility as well as to the external influence of archaeology. We can use the term '*Ersatzverzauberung*' (substitute enchantment) for this process (Ickerodt 2004a: 15, 172ff).

As part of this process, belief in a divine will or purpose has been replaced by the principle of 'progress', a principle that we see throughout the whole course of history and in which technology, science, society and the state are bound together as a single historic subject incorporating the whole human race. The descendence theory replaces the ascendance theory, which was generally accepted. Zmarzlik (1969: 150) describes this as the change 'from man-of-the-past to man-of-the-future', whose belief in tradition and the unquestioned sense of purpose in progress has more or less disappeared.

Horizontal and vertical social mobility, which has been increasing continuously since the middle of the 19th century, demands a new kind of citizen—the self-made man (Ickerodt 2004a: 169 ff). The new kind of logic that directs his behaviour can be summarized by the phrase 'outdo one's rivals'[3] and is based on oversimplified Darwinian or social Darwinian assumptions, i.e., 'survival of the fittest' in the 'struggle for life' (Ickerodt 2004a: 34-35, 43-46; 2004b).

With reference to colonial experience, the progressive mentality is applied in classifying non-European cultures (Ickerodt 2004a: 40-43), and we recognize that, since the middle of the 19th century, this classification has also been based on oversimplified Darwinian theories (Ickerodt 2003; 2006c: 440, Fig. 3). In the following sections I will examine Western progressive and competitive mentality,

2 Kant (1724-1804), in his 'Kritik der Urteilskraft', defined enlightenment as 'freedom from superstition'.

3 As an example, in a radio interview given in 2000 in connection with the World Economic Summit meeting in Davos, Switzerland, N. Walter, the Head Economist of Deutsche Bank, said that the current world economic system is not perfect, but in principle there were no grounds for complaining, since the system not only works but had also proved itself better than other systems, and therefore must be good (Ickerodt 2004b: 11).

which is based on the archaeological paradigm, and its translation into iconographic symbols. In this way, problems that originate in the fields of social science and ethics of science will be dealt with from the archaeological viewpoint.

The Archaeological Paradigm as a Component of Historic Understanding

The term 'archaeological paradigm' stands for the process of social reconstitution based on archaeology, which commenced in the middle of the 19th century. Within the course of this reconstitution, society took increasing notice of new archaeological knowledge and incorporated this into its then developing identity (Ickerodt 2004a: 218 Fig. 20). This process must be considered against the background of secularization and the disappearance of transcendental norms and values, which formerly channelled behaviour, and which in turn were gradually replaced by other norms and values. These new norms and values are claimed to have a scientific basis and in this way attained a wide-reaching social significance.

In this new form, archaeological knowledge also exerted its influence on the political ideology of the 19th century. As a form of perception, it influenced, via historic understanding, the understanding of space-time relationships in terms of '*oeconomia naturae*'.[4] At the same time archaeological research, and similarly biology, geology and the other sciences, supplied new, previously unheard of facts about natural processes and the constitution of the world, and provided a source of knowledge in terms of '*lex naturae*'. This archaeological knowledge, when it had been assimilated by society, became a significant factor influencing normative socio-economic behaviour. These processes are reflected in the changing character of time perception (Ickerodt 2004a: 76 ff) in terms of a teleological historic narrative.[5]

The time trajectory is characterized by progressive thinking based on evolutionary thinking in a society that is undergoing economic reorganization, i.e., in a developing competitive society. This requires a completely new approach to historic knowledge, which promotes individualism. The individual can and should no longer meekly accept his/her destiny, but, in order to survive in a future-oriented, competitive society, must rather break the fetters of causal patterns of behaviour. This process is interpreted as either basing environmental perception

4 According to Sieferle (1990), *oeconomia naturae* is a symbolic field that extends beyond the paradigm term and forms a 'vague concentration of basic plausibilities', covering all fields of knowledge about the way nature functions.

5 This expression, which is a scholastic term, refers, simply, to the objectivity of human activities and thus of the course of history itself. The teleological historical conception is based on one hand on the perfectioning of the material world on the way to salvation, and on the other hand on the belief that the world was made by God for man. The teleological historical concept allows events to be classified as meaningful with respect to the goal to be achieved. This classification has been extended to human society and permitted the justification of establishing social norms as if they were God's will.

on natural science or removing its supernatural basis, depending on one's outlook. Evidence of this transition in values that affects the whole spectrum of society can be seen, for example, in science-fiction literature or in art (Ickerodt 2004a: 82).

The Historization of Space Perception

Generally, Petrarcas' (1304-1374) climbing of Mont Ventoux in 1337 is taken as the beginning of modern landscape perception (Henke 2005).[6] A ruin, as a consciously perceived component of the landscape, is like a book of memories and history, which suddenly opens in front of the observer. The three dimensions of time manifest themselves in ruins: The past as viewed from the present permits one to see into the future (Ickerodt 2006c). This new, growing, understanding of space-time relationships is absolutely different from the situation we see in the Middle Ages and Renaissance times, which is far more static and largely controlled by providence. The past and the future were visualized as being—not only with respect to material culture—just the same as the present.[7]

The static mediaeval view of the past is successively modified, extended in time and increasingly differentiated. In this process, mediaeval deterministic (i.e., teleological) time perception, in which behaviour is governed by fate and providence, is gradually replaced by an environmental perception which has been developing in Europe since the Renaissance, based on chance and causality (Ickerodt 2004a).

This development has no effect, however, on the historic narratives that determine space and time perception (Ickerodt 2006c), which are still closely related to mediaeval teleological development thinking. The salvation history as understood in mediaeval times has been somewhat reshaped into a narrative of social and technological progress (Ickerodt 2004a).

A herald of this development in the Renaissance was the development in mediaeval and Early Modern times of the initial ideas that ultimately led to the concept of 'non-contemporaneous contemporaneity' (Ickerodt 2004a). It is based on the recognition that the living conditions of some non-European societies that were met with during the conquest of the world by Western society are similar to

6 In late Gothic times we see the birth of landscape painting. Konrad Witz (c. 1400/1410-1444/1446) is the first painter to portray a real, existing landscape (his painting of Lake Geneva).

7 This mode of looking at history can be exemplified by the paintings of the time. During the Middle Ages, paintings and book illustrations of biblical scenes were depicted as if they belonged to the (then) present time, i.e., the Middle Ages. In spite of the awakening of naturalism and humanism and the rediscovery of the antiquarian, no attempt was made to produce *realistic* reconstructions of historical and/or mythological events (Landgraf and Wendland 2005; Groß and Urban 2000). This persisted in biblical illustrations as late as the 19th/20th century.

historical living conditions of the Greeks or Romans for example. Study of these foreign cultures provided a better understanding of one's own past history. On the basis of images of the antiquarian past, the geographical subdivision of the Earth's surface depicts the whole history of the human race from 'wildness' to 'civilization' (Rohbeck 1990: 9). Relics of different epochs of cultural evolution therefore not only occur stratigraphically superimposed but they also exist juxtaposed, in the sense of non-contemporaneous contemporaneity, i.e., contemporary in time but not in cultural 'advancement' (Ickerodt 2004a).

This discovery, based on geological and antiquarian study, was confirmed again and again by the voyages of European explorers (Duchet 1971). The principle of non-contemporaneous contemporaneity is also supported by palaeontological discoveries, since some animals found fossil in Europe are known in other parts of the world as living examples.

This scientific discovery had some effect on the 19th century historic narratives. Even though they are still based on mediaeval teleological development thinking, they now possess a new component, a component that is used to assess and distinguish between 'wild' and 'civilized', as had been done in Roman and Greek times (Nippel 1990).

The 19th century ushered in a consequent further development of this principle. It is supported by biological discoveries, which Charles Darwin used in formulating his theories of evolution. This new principle permitted a unification of the social, political and natural-science theories, which led to the establishment of evolutionism and social Darwinism in the second half of the 19th century. This theoretical principle offered not only a unifying explanation of historic development but also allowed the hierarchic classification of all human societies from wildness through barbarism to civilization. The European explorers and conquerors saw themselves as the forefront of this progressive development (Ickerodt 2004a; 2004b). Nevertheless, the structure of the historic understanding remained in its core a teleologic narrative[8] (Ickerodt 2006b).

Development Thinking and Space Understanding as Basis of the Archaeological Paradigm

This discussion leads on to the social dimension of archaeological research, which, although scientifically accepted, is generally underestimated (Ickerodt 2005a; 2005b; 2006a; 2006b). As a component of historic understanding, archaeological discoveries and the related historical conclusions, when they have been assimilated by society, influence human space perception and thus social behaviour. Historic understanding incorporates experiences from the past

8 The consequences drawn at that time from contemporary research in biology (theory of evolution) and physics (Second Law of Thermodynamics), which showed history to be a non-directional process, were not accepted by contemporary society at large (Ickerodt 2005b).

as well as the patterns of living based on them, but also political assumptions that extend, via their function as a component of social identity, far beyond local self-definition (Ickerodt 2006a).

In order to understand these ideas more clearly, one must approach the inherent logic of Western culture (i.e., the cultural pre-understanding), which incorporates the present-day concept of space and time perception based on archaeological knowledge. This kind of study has already been undertaken elsewhere[9] and therefore can only be sketched here.

The study is based on the adoption of palaeontological and/or archaeological analogies which are used in novels to characterize a particular place or sensation (Ickerodt 2005a; 2006c).

I will begin by taking the novel *Cinq semaines en ballon* (*Five Weeks in a Balloon*) by J. Verne, published in 1864. Here Verne uses an archaeological-cum-palaeontological analogy and an archaeological-cum-geographic theme in order to illustrate the primitive nature of the African landscape on one hand and to emphasize European scientific and technological progress on the other (Verne 1864a, 19). Another, comparable device is used by Agatha Christie and by Henning Mankell. Both authors use archaeological-cum-palaeontological analogies to give a picture of the landscape. In the first case it concerns rocks on the Nile[10] and in the second case the skerries off Östergötland[11] in Sweden. This type of analogy is not only based on natural phenomena but is also extended into the technological field. The American author Kathy Reichs (2006:42-3), for example, compares the piers of the Cooper River Bridge in Charleston, South Carolina, with a triceratops.

In a similar way to Reichs, the American author Gary Goshgarian (1997: 12) describes mechanical excavators 'which crouched in the dunes like prehistoric monsters'. In this particular case the palaeontological analogy was used to produce a sinister effect. The required sinister effect is produced in a similar way by Sir A. Conan Doyle in *The Hound of the Baskervilles* (published in 1902). He succeeds in producing this effect by contrasting the moors,[12] in which there are prehistoric features, and a town-dweller, i.e., Dr Watson (Doyle 1901).

This sinister effect is based on the contrast between a world that is on the periphery of (or outside) the 'civilized' world and one in which there are other, irrational laws. He makes use of the principle of non-contemporaneous contemporaneity (Ickerodt 2004a: 41 ff), which expresses the hidden threat stemming from the past to devolution or cultural regression (Ickerodt 2004a: 125 ff). The connotive character of the archaeological-cum-palaeontological

9 Ickerodt 2004a; 2004b; 2005a; 2005b; 2005c; 2006a; 2006b).

10 'They looked down to the shining black rocks in the Nile. There was something fantastic about them in the moonlight. They were like vast prehistoric monsters lying half out of the water' (Christi 1937: 43).

11 Mankell (1998: 159) compares the rocks that stuck out of the water with fossilized prehistoric animals.

12 Conan Doyle 1901: 17, 73-75, 81, 84, 102, 111, 120.

analogy is even clearer in Mankell's work. In *The Dogs of Riga* he uses the analogy of motionless prehistoric animals to describe the silhouettes of derelict factories and thus the technological decline of Latvian industry.

These archaeological-cum-palaeontological analogies presented as a certain form of thinking and its political connotation can be understood another way. For example, in H. Ellions' (2003: 155) short science-fiction story called *Deeper than Darkness* (1957), the world consists, in the sense of non-contemporaneous contemporaneity, of developed and underdeveloped regions: 'The American continent was a modern thing, all plasteel and printed circuits, all relays and fast movement, but there had been areas of backwoods country that had never taken to civilizing'. Agatha Christie (1929: 201) has the same message when her hero Tommy Beresford comments on his butler's behaviour with 'Quite so, (…), but we're not on the prairies. We happen to be in a highly civilized city [i.e., London]'.

An archaeological-cum-palaeontological device, i.e., the simile of a journey into the past (Ickerodt 2004a: 87-91), can also put over an emotional experience to the reader of a novel. In the literature, leaving the civilized world is commonly equated with a journey into the past, or time travel. In particular, Joseph Conrad's *Heart of Darkness* (1889) can be interpreted as forming a genre. In this novel the journey into the interior of Africa is equated with a journey into a prehistoric world,[13] and, following the principle of non-contemporaneous contemporaneity, the natives of Africa are equated with primitive man.[14]

In the same way C. Holland's hero Howard Carter experiences his journey into the Valley of the Kings in Egypt as a kind of time travel into a prehistoric world (Holland 1977: 20). The writer A.K. Elkins (1993: 54) uses the same analogy. In his novel, set in Yucatán, the heroes believe they have been transposed into the past. They find themselves in the grounds of a luxury hotel surrounded by 'tamed' jungle penetrated by a smell of burnt straw.

Summarising, one can say that the archaeological-cum-palaeontological analogy is a generally understood historic analogy that is almost always used in a pejorative sense, i.e., to say that something is 'old' and 'simple' and therefore bad. Analysis of the cultural context shows that behind its apparently harmless façade, there is a social pre-understanding based on evolutionistic thinking that subconsciously influences economic, political and social behaviour.

---

13 'Going up that river was like travelling back to the earliest beginnings of the world, when vegetation rioted on the earth and the big trees were kings' (Conrad 1899: 66); 'We were wanderers on prehistoric earth, (…)'(Conrad 1899: 68).

14 'The prehistoric man was cursing us, praying to us, welcoming us—who could tell?' (Conrad 1899: 68).

The Archaeological Paradigm in Advertisements

The way a socially relevant inspiration (*Leitbild*) is created is best illustrated by another example. A useful alternative to the literature as an object of study is the advertisement. In the present context advertisements are of particular interest because they belong to a segment of the market that reacts very sensitively to the social pulse, and their message is invariably put over in a very simple and easily understood way. In this sense the advertisement can be an accurate reflection of the social treatment of the archaeological paradigm. An advertisement with an archaeological theme usually expresses the superiority of a product. The tradition and the quality of the said product are praised in comparison with the competitors' products (Ickerodt 2004a; 2004b: 15-16).

In actual practice this kind of information is put over in a very basic form. Iconographically, the structure of pictures in advertisements is suited to the standardized, up-to-date Western way of looking at things. Normally on the left is Object A, a prehistoric relic such as a pyramid or a stone artefact, in the sense of a *pars pro toto*, and the present-day product, Object B, is placed on the opposite side. The latter is a witness to biological and/or cultural evolution. The two objects are related to each other by means of linear historic concepts answering to the teleologic sense of the archaeological paradigm and in this way express the connotive message of individual and social competitiveness as a natural law (Ickerodt 2004b; 2005b; 2006b). This kind of arrangement in the picture corresponds to a form repertoire suggesting unilinear evolution, which is an object of criticism, particularly by natural scientists, biologists and palaeolontologists (Gould 1998; 1999; Wuketits 2001: 108; see also Ickerodt 2004b; 2005b; 2006b).

Alternatively, this picture used in advertisements may, however, contain intermediate stages between objects A and B, i.e., the quality of the advertised product may be illustrated by means of a typological series. This can be done in two ways: The product is shown in connection with a linear evolutionary series of the human phylogeny (Ickerodt 2004a: Section 8.6.1), or with the evolution of material culture (Ickerodt 2004a: Section 8.6.2). The latter directly reflects typological evolution in a similar way to that in which archaeologists present their typological series. Type-based representation of cultural evolution can also be treated in an abstract way or with a picture of a fossil. These kinds of pictures give the reader, implicitly or explicitly, a threat of 'dying out' as an ultimate punishment if he doesn't choose to follow the way of 'progress'. The foundation stone for this mode of thought used in advertisements is laid in one's own social environment at an early age, as has been shown in titles of children's books elsewhere (Ickerodt 2004a: 84-5, Section 7.2).

Basically one can say that the utilization of palaeontological and/or archaeological motifs is an expression of a society in which progress and competition are fundamental. A progress-oriented mentality automatically invites one to relate two objects to one another, by defining them in terms of progress and decay, and in this connection comparison of objects or cultures etc. is used in the earlier mentioned pejorative way.

The Influence of Progressive Thinking and Space-time Understanding on Political Economy

At the core of economic thinking based on the archaeological paradigm is the conviction that a natural economic system exists in which the better, or the stronger, prevails. The criteria for this are, for example, financial or social success in an apparently egalitarian society which no longer distinguishes between social classes and categories but solely between winners and losers.

This new form of political economy demands from the individual self-confidence, as well as flexibility in recognising causalities in order to see them as economic opportunities and thus to make use of them. Its social background is characterized by increasingly prominent horizontal and vertical social dynamics hand-in-hand with growing isolation and decoupling of the individual from the production and income systems. This new economic system, which established itself in the 19th century, enabled a civil society to develop, but on the other hand the considerably stronger social mobility and fragmentation mentioned above gave rise to a simultaneous increase in social isolation with all its consequences.

It would be wrong to assume that the archaeological paradigm reaches the whole of society and all social levels in the same form. The reason for this is the fact that its inherent logic is the product, as well as a component, of an adaptive process bound to individual aims. In this sense one could say that the logic of historic understanding based on the archaeological paradigm, is an elementary constituent of human space-time perception of certain groups and steers their economic and social behaviour. It structurizes behaviour and generates social cohesion. This is also the reason why, as a social Leitbild, it cannot attain the range and homogeneity of social coverage at present that it had in the late 19th and early 20th centuries. It is more true to say that, since World War II, we have to contend with an increasingly diffuse picture on account of the variety of social groupings and their historic narratives, only some of which may be based on archaeological knowledge. This means that the archaeological paradigm does not reach all social groups.

However, most social groups that are reached by the archaeological paradigm also feel increasingly strongly attracted by the scientific *ratio*, which demands a scientific verification of religious as well as alternative traditions. Science, and thus archaeological research, following the needs of a secular society to structurize and, by means of rational scientific arguments, to legitimize social behaviour. This happens with regard to the archaeological paradigm, whether it is scientifically correct or not, via teleologic historic meta-narration. This turns out to be the traditional form of understanding of space-time relationships, and therefore as an inherited way of organising social behaviour within modern and post-modern industrial and service societies, as well as in a competitive society. The seeds of the inherent logic of historic understanding are sown during socialization and constitute a subconscious process of learning.

Teleologic Space Understanding Based on the Archaeological Paradigm

Michel Foucault (1990: 34) states that the 19th century found the essence of its mythological resources in the Second Law of Thermodynamics. With the same confidence one could add that the same applies, basically to Darwin's theory of evolution. Both theories were developed in the second half of the 19th century and both theories refer to an irreversible course of historic events starting from an initial causality. Therefore in both cases we are dealing with teleonomic, i.e., non-directional, processes.

If one considers society's reception of physical and biological discoveries against the background of what has been discussed above, then the teleologic character of Western historic understanding becomes clear. It appears as if neither modernity nor post-modernity, apart from a small elite group of scientists, can accept[15] the directionless character of history that results from the above. On the contrary, one still refers to teleologic concepts, which have been tradition since the antique, and meanwhile have been supplemented by concepts based on technological positivism (Ickerodt 2007). These concepts determine not only social behaviour but also archaeological research.

But what are firstly the social and secondly the scientific consequences of this logic of teleologic and teleonomic historic models? How do they affect space and time human behaviour?

Teleologic thinking allows the reality of space to be described as an ordered structure in the sense of a world order, which, in the Christian development or ascendance sense, is the product of the ordering hand of God, i.e., *causa finalis*. In this system, time is understood as something linear, progressive and predetermined. It connects the beginning of the world (Genesis) with the end of the world (the Last Judgement). It is divided into two main epochs, i.e., the time before the birth of Christ and the time after the birth of Christ. The course of time thus forms a skeleton for the historic framework of Europe on which the historic facts are arranged (Ickerodt 2004a: 77).

In contrast, teleonomy[16] of historic processes as a scientific fact, seen in the tradition of Thomas Hobbes, is and will be perceived as a world without natural laws, which in principle can adopt any chosen form and/or condition. It offers the individual less security of life, since everyone's future is open and constitutes not only a sea of possibilities but also much uncertainty.

The teleonomic mode of thinking via the archaeological paradigm is therefore considered here to be society's attempt to compensate the uncertainty resulting from the non-directionality of living systems. A finite world picture, in which development is determined by fate, helps to overcome serious crises. The same

15 See the debate between Ickerodt (2005b; 2006b), and Porr (2005) and, in this connection, Röder (2006).

16 This term was coined in 1958 by C.S. Pittendrigh in connection with the regulation mechanisms in cells (Ickerodt 2004b: 16).

applies to the development of such pictures, which are based on the teleologic concept, that finally led in the 19th century to social Darwinism. This is where divine leadership and prophesy are abandoned in favour of the 'struggle-for-life' paradigm, in which the stronger prevails. Zmarzlik (1969: 147) refers to this false interpretation of biological and thus evolutionary 'success' as 'survival of the fittest' as the inequality dogma. This applies to the distortion of scientific facts to suit society's aims or Leitbilder and thus it supports social and inter-nation competition.

In order to understand the effect of the teleologic positivistic historic understanding[17] better, one must explain the underlying anthropological dimension.

The understanding of space-time relationships is a normal human faculty, which certainly developed during the evolution of man. It is a product of a specific cultural and historic adaptation process. The understanding of space-time relationships is permanently subject to control by the senses, which provide a constant verification of its functionality to guarantee cultural viability, in order to keep up with environmental and social changes. This is not only valid for oral and literal traditions but also to historic understanding based on archaeological knowledge. It can also be described as a category of perception, which structures and canalizes social, political and economic behaviour. It therefore steers the specifically cultural, productive use of the environment.

This learning and understanding of space-time relationships usually develops subconsciously via the natural background understanding that everyone possesses, which permits one to behave rationally in one's environment. It seems not to be valid, therefore, to claim that a logic of space-time perception inherited at an early age from one's social environment and based on the archaeological paradigm and technological positivism is an absolute space perception, even if this approach leads increasingly towards an absolute environmental understanding. It still remains a social construction, because human space-time perception will always depend on cultural pre-understanding (immanent logic), on the situation of its perception (psychic aspect) and on the situation of its utilization (specific situation context) (Ickerodt 2006c).

With reference to the teleologic, technological, and positivistic historic conception, the biological and cultural evolution of the genus Homo and the taxon Homo *sapiens sapiens* is understood as a linear increase in cultural, i.e., technological and social, complexity. In this context, this theoretical approach is certainly a child of the 19th century with respect to the politically legitimising effect of archaeological knowledge on political thinking and behaviour patterns. The resulting Leitbild, derived from archaeology and based on evolutionist and/or oversimplified social-Darwinian ideas, reflect the superiority of European culture with respect to the rest of the world and its inherent duty to civilize the

17 By historic understanding we refer to general social structures with which the past is dealt with; these structures help us to relate individual events, episodes etc. to each other and in this way to make them 'logical' and 'meaningful'.

world, as was so firmly claimed since the end of the 19th century. The place of Egypt (real) and Atlantis (fictitious) in the antiquarian world of that is taken in the world of today by Europe and the United States. This is reflected even today in the Leitbild 'scientific positivism',[18] although in a much modified form. A glance at the archaeological topics incorporated in today's belletristic shows that the dichotomy of scientific positivism and superstition, which was used as a topic by Rosny Ainé (1887) (Ickerodt 2007), is still popular (e.g., Becker 2003; Darnton 1996; Goshgarian 1997; Greannias 2006; Sigurdardottir 2006).[19]

Thus the archaeological paradigm can be understood not only as an anthropocentric view, but also as a culturally determined western view of the environment (Ickerodt 2005c). Cultural determinism and anthropocentralism are certainly the roots of chauvinism and are combined especially in modern and post-modern society with the above-mentioned historic models based on teleologic ideas.

Social and Scientific Consequences

In the previous sections, I have tried to give an account of a social environment that derives its social and political Leitbild from the field of archaeology and is referred to as the archaeological paradigm.

This describes the social assimilation of archaeological and palaeontological knowledge, as well as its influence on the culturally determined understanding of the '*oeconomia naturae*' or 'how the world goes round' as well as on the '*scala naturae*',[20] which classifies this understanding. In particular, the '*scala naturae*' is based on the attempt to investigate the position of man by means of observation, measurement and comparison (Moss 1990; Gould 1983; Ickerodt 2004a), and offers a basis of channelling social behaviour.

Thus the archaeological paradigm, as an overall system of thought based on scientific knowledge, has a direct influence on the many political concepts (liberalism, conservatism, communism etc.) and ideologies (fascism, social Darwinism etc.) that sprung up in the 19th century. It would, however, be an error to believe that this basis which steers social behaviour became obsolete as a result of World War II. Political concepts from the past are still active (neoliberalism, neofascism etc.), or they reappear in new, apparently apolitical clothing.

We must take into account that the popularization of scientific work obeys social criteria and will continue to do so. Western rationalism is after all part of

18 For criticism of the discussion on American pseudoscientific see Ickerodt (2007).

19 A significant facet in this context is that in particular the authenticity of the Christian religion is being questioned on the basis of archaeological research (e.g., Abécassis 2000; Eschenbach 2000).

20 The term relation-diminutive was introduced to express the socio-political mode of influence of the '*scala naturae*' (Ickerodt 2004a: 105-107).

the society's attempt to make a reality construction. All in all, the archaeological paradigm influences not only the social environment, but also the scientific discussion which is part of this environment.

The development of this process, in the centre of which is the change of the Western space-time perception, was discussed in the previous sections. It is characterized as a process of self-organization within the framework of an expanding society. We must recognize that this process is connected with the loss of formerly mutually shared values and the social faculty to form a broad social consensus, accelerating social self-differentiation. Against this background, the character of the archaeological paradigm can be interpreted to be a strategy of society to overcome the increasing entropy of social coherence by generating a superimposed, apparently new symbolic field. Its social relevance within this thought system has, since the 19th century, been equated with 'natural laws' derived from the understanding of the '*oeconomia naturae*' of that time. However, the social conclusions with respect to the '*scala naturae*' are, from the viewpoint of scientific ethics, highly problematic since, they are simplified and corrupted to conclusions that are rooted in the human tendency to pseudospecification,[21] thus *a priori* erring into chauvinism.

This '*scala naturae*' and the associated chauvinism, encourages one to see contrasting pairs 'civilized' and 'barbaric/wild', or in the same sense 'town' and 'country', and further, from society's viewpoint, it ranges from progressive and primitive or 'developed' to 'underdeveloped', and may go so far as to divide and classify people into of value or of no value to humanity.

In the core of the logic of historic understanding, which steers this kind of perception, lies the European progress mentality as well as competitive mentality, which developed out of progress mentality since the 19th century. Both of them have become a component of the logic of economic understanding of industrial, service and competitive societies.

From the scientific viewpoint one must consider that here social and scientific understanding are both products of a social adaptive process,[22] which determines that cultural historic developments are interpreted in the sense of an increasing

21 This term was coined by Erikson for the separation and self-definition of social groups, reminiscent of speciation. One should remember, however, that human perception of the environment structures everything that is seen tending to incorporate related cultural groups and exclude the others (Ickerodt 2004a: 105).

22 This incorporates three influencing factors, environment, historically developed environmental understanding, and structured social behaviour. All three are related to one another in their influence, which one can refer to as a feedback system. A feedback comprises the effect of a variable force on itself. In cybernetics one defines a feedback as positive when the effect increases the source effect, and as negative when the source effect is diminished or damped and eventually a homeostatic state is attained (Ickerodt 2006b: 13-14).

sociocultural and technological complexity.[23] In this sense they are related to scientific positivism as embedded in the Western rationalization process (Ickerodt 2004a: 172-81; 2007).

The biologist Riedl (1987: 261) characterizes this process as super-transcendency, which evolved from a predecessor, which he terms self-transcendency and which is based on the human faculty to recognize one's own faults and hence to improve one's own abilities.

This development takes place, however, in the different sections of society in different ways. It must be understood, with respect to Western influenced societies, as part of the two decoupling processes mentioned in the introduction. In the first phase people become decoupled from the natural environment, and in the second phase from their contemporary social environment in an over-individualized world. Society tries to oppose its own fragmentation by developing regulations and controlling mechanisms with reference to economic growth and the inherent potential economic possibilities, although at the same time the fact social structures and conventions gradually become more dynamic tends to continuously undermine these efforts.

This understanding of the teleological thinking based on the archaeological paradigm and steering scientific and historical understanding, as described above, is basic for dealing with problems in the field of the sociology and ethics of science. It is also basic to generate reliable scientific interpretations and syntheses. The historian Haydn White (1996: 67) drew attention to this problem several years ago.

References

Abécassis, E. 2000 *Die Jesus Verschwörung*. München: Diana Verlag.

Becker, W. 2003 *Missing Link*. München: Knaur Verlag.

Burmeister, S. 2000 Die ethnische Deutung in der Urgeschichtsforschung: Zum Stand der Diskussion. Auf der Suche nach Identitäten: Volk—Stamm—Ethnos. Tagung in Leipzig 8-9. December 2000. *Ethnographisch-Archäologische Zeitung*, 41, 581-95.

Christie, A. 1929 *Partners in Crime*. (10th edn 1990). London: Fontana/Collins.

—— 1937 *Death on the Nile*. (7th edn 1990). London: William Collins Sons and Co Ltd.

Conan Doyle, Sir A. 1901 *The Hound of the Baskervilles*. (1986). London: Penguin Books.

Conrad, J. 1899 *Hearth of Darkness*. (1985) London: Penguin Classics.

Darnton, J. 1996 *Neandertal*. München: Goldmann Verlag.

23 For determination of teleologic thinking on science, see Ickerodt (2004a: 135-39), and on society, see Ickerodt (2004a: 139-42).

Duchet, M. 1971 *Anthropologie et histoire au siècle des lumières*. Paris: Edition Albin Michel.
Elkins, A. 1993 *Fluch*. Ein Gideon-Oliver-Krimi. München: Haffmanns Verlag.
Eschenbach, A. 2000 *Das Jesus Video*. Augsburg: Bechtermünz Verlag.
Foucault, M. 1990 Andere Räume. In Bark, K. et al. (eds). *Aisthesis. Wahrnehmung heute oder Perspektiven einer anderen Ästhetik*. Leipzig: Merve Verlag, 34-46.
Goshgarian, G. 1997 *Der Ruf der Steine*. 1. Aufl. (2000) München: Blanvalet Verlag.
Gould, S.J. 1983 *Der falsch vermessene Mensch*. Basel, Boston, Stuttgart: Birkhäuser Verlag.
—— 1992 *Die Entdeckung der Tiefenzeit. Zeitpfeil und Zeitzyklus in der Geschichte unserer Erde*. München: Deutscher Taschenbuch Verlag.
——1998 Leitern und Kegel: Einschränkungen der Evolutionstheorie durch Kanonische Bilder. In Silvers, R.B. (ed.), *Verborgene Geschichte der Wissenschaft*. München: Knaur Verlag, 47-77.
——1999 *Illusion Fortschritt. Die vielfältigen Wege der Evolution*. Frankfurt/M.: Fischer Verlag.
Greannias, T. 2006 *Stadt unter dem Eis*. München: Heyne Verlag.
Groß, W. and Urban, W. 2000 *Dem Himmel entgegen. Niederländische Bilder des 15. bis 18. Jahrhunderts erzählen die Bibel. Sammlung Christoph Müller*. Ulm: Süddeutsche Verlagsgesellschaft.
Henke, F.A. 2005 *Topografien des Bewusstseins. Großstadtwahrnehmung, Erinnerung und Imagination in der französischen Literatur seit Baudelaire*. Inaugural-Dissertation zur Erlangung der Doktorwürde der Philologischen Fakultät der Albert-Ludwigs-Universität Freiburg i. Br. Available from: <http://freidok.uni-freiburg.de>/E-books [cited TT Monat JJJJ].
Holland, C. 1977 *Im Tal der Könige*. (2005) Augsburg: Weltbild Verlag.
Ickerodt, U.F. 2003 Primitive, Wilde und Peripatetiker—Ein Beitrag zur Wildbeuter-Ethnoarchäologie am Beispiel der Hazapi Ostafrikas. In Burdukiewicz, J.-M. et al. (eds), *Veröffentlichungen des Landesamtes für Archäologie Sachsen-Anhalt Landesmuseum für Vorgeschichte*, 57(I). Halle/Saale, 259-71.
—— 2004a Bilder von Archäologen, Bilder von Urmenschen. Ein kultur- und mentalitätsgeschichtlicher Beitrag zur Genese der prähistorischen Archäologie am Beispiel zeitgenössischer Quellen. Dissertation zur Erlangung des Grades eines Doktors der Philosophie des Fachbereichs der Kunst-, Orient- und Altertumswissenschaften der Martin-Luther-Universität Halle-Wittenberg. Bonn. Available from: <http://sundoc.bibliothek.uni-halle.de>/E-books [cited TT Monat JJJJ].
—— 2004b Die Legitimierung des Status quo: Ein Beitrag zur gesellschaftlichen und politischen Relevanz prähistorischen Forschens. *Rundbrief der Arbeitsgemeinschaft Theorie in der Archäologie*, 3(1-2), 10-23.
—— 2005a Hobsbawms erfundene Traditionen—Archäologie als Soziales Phänomen.*Archäologisches Nachrichtenblatt*, 10(2), 167-74.

—— 2005b Das Erbe der Urmenschen—Eine Anmerkung zur gesellschaftlichen Relevanz der prähistorischen Forschung. *Rundbrief der Arbeitsgemeinschaft Theorie in der Archäologie*, 4(1), 14-23.

——2005c Prähistorisch-archäologische Betrachtung zum Kulturlandschaftsbegriff. *Nachrichten zur Niedersachsen Urgeschichte*, 74, 251-63.

—— 2006a Mortui viventes obligant—Zur mentalitätsgeschichtlichen Einordnung des Völkerschlachtdenkmal am Burgdorfer Hindenburgwall. *Nachrichten aus Niedersachsens Urgeschichte*, 75, 257-65.

—— 2006b Das Erbe der Urmenschen—eine Erwiderung auf Porrs Kritik. *Rundbrief der Arbeitsgemeinschaft Theorie in der Archäologie*, 5(2), 9-19.

—— 2006c Der Kulturlandschaftsbegriff als organischer Bestandteil des historischen Verstehens. *Siedlungsforschung. Archäologie—Geschichte—Geographie*, 23, 427-64.

—— 2007 Archäologie, Pseudowissenschaft und Geschichtsvermittlung. Die gesellschaftliche Relevanz der Archäologie zwischen übertriebenem, wissenschaftlichem Positivismus und Pseudowissenschaft. *Nachrichten aus Niedersachsens Urgeschichte*, 75, (in preparation).

Landgraf, M. and Wendland, H. 2005. Biblia deutsch. Bibel und Bibelillustrationen in der Frühzeit des Buchdrucks. *Veröffentlichungen des Pfälzischen Bibelvereins*, 3. Speyer: Evangelischer Presseverlag.

Mankell, H. 1992 *Die Hunde von Riga*. (2005) München: Deutscher Taschenbuch Verlag.

—— 1998 *Die Brandmauer*. (2005) München: Deutscher Taschenbuch Verlag.

Mosse, G.L. 1990. *Die Geschichte des Rassismus in Europa*. Frankfurt/M: Fischer Verlag.

Nippel, W. 1990 *Griechen, Barbaren und 'Wilde'. Alte Geschichte und Sozialanthropologie*. Frankfurt/M.: Fischer Verlag.

Piggott, St. 1937 Prehistory and the Romantic Movement. *Antiquity*, 11, 31-8.

Porr, M. 2005 Das Erbe der Urmenschen und die Verantwortung der Archäologie—Eine Erwiderung. *Rundbrief der Arbeitsgemeinschaft Theorie in der Archäologie*, 4(1), 25-9.

Reichs, K. 2006 *Hals über Kopf*. München: Karl Blessing Verlag.

Röder, B. 2006 Erbstreitigkeiten—Ergänzungen zur Debatte über 'Das Erbe der Urmenschen' zwischen Ulf Ickerodt und Martin Porr. Rundbrief der Arbeitsgemeinschaft *Rundbrief der Arbeitsgemeinschaft Theorie in der Archäologie*, 5(2), 20-25.

Rohbeck, J. 1990 Turgot als Geschichtsphilosoph. In: Turgot, *Über die Fortschritte des menschlichen Geistes*. Rohbeck, J. and Steinbrügge, L. (eds). Frankfurt/M.: Surkamp Verlag, 7-88.

Rosny Aîné, J.H. 1887 Les Xipehuz. In: J. Baronian (ed.), *Récits de science-fiction*. Verviers (1973): Marabout.

Sieferle, R.P. 1990 *Bevölkerungswachstum und Naturhaushalt*. Frankfurt/M.: Suhrkamp Verlag.

Sigurðardóttir, Y. 2006 *Das letzte Ritual*. Frankfurt/Main: Fischer Verlag.

Steward, J. 1942 The Direct Historical Approach to Archaeology. *American Antiquity*, VII(4), 337-43.

Verne, J. 1864 *Fünf Wochen im Ballon*. (1976). Zürich: Diogenes.

Wahle, E. 1950. Geschichte der prähistorischen Forschung 1. *Anthropos*, 45, 497-538.

White, H. 1996 Literaturtheorie und Geschichtsschreibung. In: H. Nagl-Docekal (Hrsg.), *Der Sinn des Historischen. Geschichtsphilosophische Debatten*. Frankfurt/M.: Fischer Verlag, 67-106.

Wuketits, F.M. 2001 *Naturkatastrophe Mensch. Evolution ohne Fortschritt*. Frankfurt/M.: Verlag.

Zmarzlik, H.G. 1969 Zum Sozialdarwinismus in Deutschland—Ein Beispiel für den gesellschaftspolitischen Missbrauch naturwissenschaftlicher Erkenntnisse. In Altner, G. (ed.), *Kreatur Mensch. Moderne Wissenschaft auf der Suche nach dem Humanum*. München: Heinz Moss Verlag, 147-56.

Chapter 5

Art after 'Primitivism': Revisiting the 'Non' of the 'Contemporary/ Non-Western' Dichotomy

Tom Lane

Whether labelled deconstruction, post-structuralism, or re-modernisation, new orientations towards disciplinary regimes and their relations to wider historical circumstances are emerging today, which go against the grain of supposedly 'given' disciplinary objects, as well as dichotomies (such as those of experts-publics, moderns-others, subjects-objects) which variously hinge upon notions of supposedly incommensurate (if not mutually unintelligible) 'world views'. This chapter explore approaches to the tenacity of such dichotomies of Asad, Gell, Clifford, and Vogel, in light of two notorious 20th century exhibitions of 'primitive art', in order to illustrate the relevance of what some call a 'trackless thicket' (Elkins 2008) for appreciating the ways in which non-western cultural materials challenge dualist presuppositions about art.

'Trackless Thickets'?

> Since the Enlightenment, the history of science has been enlisted to show the unity and distinctiveness of Europe…Whether understood as triumph or tragedy (and there have been eloquent proposals of both views, the Scientific Revolution has been portrayed as Europe's decisive break with tradition—the first such break in world history and the model for all subsequent epics of modernisation in other cultures…[How might] a new history of science, exemplified by such works as Simon Shapin and Steven Schaffer's joint study, *Leviathan and the Vacuum Pump Hobbes, Boyle and the Experimental Life* (1985), transform the self image of Europe and conceptions of truth itself? (Daston 2006: 523).

Writing, in her book on *Crafting and the Kingly Ideal* (1993) on patterns that might be exhibited across something of the diversity of modes of objectifying linkages between 'cosmos and polis' (Toulmin 1990), Mary Helms suggests a generalised model:

> …in which the here-and-now of an organized morally informed 'cultured' social entity at the centre is surrounded on *all* sides and *all* dimensions by an outside cosmological realm which is believed to contain all manner of visible, invisible and exceptional qualities, energies, beings, and resources, some harmful, some helpful to those at the centre. We are well acquainted with characteristics of some of these outside realms as they have been explored, controlled, transformed, and exploited by traditional religious practitioners—shamans, priest, diviners, curers, midwives, etc. (Helms 1993:7).

In such a view, many cultures may have not only 'dreamed of an overall harmony between the order of the heavens and the order of human society' (Tooling 1990: 67), but also of realms 'out there' and, especially, about what the nature of such outer realms—their structures, forces, composition—including their physical, organic, social, cultural, spiritual, moral composition—suggests about the nature of the supposed centre.

For Toulmin, as for Helms and many other anthropologists and art historians nowadays, as long as there have been here have been cosmologies that situate the 'polis' at the cosmological centre of the universe, image of 'outside realms' and 'others' have been crucial motifs in 'self portraits' (Daston 2006) of the former. It is now widely recognised that hitherto predominant notions not only of so-called 'primitive art' but also 'art', in general, have been enmeshed in the history of such self-portraiture. However, many institutions structured around these notions remain as self assured as ever it has. Galleries and museums continue to maintain record numbers, whilst the markets, its auctioneers, dealers, and buyers, continue to exchange vast sums for record prices. And all this despite an economic downturn of historic significance (this essays historical present is mid 2009). In stark contrast to such stability, there is an ongoing crisis in the interdisciplinary realm of visual culture. As the art critic/historian James Elkins (2008: vii) remarked, one immediate manifestation is that 'contemporary [art historical] writing seems like a trackless thicket, tangled with unanswered questions.' This period of flux is often denounced as symptomatic of the post-modern era; a paradigm in which a widespread pandemic of indecision, and relativism has apparently halted the enlightenment engine of progress and purpose.

Without the possibility of navigating accurately through our own terrain (the post-modern thicket), perhaps some refuge, or a better understanding of art history's ethnographic present can be sought by looking towards the non west. Non-western cultural material has had a long and often tumultuous relationship with European artists, whilst its acquisition (often through disreputable means) and accumulation is representative of a far darker and now publicly transparent European history; a post colonial history. Histories, concepts, and theories surrounding the ideological production and rationalisation of the Other, as Orient, as both conceptual and tangible objects, provide new insights for understanding not only the problems of cross cultural translation, but also the structural intricacies of the west's domestic regimes of art's history and its criticism.

'Primitivism' and the Tenacity of Contemporary—Non-Western Dichotomies

In what follows I explore the relevance of what some call a 'trackless thicket' (Elkins 2008) for appreciating how non-western cultural materials challenge dualist presuppositions about art, in light of Asad, Gell, Clifford, and Vogel critiques of such presuppositions, and two 20th century exhibitions of 'primitive art'. The first exhibition I wish to examine is the infamous *Primitivism in 20th Century Art: Affinity of the Tribal and the Modern* which was held in the Museum of Modern Art (MoMA), New York, in the winter of 1984. The museum displayed African, Oceanic and American-Indian cultural material alongside 20th century European works by avant-garde artists like Picasso, Giacometti, Brancusi and Miro. An whilst the latter had already achieved the canonical status of an Ism in art's history (Primitivism), the former had yet to prove itself, therefore remaining tribal objects that had been hand picked from history.

In late Renaissance and early modern Europe, these objects were collected into 'curiosity cabinets' (or *Wunderkammer*). Beneath the banner of the mysterious, non-western objects drew together a gazing Western public and derived a monetary fortune from a lucrative mercantile economy. Trading in these increasingly global markets, wealthy merchants and their patrons accumulated vast arrays of material that was often valuable both for its material properties and its aesthetic symbolism. Importantly, neither value attributed any contextual significance to the intention, production, and reasoning behind the creation of such objects; viewing them through an Occidental prism in which the native producer was notably absent. Like collectors in the 15th, 16th, and 17th century, artists in the late 19th and early 20th century bestowed upon non-western cultural material a special reverence. Such reverence should be seen as stemming from the same natural flow of curiosity that had cultivated the preternatural speculations of earlier centuries. However, over time, such speculation had developed an accredited and institutionalised scientific basis. What these objects were, and what they represented was no longer the product of naive gawking but had become verifiable through the developing fields of anthropology and biology.

Amidst the developing sciences of man, Primitivism emerged as an art historical movement in the late 19th century in a fluid category which attempted to encapsulate European art's increasing interest with Non-European cultural material. Cultural production from as far away as Africa, Polynesia, Japan, and the Americas, had over time, come to reside among both public but predominantly private collections throughout Europe. For European artist's keen to quicken the already ferocious evolution of painterly form such collections provided an abundant source of previously untapped creative potential. Non-western cultural material therefore provided a source of both interest and novelty, but at the same time it also constituted a point for the projection of Occidental views. Limited by historically determined concepts and theories, non-western material was the subject of gross stereotypes and prejudices, perpetuated simply by a lack of understanding. As Johannes Fabian (1995: 41) remarks, 'where there is light, and where there are objects, there are shadows. On the dark side of understanding there is not-understanding and misunderstanding.'

Of course this 'dark side' of understanding is neither specific to primitivism nor to the west, even though their economic and imperial past may make them the exemplar of ideological production. Instead this lack is representative of far broader historical phenomena; one which asserts that knowledge exists within limits or totalities (ideology). This idea has evolved through many variations; from false consciousness, tendencies, and commitments, to, presence, structure, ideology, and discourse. To account for the many different views as to how this process occurs would require its own separate history. But it serves to say that the idea of there being limits to knowledge, or, to consider knowledge as consistently lacking, allows us to review moments in history as being determined by historically specific limitations. Hence in this case, it allows us the possibility to reconsider the primitive in primitivism, and the tribal in the modern, not as representative of contemporary truth, but as constituting a past truth for a historically limited set of practices.

In directing *Primitivism in 20th Century Art*, William Rubin and Kirk Varnedo both equated the extraordinary variety of non-western objects to the 'primitive' tendency discernible in modern art. Over 200 non-western objects were scattered around the gallery, in an attempt to indicate their role in the formation of a modern aesthetic. Other specific objects had actually been owned by the artists themselves and so served as a distinct reference for formal comparison. Many objects had been sourced from distinguished galleries such the Barbier-Muller, Musee de l'Homme, Hamburgisches Museum fur Volkerkunde, Friede and Guimiot. And the sheer variety served to emphasize some striking visual relationships; some 'so forceful that the casual visitor to the exhibition may leave with the impression that tribal art spawned modern art—that modernist abstraction derived directly from tribal objects' (Penny 1985: 91).

This reciprocal relationship is the subject of James Clifford's influential essay written in the year of the exhibition, 'Histories of the Tribal and the Modern'. For Clifford (1998 [1984] 191-92), there 'is *a priori* no reason to claim evidence for affinity (rather than mere resemblance of coincidence) because an exhibition of tribal works seem impressively 'modern' in style can be gathered. An equally striking collection could be made demonstrating sharp dissimilarities between tribal and modern objects.' Yet when a 'three-dimensional Eskimo mask with twelve arms and a number of holes hangs beside a canvas in which Joan Miro has painted colored shapes. The people in New York look at the two objects and see that they are alike' (Clifford 1998 [1984]: 190).

When Rubin and Varnedo contradicts this by suggesting that there does in fact lie a 'deeper' or more natural relationship than formal comparison between famous works by Picasso, and anonymous African production, it provides a source of required speculation for audiences. For beneath the breadth of the 'affinity', as an ill defined taxonomic category, the non-western and western production can co-habit in state of reciprocal flux which binds together contextually different works, and spins an open ended narrative for the viewer of some mysterious relationship surrounding 'informing principles'. Adjoined by undefined notions of formalism, abstraction, even magic and ritual, the MoMA provided an institutionalized space were 'beneath [the] generous umbrella the tribal is modern, and the modern [becomes] more richly, more diversely human' (Clifford 1998 [1984]: 191).

As a consequence of art's generosity, it was works by Picasso, such as *Girl before a Mirror* (1932), or *Les Demoiselles d'Avignon* (1907), and Giacometti's *Spoon Women* (1915), as well as works by Miro and Brancusi that have been attributed with 'striking similarities'. Non-western material in turn simply fluctuated in and out of taxonomic categories, from anthropological specimen to informing references for canonised artworks. Such appropriation and assimilation by the institution reveals the ideological commitments of the exhibition, and in doing so reveals the structure of western modernism as a sympathetic category which privileges form over context. Clifford also suggests that the ideological strategy of the exhibition can be revealed not simply by identifying coincidental resemblances (what is included), but also by discerning dissimilarities (that which is excluded).

This dialectical version of history, as process of inclusion/exclusion, inside/outside, represents the hallmark for the advent of structuralism/deconstruction. Reconstituting events in these terms allows us to better identify ideology, but also to perhaps more accurately grasp Foucault's notion of Power relations. Two points were distinctly lacking from the exhibition take the form of two absences, two points of exclusion. Firstly, there is the absence of the Other, the non-western individual who produced such material. And the second absence regards the exclusion of the primitive in primitivism; the exclusion of what is supposed to represent the tribal in the modern. Such exclusions portray more clearly the Power relations implicit in western culture, specifically here the Power authorised and legitimated by the art institution.

Regarding the first instance of the absent Other, such a move is required if the formal motifs are to act as the harmonizing affinity. If the alternative productions present in the gallery space were to be documented amidst their own specific cultural relations, in short, on their own terms, the idea of an affinity would evaporate. In terms of the material of the non-west the exhibition was context free; of the two-volume, 690-page exhibition catalogue, that presented 19 essays by 16 authors and 1,087 illustrations, very little space was given over to their constitution. Susan Vogel explains that, 'in an art museum, works are shown for their intrinsic merits and not for what they tell us about particular ideas or cultures' (Vogel 1982: 44). So by maintaining a critical distance between the museums representation and the cultural realities of the non-western objects, western institutions, and in this case the MoMA, can subtly appropriate object's effectively 'culture free'. The majority of the non-western objects were displayed without accompanying wall texts which prevented discussion as to the use, creation, consumption and meaning that alternate cultures place, construct, and imbue in their work. The public were unknowingly denied access to this illusive Other's cultural systems, with the desired outcome of consigning the primitive objects to some former history, and the notion of a primitive world view to the past.

The historical emergence of the primitive as an anthropological and scientific concept undoubtedly taints the history of primitivism in European art, this is exemplified in its exclusion from the 1984 MoMA exhibition. Whilst considered to be inexcusable today, for the early 20th century the existence of primitive minds and alternative worldviews was common currency. From Lévy-Bruhl's *mentalité primitive*, to, Lévi-Strauss's *pensée sauvage*, along with scientific questions concerning eugenics, such as the dangers of race mixing, encounters with alternate cultures saw western society project genuine concerns upon muted peoples. The Occident's Other was indisputably real, but importantly it was also historically recent.

Quite simply, the primitive as a 'worldview' did not always exist, and could not have done until the idea of the has become saturated inside the western psyche. As Said (1993) points out, the west developed the idea of barbarism in order to convince the barbarians or their own barbarity to further justify their own imperial actions; the idea that these Other cultures viewed themselves as barbaric; that they we fully transparent unto themselves in highly improbable. As Mary Douglas (1966: 91) remarked, 'The anthropologist who draws out the whole scheme of the cosmos which is implied in [the observed] practices does the primitive culture great violence if he seems to present the cosmology as a systematic philosophy subscribed to consciously by individuals.'

In attempting to divert attention away from this volatile subject, *Primitivism in the 20th Century Art* excluded any detailed historical information regarding the specific relation between canonised European works and the emerging human sciences (Biology, anthropology) that formed there historical surroundings. By placing the emphasis purely upon formal resemblance the MoMA manage to avoid the possible pitfalls of interpreting the works in relation to theories that had been designed to explain the existence of a primitive, or alternative world views. Clifford confirms this regard to the exhibitions take on *art nègre*,

> Overall one would be hard pressed to deduce from the exhibition that all the enthusiasm for things *nègre*, for the 'magic' of African art, had anything to do with race. Art in this focused history has no essential link with coded perceptions of black bodies—their vitalism, rhythm, magic, erotic power, etc.—as seen by whites. The modernism represented here is concerned only with artistic invention, a positive category separable from a negative primitivism of the irrational, the savage, the base, the flight of civilization (1998 [1984]: 197).

Questions regarding the methodological problems inherent in the process of writing ethnography have become a focal point for an anthropology keen to untangle itself from any western ideological commitment. In his 'The Concept of Cultural Translation in British Social Anthropology' (1986), Talal Asad questions the ethnographer/anthropologist's preoccupation with cross-cultural translation. Whilst William Rubin and Kirk Varnedo did not claim to be providing an ethnographic comparison (this comparison was deliberately excluded) between the non-western objects and European art, their notion of 'affinities' envisages formal comparisons as evidence of commensurate motivations.

Four years later an exhibition called *ART/Artifact: African Art in Anthropology Collections* (1988) was held at the Centre for African Arts, New York, reconsidered non-western cultural material and argued that 'our vision is conditioned by our own culture' (Vogel 2006 [1988]: 209). This influential exhibition openly challenged the manner in which alternate cultures had come to be represented and assimilated by both western categories and western concerns, even looking into the idea of excessive charity. Susan Vogel, the curator and director of the exhibition claimed that 'unless we realize that the image of African art we have made a place for in our world has been shaped by us as much as by Africans—we may become misled into believing that we see African art for what it is' (2006 [1988]: 209). This exhibition of African objects and cultural production attempted to be transparent about the contact with ideology and discursive values that museums have to negotiate. In doing so it once again highlighted the structure of western art and institutions and their implicit political stances.

Speculating about the epistemic distance between the art object and the artefact, Vogel's collection stimulated a dialogue that no longer tried to find formal resemblances amongst objects, but instead tackled head on the problematic confrontation of non-western objects with institutionalised ways of viewing; does art become art in the institution? Can non-western objects be seen to be on par with their western counterparts? And what happens when alternative cultural material becomes trapped behind western glass windows in order to be venerated upon podiums inside art, anthropological, and natural history museums? Pushing these ideas into the foreground forced art historians to justify artistic classifications, which in doing so exposed some internal regularities similar to the strategies undertaken by the MoMA in promoting formal affinities.

'Traps as Artworks and Artworks as Traps'[1]

Vogel's exhibition instead encouraged contextual contradictions, and a reconsideration of the 'ways of knowing' about classifications and the authority behind 'ways of looking'. Part art historical/part ethnographic, *ART/Artefact* was considered by Alfred Gell to be a 'master-stroke' of curatorial self reflection. Gell's influential essay 'Vogel's Net: Traps as Artworks and Artworks as Traps' (1996) focused upon the presence of a Zande hunting net. The Zande hunting net 'is a large, interesting looking, honey coloured bundle of rope with regular knots visible beneath the binding. [...] 'For the Zande its purpose and meaning were

1 Gell 1996.

straightforward—to catch animals in communal hunts that brought meat to the village' (Vogel 2006 [1988]: 215-6). But for Gell, the net is an exemplar of the crises in cross cultural translation, but also more broadly representative of the problem with interpretation inside western art history.

Arthur Danto, the exhibitions catalogue essayist, contributed to the debate by suggesting that such non-western cultural material should not be conceived as art even though, by his own institutional and interpretive standards, some of the works provides the same complexities as western production. Gell (Gell 1996: 19) identifies that for non-western artefacts to become art 'the interpretation must relate to a tradition of art-making that has internalized, reflects on and develops from its own history, as western art has done since Vasari, and maybe before.' This notion of an internal coherence, and self reproduction through which the new relates to the past forms part of a structural view of history, consistent with Derrida's (1967) concern with Presence or Foucault's (1969) understanding of the proliferation of discourse. Each attempts to account for what appear to be regimes or narratives that are specific to a subject disciplines operation. Gell (1996: 20) counters Danto by suggesting that his own disciplinary commitments are responsible for his unwillingness to accept these non-western objects as art; 'Danto's dilemma is, essentially, that his interpretive theory of art is constructed within the implicit historical frame of western art, as was its Hegelian prototype.' As Gell goes on to remark that Danto himself becomes 'ensnared' in the hunting net, because he is so thoroughly enmeshed in his own disciplinary regime and his own ideological tendencies.

What becomes difficult is the ability to differentiate the positive from the negative connotations; whether to affirmative non-western objects as art and risk cultural assimilation, or be repressive and risk exclusion, both of which can still be viewed as two sides of cultural imperialism. In denying the non-western cultural production the status of an art object, Gell is quick to point out Danto's ideological commitments to a Hegelian art history which privileges a certain aesthetic function over other functions such as immediate use value, or even process value created during production. But Gell (1996: 37) himself remained divided on the issue claiming instead that 'Vogel's net was set with care, and in it she captured, besides sundry philosophers and anthropologists—including this one—a large part of the question 'what is art?' 'Contemporary art historians continue to remain enmeshed in Elkins' 'trackless thicket' because of the traps set by figures like Vogel, and misplaced actions by curators and critics.

ART/Artefact after 'Primitivisms'?

At what point then can a line be drawn that separates positive and negative interpretation; sympathetic and repressive translation? How can we escape an era obsessed with denouncing normalisation when actions are normalising to the extent that they one of many possible actions? How is it possible to be inclusive without being denounced as exclusive? Art's history, like any other discipline, was set up to encapsulate and orchestrate and great multiplicity of creative production under one category, but like any other affirmation, 'art' simultaneously differentiates itself from not-art, or Other objects; such binary opposition inhibits our capacity to move past such complexities. In a multi cultural society the affirmation of minority rights is one clear example of positive discrimination, set up in reaction against discrimination proper (always negative, repressive) by dominant groups. *ART/Artefact* performed a similar kind of positive discrimination by blurring classifications, and providing a public forum for institutional comparison.

The Other remains a problem for all cultures which seek to distinguish themselves, or whose cultural categories lack the breadth to encompass internal transformations or confrontations from the outside. Primitivism was representative of a misplaced faith in the historic validity in science. Now disproved, and even repressed, the idea of the primitive is notably absent from works which remain examples of Primitivism; they are called by that which they no longer express. Hollowed of negative assumptions Primitivism, in the hands of a western institution, then became subjected to yet another barrage of criticism both for its positive interpretation and its subsequent assimilation of non-western objects; the very objects through which it had hoped to cast a new narrative to bolster the status of the already canonical works. The products of alternative cultures then reappeared to once again challenge the very structure of western art, its history and its institutional role. Since the advent of a post colonial methodology western art has seen the emergence of positive discrimination in favour of non-western works. Such positive discrimination whilst seemingly necessary for those on the side of the repressed unfortunately only harms further those it seeks to protect. For in discriminating positively non-western objects lose their individuality and become part of the very Same they should be distinguished from: being equally legitimate does not necessarily entail equality.

References

Asad, T. 1986 The Concept of Cultural Translation in British Social Anthropology, in J. Clifford (ed.) *Writing Culture: The Poetics And Politics Of Ethnography*, University of California Press: 141-64.

Clifford, J. 1998 [1984] Histories of the Tribal and the Modern, in J. Clifford, *The Predicament of Culture*, Harvard University Press: 188-214.

Daston, L. 2006 The History of Science as European Self Portraiture. *European Review* 14: 523-36.

Douglas, M. 1966 *Purity and Danger: an analysis of concepts of pollution and taboo*. Routledge and Kegan Paul, London.

Elkins, J. 2008 *Six Stories from the End of Painting: photography, astronomy, microscopy, particle physics and quantum mechanics*. Stanford: Stanford University Press.

Gell, A. 1996 Vogel's Net: Traps as Artworks and Artworks as Traps, *Journal of Material Culture*, Vol. 1 (1): 15-38.

Helms, M. 1993 *Craft and the Kingly Ideal. Art, trade, and power*. Austin: University of Texas Press.

Penney, D.W. 1985 Review: *Primitivism in 20th Century Art: Affinity of the Tribal and the Modern*, by William Rubin, in *African Arts*, Vol. 18, No. 4 (August): 27-91.

Shapin, S. and Schaffer, S. 1985 *Leviathan and the Vacuum Pump. Hobbes, Boyle and the experimental life*. Princeton: Princeton University Press.

Toulmin, S. 1990 *Cosmopolis. The hidden agenda of modernity*. Chicago: University of Chicago.

Vogel, S. 2006 [1988] ART/Artifact: African Art in Anthropology Collections, in H. Morphy and M. Perkins (eds) *The Anthropology of Art: A Reader*, Oxford: Blackwell Publishing Ltd: 209-218.

Chapter 6
Doku-porn: Visualising Stratigraphy

Geoff Carver

Stephanie Koerner's session on iconoclasm held at the EAA conference in Cork in 2005 got me thinking about imagery in archaeology, especially the graphics we use in both documentation and publication. I had already been working on various ideas about the iconography of stratigraphic profiles and their schematic representation in the form of the Harris matrix, but I've also been vaguely concerned about the accuracy, precision and—ultimately—objectivity implied by the use of computer graphics. In the following I really want to talk about the power of imagery, specifically the transformation into icon; hence iconography and—as in Cork—iconoclasm.

I will start by referring to a few photos which have so deeply infiltrated our collective conscience that we may label them as 'icons'; images which, despite being familiar, somehow manage to carry such strong emotive associations after all these years (especially, perhaps, in the unfamiliar contexts of a lecture on archaeology). The first shows a Jewish boy captured by German troops during the Warsaw uprising in 1943 (Figure 6.1); the second shows a naked girl (Kim Phúc) running down a road in Vietnam after a napalm attack, and the third is one of the pictures of the 'Tank Man' at Tiannanmen Square. For reasons of copyright I won't reprint the last two, but readers should be familiar with them already.

And I will also tentatively suggest that what unites these images—besides the theme of conflict in this somewhat arbitrary sample—is a sense of *humanity*. I'll get back to this, but please keep in mind, as I contrast this form of iconography with another: *pornography*. And again: you need only think of the iconic image of the centrefold; I don't think I have to reproduce one.

The feature common to all these images is the human form. The difference, though, is that, although the centrefold itself, as an *image—is* iconographic—no individual centrefold—as a *person*—is. Essentially the women in these magazines are anonymous and interchangeable. As images, they are *neither* as memorable *nor* as powerful as those I began with because they lack this sense of *humanity*.

I'd like to suggest that there are two main reasons for this. The first is that pornography does not depict *people*, it represents a consumer commodity. Specifically: it is trying to sell more *magazines*. And with all consumer commodities, the goal—ultimately—is not the *fulfilment* but rather the *creation* of desires, or needs. If you went out and bought a pornographic magazine, and it really left you fulfilled—if you really had been satisfied—then you would have no reason to buy the next issue, and the magazine would soon go out of business.

Figure 6.1 Photo from Jürgen Stroop's report to Heinrich Himmler from May 1943 showing Tsvi Nussbaum and other Jewish survivors of the Warsaw Ghetto Uprising

Like many consumer products, pornography is designed to leave you unfulfilled: to make you buy something you don't need (i.e., the next issue); to make you dissatisfied with yourself and with the real people around you, who aren't quite as perfect as the artificial creations splashed across the glossy pages of some cheap magazine. So you'll be dissatisfied with your car, the music you listen to, your clothing, weight, breath, toothpaste, deodorant, aftershave, and everything else in your life that keeps you from meeting the girls of your dreams, and that the advertisers want you to *buy, buy, buy…!*

The second reason—related to the first—is that pornography is *fantasy*. Take away that sense of anonymity—allow imperfection in the form of character, personality, *humanity*—and you burst the bubble, bringing you back into a world where you *don't* drive a Porsche, you *don't* look like James Bond, and where beautiful women *aren't* quite as open and available as the magazines depict them as being. Because after all: if life really was like that, you wouldn't be buying porn in the first place. So to fulfil this fantasy these girls are inflated, mass-produced and digitally—quite possibly surgically—enhanced; all aspects of personality—*humanity*—air-brushed away.

And—at the risk of making myself really unpopular—I would like to suggest that something similar happens in archaeology. Because of my larger research interests—and for the sake of clarity—I'll use what I hope will be a relatively simple example to explain what I mean.

In *The Archaeological Process*, Ian Hodder describes something like this fantasy as one of the stages of initiation into the secrets of archaeological science:

> If an untrained crew member or student is placed in front of a section and told 'draw what you see', the product is often gobbledygook! There are limitless numbers of lines, lenses, changes in colour and texture than can be observed on any complex section. The untrained eye has no way of discerning the relevant patterning in the soil (Hodder 1999: 53).

This suggests any number of problems dealing with perception, reification and classification, but I will concentrate on the idea of 'discerning the relevant patterning in the soil,' focusing on the good stuff as though nothing else is important. And I'd like to argue that this is the same sort of fantasy that we see in pornography.

Among other things, by drawing this way we perpetuate the fantasy of the closed context: the fantasy that each context is closed, sealed and impregnable, unaffected by post-depositional transformations, intrusives, etc. If I wanted to be crude about it I could almost say that this archaeological fantasy of the closed context is the *inverse* of pornography where everything is open and freely available, but the idea of being closed is a direct consequence the argument that archaeological stratigraphy is analogous to geological stratigraphy; that archaeological stratigraphy is like—but *not* identical to—geological stratigraphy. The qualification is worth emphasising: if archaeological stratigraphy was *identical* to geological stratigraphy, then there would be no need to distinguish between archaeological and geological stratigraphy, and we could simply refer to stratigraphy.

Archaeological stratigraphy is said to be analogous to geological stratigraphy in order to make the claim that archaeological stratigraphy—like geological stratigraphy—is subject to a number of 'laws':

1. Superposition,
2. Original Horizontality, and
3. Original Continuity.

Harris (1989: xii) argues that this analogy:

> fails to account for the fact that most of the stratigraphic problems in archaeology today stem from the fact that we did not divorce ourselves long ago from geological notions of stratigraphy, which are entirely useless in many archaeological contexts.

But we continue to make this analogy so we can argue that—despite all we know of post-depositional transformations and intrusives—'contexts are defined as homogeneous entities' (Carver 1990: 90), and that 'no matter how extensive the upper and lower boundaries or interfaces of a stratum may be, the material enclosed by these boundaries is broadly the same age' (Dirkmaat and Adovasio 1997: 45), and that 'all material or debris contained within a layer or deposit must be the same age or older than the formation of the deposit. While it is possible that material from an older deposit can be incorporated in a younger deposit the opposite cannot be true' (Darvill 2003: 410), etc.

And the styles we use in the drawings we make while documenting our sites reinforce this fantasy of a fixed, closed context. I'll try to illustrate this process—and humanise it—by outlining a minor conflict over the respective drawing styles of Sir Mortimer Wheeler and the German archaeologist Gerhard Bersu (cf. Carver 2005: 21-24). In British archaeology, at least, Wheeler's style is often contrasted with that Bersu's (cf. Webster 1963: 147-50; Barker 1998: 41, 169, 173; etc.):

> Wheeler's type of code has appeared to very many of us to be at once honest to scholarship and conformable to the 'requirement' of our study today. It does not pretend not to be the result of the thought and experience of the excavator applied to the problems of his excavation, and presented in visual form. Alternative conventions have been proposed, the late Gerhard Bersu in his Little Woodbury report being their main proponent in this country, in which an attempt to produce a black-and-white rendering of the uninterpreted complexities of the original section is made (Piggott 1965: 176).

This is because British archaeologists were familiar with both men's work.

In Wheeler's drawings:

> the strata were firmly outlined, filled with symbolic representations of the various types of soil, stones and the like, numbered, named and interpreted…the Wheeler method produced a highly schematic rendering of actuality which was also a definitive interpretation of its meaning…[Wheeler's] sections, equally with his writing style, were clear-cut, positive, authoritative (Hawkes 1982: 86-7).

Among other things—and one of the main reasons why Wheeler's style will be so familiar—is because it readily lends itself to the construction of Harris matrices (Figure 6.2). These clear outlines are visual representations of Harris' interfaces, reinforcing the idea of superposition, and the general belief that contexts are closed—sealed—mutually exclusive.

Bersu, on the other hand, did not draw 'clear outlines' (Figure 6.3). In contrast to Wheeler's interpretive—almost schematic—representation, he developed what Wheeler called a "pictorial' method' of section drawing:

> Reference must be made to an alternative method which was occasionally (though not normally) used long ago by Pitt Rivers, and has from time to time found favour with others, amongst whom Dr Gerhard Bersu is notable...This may be called the 'pictorial' method...The strata are not outlined...but are, to a greater or less extent, differentiated from one another by what may be called a sort of chromatic shading. The result is an impression of the section, more akin to a photograph than to a diagram. The different 'tones' are not consistently framed or demarcated, nor are they numbered.

Although he praised Bersu's innovative fieldwork (Wheeler 1954a: 18) and his excavation at Little Woodbury (Wheeler 1954a: 128-29), Wheeler was critical of Bersu's drawing style. His criticism is somewhat subdued in this instance:

> Granted that, in the hands of an artist, this impressionistic technique has merits, it is equally plain that in the hands of one who is a mere delineator rather than an artist chaos will prevail. A technique which is beyond the reach of most excavators is, on that ground alone, ruled out where a workable alternative exists. The impressionistic technique cannot be defended on the grounds that it is 'less conventional'; in its own fashion, it is no less conventional than the linear technique, only, it is a different convention. And it has this crowning disadvantage: it lends itself to nebulousness, to a blurring of detail and a lack of precision in diagnosis. Even some of Dr Bersu's skilful drawings show this, and a

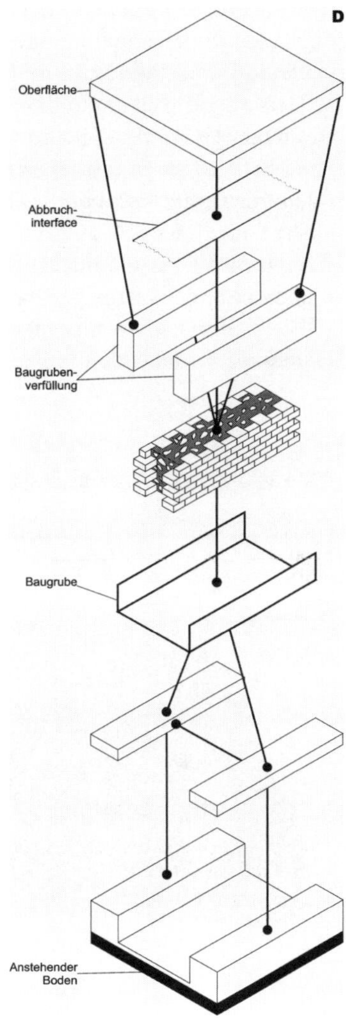

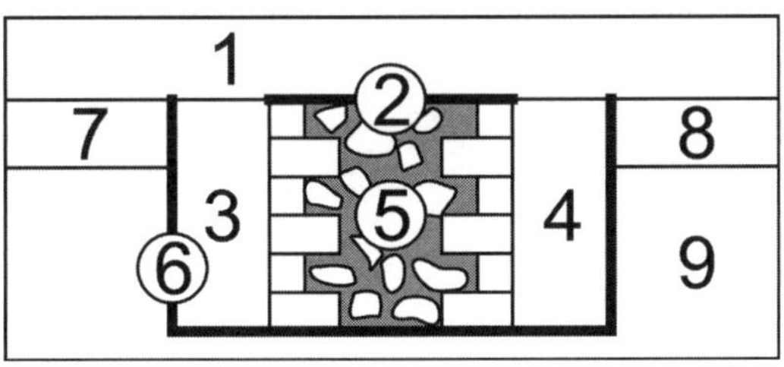

Figure 6.2 Exploded view of stratigraphy
Source: After Harris 1989: 39.

> less skilful performer is liable to be utterly lost. This is a grave defect. If I have tried to emphasize one thing more than another it is the need for considered precision in our work. Any medium or convention which is likely to encourage woolly thinking is to be deprecated: and, save at its rare best, the impressionistic technique is woolly. I do not commend it to the average worker (Wheeler 1954a: 59-60).

Elsewhere, however, the gloves come off. Wheeler (1968) once criticised some profile drawings for being "of the obsolete Bersu type in which pictorial smudgery was substituted for hard-headed analysis' and had, moreover (another deadly sin), 'neglected the essential number of the layers" (Hawkes 1982: 354).

Since I was trained in the Wheelerian tradition, when I first encountered it, the Bersu style was unfamiliar, and before dealing with Hodder's problem of 'the untrained eye [having] no way of discerning the relevant patterning in the soil,' I first had to get over the 'shock of the new.' I still have trouble with Bersu's style of drawing; I still don't know how I'd use his 'pointillist' style in the digital illustrations I work with; how to import this 'gobbledygook' into CAD, for example, nor how to transform this into 3D imagery (for an interesting example of an attempt to convert paper-based records into 3D GIS, see Lieberwirth 2006).

But I didn't think this mattered because I thought 'the obsolete Bersu' style was obsolete—an exception—and this seemed to explain why this style died out.

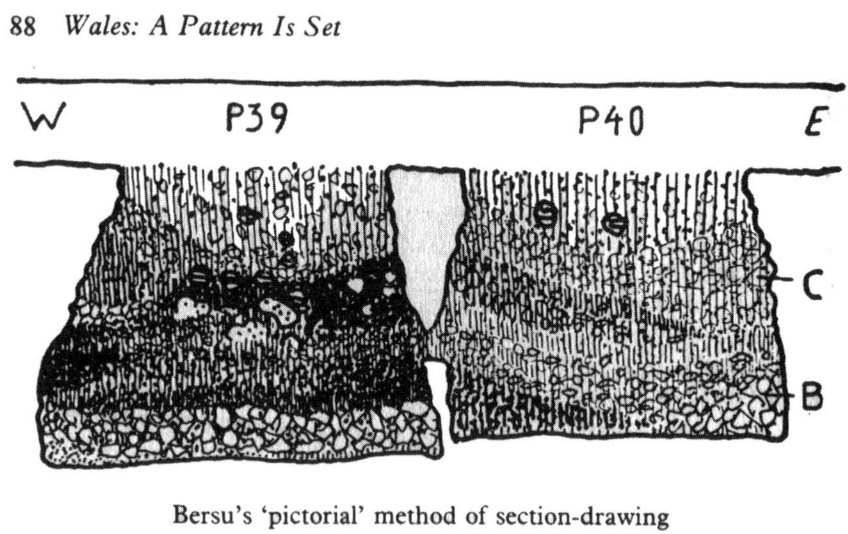

Figure 6.3 Bersu's 'pictorial' method of section-drawing

Source: From R.E.M. Wheeler (1954a), *Archaeology from the Earth*. By permission of Oxford University Press.

If you believe the British, that is. Because Bersu's style isn't dead, after all, it's just not very common within the Anglo-American tradition, the tradition with which I am still most familiar. This may be due to Wheeler's criticism, but also to an underlying ideology—or superstructure—governing the wider aims of Anglo-American archaeology.

Bersu's style actually has a wider distribution than I would have suspected, had I only read the British sources; some randomly selected examples include: Smith 1926: 28 (fig. 15); Cunnington 1926: plates 6-7, 9-13, 46; ÒÀÐÀÊÀÍÎÂÎÉ and ÒÅÐÅÍÒüÅÂÎÉ 1959; Sjövold 1962: 49 (fig. 7); Solecki 1960: facing pg. 620 (fig. 8); Stjernquist 1961: plates X-XIII; Stekelis 1972: plate 4; Ament 1976: 28 (fig. 2); Sugihara 1977: 17-19 (figs. 4-9); Smith 1979: 114 (fig. 22); Banks et al. 1979: 131 (fig. 26), 133 [fig. 27], 135 [fig. 28], etc.; Siiriänen 1984: 36 [fig. 32]; Svoboda et al. 1996: 211 (fig. 9.3); Antoniewicz: 255-263 [figs. 11-12], etc.

There is also any number of variations to be seen in the literature. Bersu- (Scollar et al. 1970: 19 [fig. 4], 20 [fig. 6], 21 [fig. 7]) and Wheeler-style drawings (Scollar et al. 1970: 21 [fig. 8]) in the same publication, for example. Or Bersu-style with Wheelerian outlines (Banks et al. 1979: 117 [fig. 23], 121 [fig. 24], 127 [fig. 25]), or Bersu mixed with accurate outlines of the shape of a cave (Sedlmeier 1982: 20-21 [figs. 3-4], 23 [fig. 6], 25-26 [figs. 8-10]). Then there is the so-called Polish 'hyperrealism' (Kobyliñski 1993: 58), etc.

The point is that there is a wide variety—a spectrum—of drawing styles, not just the binary opposition of Wheeler vs. Bersu.

And the fact is, despite the mythology and whatever he may have written on the subject in *Archaeology from the Earth* and elsewhere, Wheeler himself does not seem to have followed any hard and fast rules. Some of his drawings had outlines (Wheeler and Wheeler 1936: plate XVII; Wheeler 1943: 52 [fig. 10], 53 [fig. 11]), and some did not (Wheeler 1954b: 17 [fig. 5], plates VII, XIV, XVII). Some showed a mix of layers with and without outlines (Wheeler 1943: 39 [fig. 7], 54 [fig. 12A], 87 [fig. 15], 95 [fig. 18], plates V-VI, VIII-IX, XI-XIII, XIX), sometimes with find-spots indicated (Wheeler 1943: 85 [fig. 14]).

Many of his drawings indicated textures using some form of hatching (cf. Wheeler 1954a: 60 [fig. 13])—sometimes with lines indicating layer boundaries or discontinuities within a single hatching pattern (Wheeler and Wheeler 1936: 55 [fig. 3])—while others were mostly empty (without hatching) but had the locations of finds marked between borders (Wheeler and Wheeler 1932: plate IV; Wheeler and Wheeler 1936: plate XVIII). Sometimes there was hatching but no outlines, especially in the layers immediately under the symbol indicating humus. And this seems to offer a sort of explanation, a key to understanding the iconography of Wheeler's graphic language: the layer outlines seem to represent archaeologically significant boundaries, not simple differences between natural soils of no archaeological significance; as indicated, for example, in Wheeler and Wheeler 1936: 55 [fig. 3], plates VII, XIII-XIV, XIX-XX, Wheeler and Wheeler 1936: plates VII, XIII-XIV, XIX-XX; Wheeler and Richardson 1957: 29 [figs. 4 and 4A], plates III-IV, XVII, XXIX XXXI, XXXIX), etc.

My own analysis has not yet allowed me to judge Bersu's work on an equal footing with Wheeler. I will note that—from what I have seen so far—his 'smudgery' (or 'gobbledygook') reflects a strong sense of the real-world difficulties of preparing and drawing profiles disturbed by stones and roots (Bersu 1945: 8), post-depositional transformations (Bersu and Wilson 1966: 4, 7-9), etc.

And then there is the suggestion that the story might actually be much more complicated:

> One very definite error, and one that had already provoked Gerhard Bersu's disapproval during his inspection, concerned the numerous ovoid pits sunk deep into the chalk among the huts. [Wheeler] had been writing of pit-dwellers as a small boy and was reluctant quite to abandon the idea now, when it had become generally discredited. Although in Maiden Castle he conceded that most were storage pits later used to dump rubbish, he stubbornly maintained that some had been used as sheltered rooms, and that in one the occupants had crouched round a fire, eating mutton and throwing the bones over their shoulders. So strongly had this picture taken possession of his imagination that he ignored both informed opinion and the probability that his mutton-eaters would have been both roasted and suffocated (Hawkes 1982: 177).

Certainly, whatever the truth in this particular matter may be, I would at least like to suggest the possibility that the theoretical—or even ideological—foundations to a widely influential drawing style may be attributable to a minor squabble between two eminent archaeologists over the interpretation of some pits.

Then there's also the problem that Bersu's too *artistic*. His style might be OK for 'Continentals,' but Anglo-Americans want to buy into the fantasy that we're *scientists*, and that what we're doing is *science*.

So we try to make our illustrations and our documentation *look* scientific, in much the same way that we try to adopt what we think of as being a 'scientific style' in our writing (cf. Selzer 1993). Originally we modelled everything—including the idea of the 'closed context' (which works with dinosaurs encased in stone but not necessarily for artifacts in soil, by the way)—on geology, because Charles Lyell had transformed geology from an out-of-fashion hobby—fossil-hunting—into a science worth emulating. A little later—when we switched loyalties and strove to mimic physics—a lot of the processualists thought systems analysis was cool. So now we use flowcharts, and even try to reduce—or transform—archaeological sites into the flowchart format of the Harris Matrix (it may perhaps also be worth noting that even Lyell [1991: 243 figs. 58 and 59] *re*printed something that looks remarkably like a Harris matrix).

And there's nothing really wrong with this, unless you make the mistake of thinking that—since it *looks* scientific—then it must be so. This surprisingly easy mistake manifests itself in any number of ways: the 'expressiveness and integrity' of Wheeler's numbers and 'clear outlines,' for example, but there was also an older tradition that images themselves constituted 'evidence' antiquaries and early archaeologists (Burke 2003). Which brings us to the related belief that the camera doesn't lie, a mistake no archaeologist would ever make, especially since there's obviously a world of difference between a commercial pornographic image intended merely to titillate and a scholarly work of scientific documentation, such as that shown in Figure 6.4. Figure 6.4 shows one of Edward Muybridge's studies of 'human locomotion.' I often use this as an example of the way archaeologists compose narratives by arranging events or contexts along a temporal axis; Muybridge and filmmakers put individual images together we can get at least the illusion of continuity and the passage of time in the same way. I especially find this useful as an example of multiple perspectives when trying to explain post-processualism—or comparing the plans and profiles that we document our sites with. This metaphor becomes even stronger when you remember that archaeological documentation derives from architecture: showing the plan, perspective and section of a building, reducing 3D or even 4D reality to the two dimensions of the printed page.

I'd like to look at this as an image in itself, though, and try to suggest another sort of equivalency between Muybridge and a Harris matrix, by asking what this photo really shows. Muybridge has been subject to some fairly substantial criticism:

> In spite of the anonymity of the gridlike background in these pictures, the seeming objectivity of the camera and authenticity of results, and even the evident seriousness of the direction under which the work was carried out, these pictures are inconsistent with what we understand to be a scientific analysis of locomotion. The unsystematic and incongruent aspects of Muybridge's photographs effectively obscure any knowledge of the underlying laws governing the mechanics of movement. In many cases the moving limbs are hidden by drapery (Braun 1992: 247).

More important in terms of content, however:

> The very movements chosen for scrutiny, in the human realm at least, are often unrepresentative. The scientific purpose served by a study of one woman chasing another with a broom or a girl falling into a pile of hay…

or (as in the example shown in Figure 6.4), of one woman pouring a bucket of water over another—

> …is impossible to fathom without directions. Are we to understand that there are constant factors that govern 'Fancy Dancing' (the subject of Muybridge's plates—187-89 and 191-94) or numerically expressed laws that can be derived from hickens Being Scared by a Torpedo,' the final plate in *Animal Locomotion*? (Ibid.).

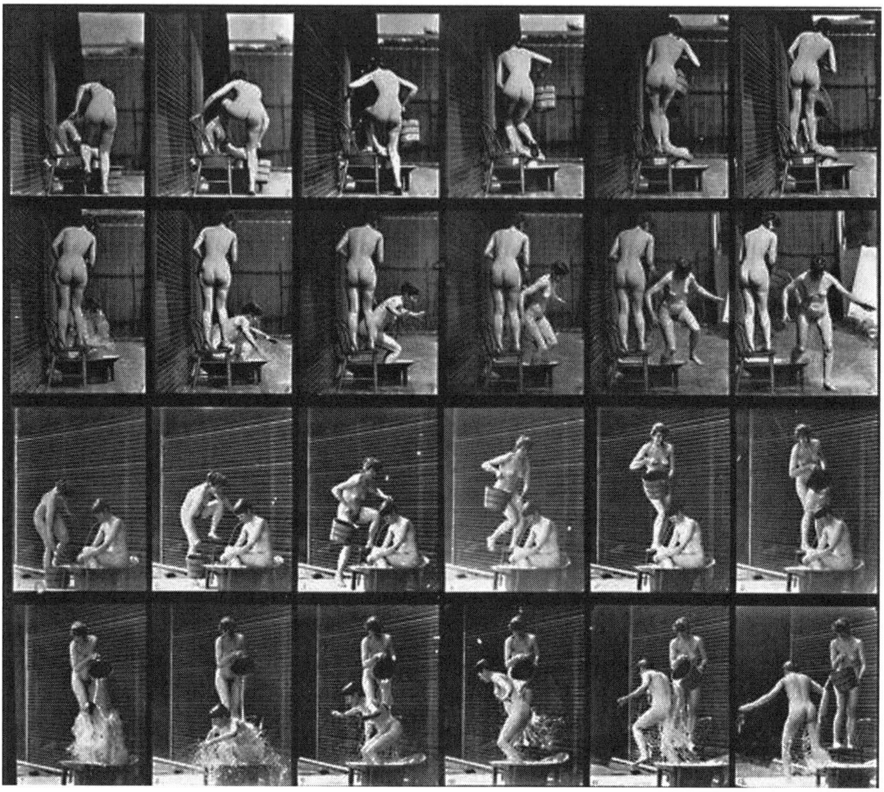

Figure 6.4 A photo by Eadweard Muybridge, of a 'Woman pouring a bucket of water over another woman'

Source: Plate 407 from his *Human Locomotion* series taken from multiple perspectives.

What I'm ultimately concerned with is this process of reification, this self-fulfilling prophecy that if it *looks* scientific, then it *is* scientific. We find this especially and increasingly in computer graphics. They look so real that you *want* to believe them. And for all that they have taken from studies of textual analysis (i.e., hermeneutics; cf. Hodder 1999: 32, Holtorf 2000: 167, Hacking 2002: 92-3, etc.), I have yet to see any post-processualists address this problem of the willing suspension of disbelief.

But that's not really what I wanted to talk about, anyway, because—ultimately—I'm not really as interested in archaeological imagery as I am in what we do with it, and why. And it's not so much that we've been so seduced by our imagery that we take it for the reality it is meant to represent; I'm more concerned with what influence it has in reinforcing our stated aims and goals as archaeologists.

This is why I'm trying to equate Muybridge's *constant factors* with the 'universal laws of human behaviour' processualists claimed is what we are supposed to uncover. I reject that aim for a number of reasons. First off, because we can't excavate 'universal laws of human behaviour,' we can only dig up artifacts. At best we dig up—in the words of David Clarke (1973: 17)—'bad samples' of 'indirect traces' of 'unobservable hominid behaviour patterns.'

And second, I reject this because archaeology is and should be *about* people. When has anyone ever been as impressed by a flowchart outlining social hierarchies, or a kinship system, or a typology or a stratigraphic profile—as they were by what Joyce (1993: 204) described as 'the feeling which arrests the mind in the presence of whatsoever is grave and constant in human sufferings'?

In the end, I fear—and fight against—being seduced by the fantasy of objective science, away from the reality of what we really are.

References

Ament, H. 1976 *Die fränkischen Grabfunde aus Mayen und der Pellenz*. Germanische Denkmäler der Völkerwanderungszeit; Serie B: Die fränkischen Altertümer des Rheinlandes Gebr. Mann, Berlin.

Antoniewicz, W. Eneolityczne groby szkieletowe i ziemianki mieszkalne w Nowym Darominie (pow. sandomierski, Ma³opolska). *Niederlův Sborník* 243-74.

Banks, P.J., Jones, R.J.A. and Smith, C. 1979 Site SK187082. In *Fisherwick: The Reconstruction of an Iron Age Landscape*, edited by Smith, C., pp. 115-49. BAR, Oxford.

Barker, P. 1998 *Techniques of archaeological excavation* 3rd edn. Routledge, London.

Bersu, G. 1945 *Das Wittnauer Horn*. Monographien zur Ur- und Frühgeshichte der Schweiz Birkhäuser, Basel.

Bersu, G. and Wilson, D.M. 1966 *Three Viking Graves in the Isle of Man*. Monograph Series Society for Medieval Archaeology, London.

Braun, M. 1992 *Picturing Time: the work of Etienne-Jules Marey (1830-1904)*. University of Chicago Press, Chicago.

Burke, P. 2003 Images as Evidence in Seventeenth-Century Europe. *Journal of the History of Ideas* 64(2):273-96.

Carver, G. 2005 Archaeological Information Systems (AIS): Adapting GIS to archaeological contexts. In *Workshop 9: Archäologie und Computer 3-5. November 2004*, edited by Börner, W. and Uhlirz, S. Magistrat der Stadt Wien, Wien.

Carver, M.O.H. 1990 Digging for data: archaeological approaches to data definition, acquisition and analysis. In *Lo Scavo archeologico: dalla diagnosi all'edizione: III Ciclo di lezioni sulla ricerca applicata in archeologia*, edited by Francovich, R. and Manacorda, D., pp. 45-120. All'insegna del giglio, Firenze.

Clarke, D. 1973 Archaeology—Loss of Innocence. *Antiquity* 47(185):6-18.

Cunnington, M.E. 1926 *Woodhenge: A Description of the Site as revealed by Excavations carried out there by Mr. and Mrs. B.H. Cunnington, 1926-7-8. Also of Four Circles and an Earthwork Enclosure south of Woodhenge*. George Simpson and Co., Devizes.

Darvill, T.C. 2003 *Oxford Concise Dictionary of Archaeology*. Oxford University Press, Oxford.

Dirkmaat, D.C. and Adovasio, J.M. 1997 The Role of Archaeology in the Recovery and Interpretation of Human Remains from an Outdoor Forensic Setting. In *Forensic taphonomy: the postmortem fate of human remains*, edited by Haglund, W.D. and Sorg, M.H., pp. 39-64. CRC Press, Boca Raton.

Hacking, I. 2002 *Historical Ontology*. Harvard University Press, Cambridge, Mass.

Harris, E.C. 1989 *Principles of Archaeological Stratigraphy*. 2nd edn. Academic Press, London.

Hawkes, J. 1982 *Mortimer Wheeler*. Weidenfeld and Nicolson, London.

Hodder, I. 1999 *The Archaeological Process: an introduction*. Blackwell, Oxford.

Holtorf, C. 2000 Making sense of the past beyond analogies. In *Vergleichen als archäologische Methode: Analogien in den Archäologien; Mit Beiträgen einer Tagung der Arbeitsgemeinschaft Theorie (T-AG) und einer Kommentierten Bibliographie*, edited by Gramsch, A., pp. 165-75. B.A.R., Oxford.

Joyce, J. 1993 *A Portrait of the Artist as a Young Man*. Wordsworth, Ware, Hertfordshire.

Kobyliński, Z. 1993 Polish medieval excavations and the Harris Matrix: applications and developments. In *Practices of Archaeological Stratigraphy*, edited by Harris, E.C., Brown, M.R.I. and Brown, G.J., pp. 57-67. Academic Press, London.

Lieberwirth, U. 2006 *Landscape and archaeological deposits. 3D modelling and analysis of archaeological stratigraphy using Geographical Information Systems in archaeology—it's possibilities and advantageous*. Humboldt-Universität zu Berlin.

Lyell, C. 1991 *Principles of Geology*. Facsimile of 1st edn Vol. 3. University of Chicago Press, Chicago.

Piggott, S. 1965 Archaeological Draughtsmanship: Part I. *Antiquity* 39(155): 165-76.

Scollar, I., Spitaels, P. and Goff, C. 1970 The Excavation Report. In *A Medieval Site (14th Century) at Lampernisse (West Flanders, Belgium)*, edited by Scollar, I., Verhaeghe, F. and Gautier, A., pp. 11-29. Series edited by Laet, S. J. d. De Tempel, Brugge.

Sedlmeier, J. 1982 *Die Hollenberg-Höhle 3: Eine Magdalénien-Fundstelle bei Arlesheim, Kanton Basel-Landschaft*. Basler Beiträge zur Ur- und Frühgeschichte Habegger, Derendingen-Solothurn.

Selzer, J. 1993 *Understanding Scientific Prose*. Rhetoric of the Human Sciences University of Wisconsin Press, Madison.

Siiriänen, A. 1984 *Excavations in Laikipia: An Archaeological Study of the Recent Prehistory in the Eastern Highlands of Kenya*. Finska Fornminnesföreningens Tidskrift Suomen Muinaismuistoyhdistyksen Aikakauskirja, Helsinki.

Sjövold, T. 1962 *The Iron Age Settlement of Arctic Norway: A Study in the Expansion of European Iron Age Culture within the Arctic Circle*. Tromsö Museums Skrifter I: Early Iron Age (Roman and Migration Periods). Norwegian Universities Press, Tromsö.

Smith, C. 1979 Site SK1 83098. In *Fisherwick: The Reconstruction of an Iron Age Landscape*, edited by Smith, C., pp. 113-14. BAR, Oxford.

Smith, R.A. 1926 *A Guide to Antiquities of the Stone Age in the Department of British and Mediaeval Antiquities*. British Museum, London.

Solecki, R.S. 1960 Three Adult Neanderthal Skeletons from Shandihar Cave, Northern Iraq. In *Annual Report of the Board of Regents of the Smithsonian Institution*, pp. 603-635. United States Government Printing Office, Washington DC.

Stekelis, M. 1972 *The Yarmukian Culture of the Neolithic Period*. The Magnes Press—The Hebrew University, Jerusalem.

Stjernquist, B. 1961 *Simris II: Bronze Age Problems in the Light of the Simris Excavation*. Acta Archaeologica Lundensia (Papers of the Lunds Universitets Historiska Museum) Rudolf Habelt/CWK Gleerups, Bonn/Lund.

Sugihara, S. 1977 *Two Lithic Cultures of Take, Gumma Pref., Japan*. Reports on the Research by the Faculty of Literature, Meiji University Meiji University, Tokyo.

Svoboda, J., Ložek, V. and Vlček, E. 1996 *Hunters between East and West: The Paleolithic of Moravia*. Interdisciplinary Contributions to Archaeology Plenum, New York.

Webster, G. 1963 *Practical Archaeology: An Introduction to Archaeological Field-work and Excavation*. Adam and Charles Black, London.

Wheeler, R.E.M. 1968 Review of *Hod Hill, Vol. 2: Excavations carried out between 1951 and 1958*; Ian Richmond and others. The British Museum, London, 1968. *Antiquity* 42(166):149-50.

—— 1943 *Maiden Castle, Dorset*. Reports of the Research Committee of the Society of Antiquaries of London Society of Antiquaries, Oxford.

—— 1954a *Archaeology from the Earth*. Oxford University Press, Oxford.

—— 1954b *The Stanwick Fortifications: North Riding of Yorkshire*. Reports of the Research Committee of the Society of Antiquaries of London Society of Antiquaries, Oxford.

Wheeler, R.E.M. and Richardson, K.M. 1957 *Hill-Forts of Northern France*. Reports of the Research Committee of the Society of Antiquaries of London Society of Antiquaries, Oxford.

Wheeler, R.E.M. and Wheeler, T.V. 1932 *Report on the Excavation of the Prehistoric, Roman, and Post-Roman Site in Lydney Park, Gloucestershire*. Reports of the Research Committee of the Society of Antiquaries of London Society of Antiquaries, Oxford.

Wheeler, R.E.M. and Wheeler, T.V. 1936 *Verulamium: A Belgic and two Roman Cities*. Reports of the Research Committee of the Society of Antiquaries of London Society of Antiquaries, Oxford.

ТАРАКАНОВОЙ, С.А. and ТЕРЕНТьЕВОЙ, Л.Н. 1959 *ВОПРОы ЗТНИУЕСКОЙ ИСТОРИИ НАРОДОВ ПРИБАЛТИКИ*. Moscow.

Chapter 7

Doing Archaeology in the Risk Environment: A Theoretical Sketch, Some Observations and Propositions

Koji Mizoguchi

I

'Risk' is a uniquely modern concept. The concept differs from that of danger or fate. The former is the possibility of damage or harm that is inflicted by known but uncontrollable forces. The latter is the way to make sense of damage or harm inflicted or feared to be inflicted by uncontrollable forces by attributing their causes to God's will or to predetermined programmes designed by the transcendental. In contrast to them, risk can not only be foreseen but also reduced or 'managed' to a degree. Or, risk is the possibility of damage or harm which is *perceived* to be controllable and manageable. In other words, as danger and fate are communicated, and made sense of, as something uncontrollable, risk is communicated, and perceived, to be manageable.

II

This difference between the concepts derives from that between the pre-modern, or hierarchical, and the modern, or functionally-differentiated, modes of communication in which these concepts are embedded.

Let me explain. From the emergence of ancient/early states until the dawn of modernity, the reference to transcendental entities functioned to solve or conceal issues concerning the intrinsic uncertainty of the world. By making '*being* or *not being faithful* to God' the ultimate referential point, contingent occurrences were classified into the categories of good and bad events, and were made sense of by attributing their causes to God's will, approval and anger: good events were the sign of God's approval and bad events God's disapproval and anger. Their classification in social discourse was conducted by higher-ranked people such as kings, monarchs and monks as representatives of God's will, and no one questioned the righteousness of the hierarchical order.

Modernity replaced God with Science and Technology. Vertically divided societies transformed to horizontally or functionally divided and differentiated ones, as the representatives of God, such as kings, monks and monarchs, disappeared, and horizontally structured sectors such as economy, politics, science, culture, and so on began to function and reproduce themselves by drawing upon mutually distinct codes/systems of value, e.g., payment and non-payment in economy, wining and losing elections in politics, truth and untruth in science, and so on. And the relationship between those sectors is one of mutual commentary and relativisation; politics tries to make sense of and control economy by referring to its own code, or a 'truth', and vice versa, for instance, and they do so by knowing that the codes upon which they operate are different and unshareable and that the complete control by one sector of another is impossible. To put it simply, Modernity lost such universally-sharable vale systems as 'traditions' and God. Accordingly, the cause of dangers and disasters, i.e., realised dangers, cannot be attributed to God's will any longer, nor can be made sense of by referring to higher-ranked individuals' explanation such as monks or kings.

On the other hand, the coming of horizontally-situated, mutually commenting communication fields or value systems opened up a new possibility; a danger generated or a disaster occurred in one communication field can be commented upon, and hence made sense of, by referring to the value systems of other communication fields. For instance, an economic disaster can be explained as a failure of political system, a disaster in heritage management can be blamed as a failure of science system, and so on. And this newly generated potential of mutual explanation and attributing causality to the operation of other social systems than that of the system itself in which disaster occurs led to generate the notion of calculable, and hence manageable, danger. It marked the emergence of the concept of risk: potential damage caused by a foreseeable danger is now calculated in terms of how that would disrupt the operation of other social systems, and the potential cause of a foreseeable danger is attributed to the malfunction of other social systems.

III

However, the foundation upon which the calculation of danger or risk assessment is made is bound to be relativised, because that foundation itself is a *chosen* position. Economic risk assessment in heritage industry made from an political-administrative viewpoint can be criticised from a scientific viewpoint, for instance. Without some kind of transcendental referential points, this endless multiplication of relativisation cannot be halted.

In other words, any type of risk calculation and risk management has to take this type of *indeterminacy*, intrinsically generated by the multiplication of mutual relativisation between functionally differentiated systems, for granted, and hence have to develop the ways in which to cope with it. And, because the concept of risk itself is a product of horizontally-situated, functionally-differentiated, communication fields, the way to cope with the indeterminacy can only be found in the way we manage our communication.

The following seem to me two possible devices that can be adopted. One is to *heuristically assume* that a particular value system can be prioritised and applied as long as the objective of a risk assessment and the stances of the stakeholders involved are clear, and to try to *tame* the indeterminacy, i.e., to make it feel controllable, whereby to actually control it to some extent. The other is to give up hoping to apply a single value system to the practice, and instead to concentrate on just *getting on with* ever-changing communicative situation in risk management. In practice, the former can take the following forms:

- To choose a particular value system as the just framework for the assessment and management of a risk by examining the positions of the stakeholders and the matrix of their interests and potential loss and gains. Archaeological interventions to post colonial issues tend to adopt this strategy.
- To heuristically construct a hierarchical order of values and priorities for the assessment and consultation, and progressively reduces the range of permissible potential outcomes as the process of the assessment and consultation ascends the hierarchical structure. A typical example is the judicial system constituted by hierarchically-situated courts, and archaeological heritage management involving both governmental and non-governmental institutions tend to adopt this strategy.

The latter, in contrast to that, tends to take the following form:

- To focus on how to avoid altogether introducing a hierarchical order or hierarchical referents to the practice, and to hope a shareable value system is naturally to emerge amongst the stakeholders.

Logically speaking, none of these strategies can overcome the indeterminacy problem completely, simply because they cannot modify the intrinsic nature of the conception of risk and communication about it; a particular way to assess and manage a risk is bound to be relativised by different perspectives based upon the operation of different social systems and different interests. In that sense, what matters, or what we can meaningfully be concerned about in choosing one out of these possible strategies is: what *advantages* and *risks* the adoption of each of them generate.

IV

Western academic archaeology tends to adopt our second strategy. The disciplinisation of archaeology in the West has progressed by drawing upon the model of natural scientific cycle, and the ethos of research post-oriented career pursuit has enhanced hierarchisation in the invention and innovation of methodology and theory, a form of archaeological risk management, and that in the publication of research outcomes.

The situation is, however, rapidly changing. The reflection upon post-colonial conditions, negative effects of globalisation, and the involvement of 'Western' archaeological practices to their spread leads archaeologists to place high hope upon the autonomous generation of shared agreements through communication with the minimum imposition of hierarchical orders. The trend also actively endorses position-taking as stakeholders, and heuristically assumes that making explicit the interests each stakeholder has enables to reach a context-sensitive risk assessment and risk management.

This strategy appears to be close to our third strategy; risk assessment through non-hierarchical discourse without prioritising any specific value system naturally generates a middle ground and a sharable value system with which to calculate and minimise risks. However, in practice, this strategy tend to tacitly imply the prioritisation of a specific *moral* ground, that is, minority value systems always have to come first. This prioritisation strategy would work as long as the strategy encourages those who have been silenced to come out to have their say and as long as that gets previously stifled communication going smoothly and productively. However, if the strategy were perceived by sectors of the participants to be a forcible imposition of a specific value system upon them, it would result in the division between those who are for and against the prioritised value system, and the issue to be solved would shift from the assessment and management of a specific risk to the management of dispute over the way in which how to assess and manage the risk is discussed. In other words, the prioritisation of a specific value system in risk assessment and management can generate new risks, i.e., risks of disrupting the smooth and fruitful continuation of communication concerning better risk assessment and management.

This stance might also make impossible sufficiently narrowing down the range of variance in the perception of a risk to be assessed and managed amongst different stakeholders. Behind the prioritisation of minority value systems lies the belief that the minorities have been excluded from the decision making which might cause dangers, i.e., uncontrollable risks, to them. That feeling, the feeling of powerlessness, makes their fear and the perception of dangers uncontrollable and impossible to define. That would make almost impossible to agree amongst the stakeholders what dangers they need to recognise as risks. Without the shared understanding of what risks they have to assess and manage, we cannot communicate risk communication.

Another possible negative consequence of prioritising minority value systems, which is getting perceived as a taken-for-granted position by an increasing number of archaeologists, may be felt particularly strongly when the risk assessment made in the archaeological communication field is assessed and implemented by experts in other social communication fields. As a scientific communication system, a dominant expectation assigned to the working of the archaeological communication system is: it is reproduced by drawing upon the binary distinction between truth and untruth. If this expectation were betrayed and instead the impression that the decisions are made by drawing upon a specific moral stance, not only would the credibility of the archaeological communication system dismissed but also that would lead to negative, at times even hostile, reactions to the outcomes of archaeological discourse by other communication systems such as the economic and the political.

V

The conclusion of this short essay, as you would guess, has to be ambiguous and negative. In order for us to better handle difficulties involved in archaeological risk assessment and management we had better concentrate on devising the ways to continue risk communication smoothly and fruitfully. By this I mean that we had better abandon the illusory hope that the sharing of a value system with which every stake holder can agree with in assessing and managing risks is possible.

The recently gained recognition that modern scientific archaeology is intrinsically a colonial practice encourages an increasing number of archaeologists to prioritise minority value systems in assessing and manage risks which their practice generates, but that strategy, the above consideration suggests, might be met with multi-layered problems. One of these would be that the strategy, even if adopted in the best intention of making previously silenced voices to be heard, if not associated with the attitude of being open to dialogue concerning the validity of the prioritisation, might be perceived as the imposition of a particular moral stance which is to be countered with different moral stances based upon different value systems and understandings of history. By a 'moral stance' hear I mean a specific value system the validity of which is perceived by its advocates as 'non-negotiable'. In the worst cases, the strategy might result in the stoppage of the communication altogether, or lead to the shift of the objective of the communication from reaching a good risk assessment and risk management to negotiating over the way how to discuss about the way a better risk assessment and management can be conducted. This might better be recognised as waste of time and our precious energy.

For the sake of the smooth and fruitful continuation of archaeological risk communication, we had better be prepared to abandon our taken-for-granteds, including attractive stances of prioritising minority or formerly oppressed value systems, and strive to narrow down, step by step, the mutually permissible range of variants in the recognition of what risks we have to deal with and how to assess and manage ttem. To add, we have to remind ourselves that archaeological communication system is perceived by other communication systems to be a subsystem of the general science communication system. Portraying ourselves to be other than that would disrupt risk communication between ourselves and those who identify themselves to be the members of other communication fields.

Acknowledgements

The argumentation derives heavily upon the thoughts of the late Niklas Luhmann, the German Sociologist, and the essence of his social systemic thinking with fundamental epistemological implications can be found in (Luhmann 1995). My stance as to how to confront with the risk environment as a fundamental trait of the post-modern condition, please see (Mizoguchi 2006). I am indebted to Staphanie Korner for encouraging me to go deep into the relationship between archaeology and contemporary epistemic-ontological issues and for continuously providing me with opportunities to talk about my on-going thoughts on the issue. I also thank Ian Russell for encouragements at the right moments.

References

Luhmann, N. 1995 *Social Systems.* Stanford, CA: Stanford University Press.

Mizoguchi, K. 2006 *Archaeology, Society and Identity in Modern Japan.* Cambridge: Cambridge University Press.

Interface 1

Re-designing Mobilities

Aspects of John Urry and Stephanie Koerner's Conversation, April 2009

John Urry (JU): In 1990 when I first published the *Tourist Gaze* it was much less clear just how significant the processes we now can 'globalisation' were to become. Indeed the 'internet' had only just been invented and there was no indication how it would transform countless aspects of social life, being taken up more rapidly than any other previous technology. And no sooner than the internet had begun to impact, than another 'mobile technology', the mobile phone, transformed communications practices 'on the move'. Overall the 1990s have seen remarkable 'time space compression' as people across the globe have been brought closer through various technologically assisted developments… And part of this sense of compression has stemmed from the rapid flows of travellers and tourists physically moving from place to place [and] there is no evidence that virtual and imaginative travel is replacing corporeal travel, but there are complex intersections between these different modes of travel that are increasingly de-differentiated from one another…Of course not all members of the world community are equal participants…[in emerging] countless mobilities, physical, imaginative and virtual, voluntary and coerced (Urry 2001: 1-2, 8).

Stephanie Koerner (SK): I very much appreciate that you take the time to talk with me. Over the last decade insights drawn from your works have contributed greatly to awareness of the embeddedness (in a number of contradictory trends discussed in this volume) of changes taking place in the dynamics of pedagogical institutions—such as departments of anthropology and archaeology in universities, museums, educational television—and wider public affairs. Especially influential have been not only your writings and publications together with Scott Lash, Cris Rojek and others on the social and ecological implications of mobility and visualisation technologies, but also on policy implications of the indeterminancy of human-environmental relationships. Thank you very much for a copy of the book that you and Kinsley Dennis wrote, *After the Car* (2009) and for suggesting that I attempt to integrate some of its themes into the writing up of aspects of our discussion.

JU: You are welcome, and I hope that some of the themes of our discussion might be further pursued in relation to the series of workshops that you are organising together with Bron Szersynski and Brian Wynne for the 2010

Annual Research Programme (IAS) of Lancaster University's Institute for Advanced Studies around the theme 'Experimentality'. What themes should our conversation stress?

SK: I would like to ask about the backgrounds of the directions your research has taken, themes that may run through a number of your publications, and about what you think might be shaping the future of 'transformations of tourism, travel and theory' (Rojeck and Urry 1997).

JU: That sounds more like another book on its own! But we can give it a try.

SK: Do not worry. We do not need a great deal of time…I'll integrate background materials into our discussion's write up from publications—for instance the quote at the onset from your 2001 paper. I recall that we started talking about some of these materials at the final workshop in 2008 IAS Research Programme around the theme 'New Sciences of Protection'. One of the most interesting questions explored in that Programme was that of whether 'safety' based on the various technologies discussed was achievable and desirable—as foundations of politics and as ethics. You discussed a number of examples to illustrate the relevance of questions of 'achievable' and 'desirable' by and for whom, which compares in very important ways to themes of the 'unquiet pasts' volume, such as the question of for and by whom heritage is 'managed'. One issue at stake in both cases—as with mobility and visualisation technologies—is that means of protection, inclusion, and so on can and have also been hazards, threats, means of control and exploitation, exclusion, etc. Does that sound like a theme that has long motivated your research or place to start?

JU: Yes, but we should stress that any analogies we might want to draw must depart—not from a universalising generalisation and or set of assumptions about history, modernity and so on—but from considerations of comparisons that might be made between concrete situations. I cannot remember ever not being concerned to stress such approaches. In the 1980s I was involved in numerous discussion and events with socialists and what are now called specialists in science studies (STS), which stressed the importance of such approaches to relationships between technological and social change. Much discussion developed about the relevance of such approaches for research on relationships between processes that Harvey (1989) and Giddens (1990) described as 'time-space distanciation and compression.' I was also very interested in the relevance for 'contextual approaches' of insights developed by Sheila Jasanoff (1995) and Brian Wynne (1992, 1996), for instance, concerning the importance of sociological 'analyses of 'local knowledge and practices'… can challenges the ways in which…standardising discourses of science can conceal unwarranted political assumptions' (Macnaghten and Urry 1998: 18). Considerable opportunities were provided for developing such approaches not only because of directions research was taking in Lancaster University's social science programmes, but also because of the university's geographic location. My first inquiries into the history and transformation of tourism and social impacts of mobility and visualisation technologies centred on Morecambe Bay—once a core centre of industrial northwest England—and the Lake District. The area has not only a long history as source of raw materials and labour. It has also a long history as tourist destination. For example, today a favourite image of the Lake District is a photograph of Ullswater taken from the lakeshore with daffodils slightly offset in the foreground. According to literary historians, this is the scene that inspired William Wordsworth to write his poem 'The Daffodils'. Various views of the spring flowers at Ullswater are photographed and reproduced each year in calendars, books, travel guides, and as postcards. The apparently realistic images are metonymic of the District. Without descriptive texts or explanation, these photographs promise the experience of 'wandering lonely as a cloud' over the hills, in the traces of one of England's most famous poets (Crawshaw and Urry 1997: 184-185).

SK: You have mentioned that research on dynamics of mobilities and visualisation technologies centring on tourism has played important roles in shaping your approaches not only to globalisation but also to associated change in orientations towards 'nature' and 'society' such as those you discussed in *Sociology beyond Society* (2000) and those, which you and Phil Macnaghten discussed in *Contested Natures* (1998)?

JU: Yes, we could say something about that and then return to 'tourism'.

SK: That also sounds like a useful way to lead into topics of your recent work which relates to factors shaping the future of 'transformations of tourism, travel and theory' (Rojeck and Urry 1997).

JU: In *Sociology beyond Societies: mobilities for the twenty-first century* (2000: 1), I wanted to develop categories that might be relevant for sociology as a discipline as we enter into this new century, which centres on the 'diverse mobilities of peoples, objects, images, information and wastes; and of the complex interdependencies between, and social consequences of, these diverse mobilities'. These mobilities call into question not only assumptions about the 'givenness' of 'society' but also the notions of nature that motivate the notion of 'timeless' time (2000: 107) which has been critiqued by Ilya Prigogene, Isabelle Stengers (1984, 1997) and others mentioned in the 'unquiet' volume. Amongst other things, I (2000: 2-3) argued that these 'mobilities crisscrossing societal borders in strikingly new temporal and spatial patterns hold out a possibility for a major new agenda for sociology'.

SK: For localising the global and exploring patterns of instantiation?

JU: I presented a long list of methodological requirements for such approaches, including that of interpreting how 'chaotic unintended and non-linear social consequences can be generated which are distant in time/space from where they originate and are of quiet different and unprecedented scale' (Urry 2000: 20).

SK: Such an approach is likely to have been crucial for pursuing the aim of *The Tourist Gaze* (Urry 1990) to explore something of the variety and changes taking place in forms taken by the dynamics of mobility and visualisation technologies involved in tourism practices?

JU: Much attention focused in that work on the visual tourism experience and practices—in senses that drew upon Foucault's emphasis on the embodied and materially embedded nature of visualisation instruments and practices (see also, Royek 1993). More recently, Carol Crawshaw and I have revisited a number of the issues such a focus poses, noting that 'photographs help not only to panopticalise'—control and survey—'...photography as a mirror, as ritual, as language, as dominant ideology, and as resistance [are] examined in relationship to research we conducted in the English Lake District' (Crawshaw and Urry 1997: 184)

SK: Might that relate to the ways in which photographs and other visualisation techniques can help challenge claims to authority grounded in the notion of 'timeless' time we mentioned before?

JU: Maybe. You have mentioned something similar a bit ago—about the relevance for concerns with political and ethical reflexivity of the indeterminacy of nature and histories.

SK: And also memory. And maybe focusing on things summarised by the expression 'memory' might help to illuminate such concerns in new ways. I am thinking about similarities between the approaches to time that you developed, for instance, with Macnaghten in *Contested Natures* (1998) and the approaches discussed by several contributors to the 'unquiet volume' that stress political and ethical implications. Interestingly, you and Dennis stress the significance of approaches to time, the emergence of order and to how things might endure and transform in states far from equilibrium not only for understanding but also to address contemporary problems. Might that be a theme that runs through many of your works?

JU: I am not sure. But Dennis and I devote much of the chapter on 'system's to related themes, and it is likely to inform how I would envisage an approach to forces impacting the future of tourism.

SK: Including forces undermining the 'car system'?

JU: For Dennis and I:

> …powerful forces around the world are undermining [the] car system and will usher in a new system at some point in this century. The car system is based upon nineteenth century technologies, of steel bodies and internal combustion engines, incidentally showing how old technologies can remarkably endure. We believe that this mass system of individualised, flexible mobility will be 'redesigned' and reengineered' before the end of this century. This book argues that a new system is coming into being. It is a bit like the period around 1900 when the car system was being formed: it was emergent, although no one at that time could imagine exactly what it was going to be like (Dennis and Urry 2009: 2).

SK: All this will surely impact tourism, and wider mobility technologies, as well as conceptions of 'nature' in the humanities and social sciences. I think it might help tie together themes of the volume if we concluded with reference to a passage form your book with Macnaghten—also because it mentions Ulrich Beck and Tim Ingold, whose work has provided many contributors to this volume with much inspiration.

JU: I hear that they may be taking part in the IAS Annual Research Programme you, Szersynski and Wynne are organising…

> In the past the strength of 'nature' lay in the ways its cultural construction was hidden from view (see Latour 1993). But in the contemporary world of uncertainty and reflexive modernity, this no longer seems to be true. A major task for the social sciences will be to decipher the implications of what has always been the case, namely, a nature elaborately entangled and fundamentally bound up with social practices and their characteristic modes of cultural representation (see Beck 1996). A bold and convincing attempt to dissolve the nature/culture divide recently has been provided by Ingold (1996) in which he sets out a framework where embodied human relationships can be embedded with the continuum of organic life (Macnaghten and Urry 1998: 30).

References

Crawshaw, C. and Urry, J. 1997 Tourism and the Photographic Eye, in C. Rojek and J. Urry (eds) *Touring Cultures. Transformations of Travel and Theory*, 176-195. London: Routledge.

Dennis, K. and Urry, J. 2009 *After the Car*. Cambridge: Polity Press.

Giddens, A. 1990 *The Consequences of Modernity*. Stanford, CA: Stanford University Press.

Ingold, T. 1996 Situating Action: The History and Evolution of Bodily Skills. *Ecological History* 8: 171-182.

Jasanoff, S. 1995 *The Fifth Branch: Science Advisors as Policy Makers*. Cambridge: Harvard University Press.

Latour, B. 1993 *We Have Never Been Modern*. Cambridge, MA: Harvard University Press.

Macnaghten, R. and Urry, J. 1998 *Contested Natures*. London: Sage Publications.

Prigogine, I. 1997 *The End of Certainty. Time, chaos and the new laws of nature*. London: Free Press.

Prigogine, I. and Stengers, I. 1984 *Order Out of Chaos*. New York: Bantam.

Royek, C. 1993 *Ways of Seeing—Modern Transformations of Leisure and Travel*. London: Macmillan.

Rojek, C. and Urry, J. (eds) 1997 *Touring Cultures. Transformations of Travel and Theory*. London: Routledge.

Urry, J. 1990 *The Tourist Gaze*. London: Routledge.

—— 1995 *Consuming Places*. London: Routledge.

—— 2000 *Sociology beyond Societies: mobilities for the twenty-first century*. London: Routledge.

—— 2001 *Globalising the Tourist Gaze*. Lancaster: Lancaster University Papers in Sociology.

PART 2
Lived Cultural Heritage

Introduction to Part 2
Contextualising Changing Approaches to Heritage

Stephanie Koerner

Few—if any—areas of specialisation in archaeology are likely to be more deeply affected by complex and contradictory forces of 'globalisation' and 'risk society' than those relating to the expression 'lived cultural heritage'. This is especially the case in contexts where controversies are enmeshed in deepening inequalities with regards to exposure to techno-science hazard, unsustainable development and political conflict (e.g., Friedman 2001). Writing on challenges posed by these contexts, in his contribution to an edited volume on *A Future of Archaeology* (Layton et al. 2006), Henry Cleere (2006: 65) notes that for many researchers, the UNESCO World Heritage Convention represents both 'the best and the worst of contemporary perceptions of and approaches to the tangible and intangible legacy of countless generations of ancestors. One of the most serious charges levelled against the Convention (or those who implement it) is the way in which it is often seen to ride rough shod over the rights and aspirations of local and indigenous communities.' Different categories of World Heritage have different implications for the question 'management for and by whom' (Cleere 2006). This is especially the case for the three main categories of 'cultural landscapes':

- 'the most easily identifiable is the clearly defined landscape designed and created' as such intentionally, such as 'garden and parkland landscapes,'
- 'the organically evolved landscape' that 'results from an initial social, economic, administrative and/or religious imperative and has developed its present form by association with and in response to its natural environment,' the two major sub-categories of these are: 'a relic or fossil landscape, in which an evolutionary process came to an end at some time it the past' but distinguishing features have remained, and 'a continuing landscape that has an active role in contemporary society closely associated with the traditional way of life, and in which the evolutionary process is still in progress,'
- 'the associative cultural landscape' frequently defined in terms of 'the powerful religious, artistic or cultural associations of the natural element rather than material or cultural evidence, which may be insignificant or even absent' (Cleere 2006: 68-69).

Having worked more than a decade in World Heritage policy processes, Cleere (2006: 69) notes that it has been category (2) and, especially those where the evolutionary process are currently underway, that pose the most difficult challenges with regards to questions of 'management by and for whom'.

The contributions to this section variously relate to developments bearing upon this question. For heuristic purposes, my aim in this introduction is to highlight some of the reasons for arguments that the World Heritage Convention represents both the best and the worst of contemporary perceptions of and approaches to the tangible and intangible legacy of countless generations of ancestors (Cleere 2006). I will conclude these comments with reference to examples of projects, which not only illustrate several features shared by promising approaches, but also bring theoretical and methodological implications into sharp relief.

Heritage as a Public Policy Issue

The emergence of heritage and/or Cultural Resource Management (CRM) as a public policy issue has likewise occurred in tandem with other changes that have taken place in policy processes (Holtorf 2001; Carman 2002). For example, changes in the scale, normative roles, and conceptions of the sources of the objectivity of scientific expertise have contributed to increased expectations, on the part of government and commercial policy authoritiesm that super computer technology provide a 'sound scientific basis' for adjudicating:

- investments in research, methodologies, relevant findings, and so on,
- classification of cultural resources and or heritage, as well as the issues different categories pose,
- relationships between academic, government, commercial and public sectors.

In tandem with this development, there has been considerable growth of interest in the implications of the complex history of heritage institutions and practices for challenges facing those working in the field today. For example especially amongst those working in contexts where controversies over heritage are embedded in long histories of social and environmental problems, new light is being thrown on complex historical connections between human rights and cultural heritage which go back to 1948 when the United Nations (UN) included 'culture' in its Declaration of Human Rights (Silverman and Fairchild Ruggles 2007: 1-7). Researchers are only beginning to grapple with the extraordinary complexity of relationships between that Declaration and heritage conservation as a normative issue. It bears noting that in 1954 UNESCO passed the Hague Convention for the Protection of Cultural Property in the event of Armed Conflict. Like the Universal Declaration of Human Rights, this convention was written in response to the terrible destruction of World War II (Silverman and Fairchild 2007: 8). Today much controversy over heritage and human rights amongst human rights centres on the UNESCO classification of the category 'intangible heritage' and or 'lived' heritage. The category was introduced for a variety of reasons, including to:

- 'balance the list of world heritage' and to make it 'multi-cultural,'
- target heritage 'in need of urgent safeguarding',
- include in the list sites that preserve memory of things that should never have happened and should never be allowed to occur again.

William Logan (2007: 33) notes that, according to the UNESCO's Convention for the Safeguarding of Intangible Heritage, while:

> tangible heritage is conceived as 'inanimate' forms of nature and culture (places and artifacts), 'intangible heritage' includes 'practices, representations, expressions, knowledge and skills'—in other words, heritage that is embodied in people rather than inanimate objects.

Much controversy around divisions between 'academic research' and 'rescue archaeology' revolves around archaeology and anthropology's governance roles. Laurajane Smith (2007) argues that:

> Conflicts over the control of cultural heritage must be understood as existing within the parameters of political negotiations between the state and a range of interests over the political and cultural legitimacy of claims to identity [Smith 2004]. Drawing upon Foucault's thesis of 'governmentality', that is a body of knowledge that the state deploys to help policy makers and legislators understand, make sense of, regulate and govern demands and claims based on appeals to the past...Certain 'truths' about representations can be more easily entered into political calculations than other forms of knowledge...Thus intellectual fields become part of the mechanics or technologies of government... (Smith 2007: 161; cf Foucault 1970).

For Smith and many others, a crucial issue is that in contexts where heritage controversies are embedded in far reaching socio-political strife—including threats to human rights—it is dangerous to place commensurate value on people and things and to couch [loss of heritage] in a language reserved for genocide, since these do not inhabit the same order of existence (Meskell 2002: 564). There is also much concern that archaeologists and anthropologists ensure that heritage recognition and conservation is not an 'empty gesture' with little material significance for those whose heritage is at issue (Smith 2007: 160).

Smith (2007: 162) notes that the mobilisation of archaeology as a technology of government does not always mean that archaeological wishes are upheld. As is the case for contexts where social sciences have been assigned advisory roles on environmental issues, archaeological expertise and concerns can and have been marginalised by the interests of more powerful knowledge based political economy agencies. Fortunately, especially amongst those working in contexts where controversies over 'tangible and intangible heritage' are embedded in long histories of jointly social and environmental crises, there has been considerable growth in awareness of the ways in which gaps between what are useful to describe as 'two cultures' perpetuate tendencies of much heritage policy to:

- presuppose distinctions between realms of nature, science and technology versus realms of trust, social ends, and culture, and between truth claims, politics and ethics,
- overlook the historical contingency of the meanings and functions of 'tangible heritage' that is, the extent to which not just 'intangible heritage' but certainly 'tangible heritage' is lived, embodied, and performed,
- ignore the extent to which the significance of scientific objects, practices and instruments (including objects of expanding 'risk sciences' hinges upon complex 'lived heritage' too).

This awareness may be contributing to growing interest in the importance of 'critical inquiry' into 'the academic cultural production of and representation (of both past and present people)' for politically and ethically reflexive archaeology and anthropology (cf. Lazzari 2008). But it also bears underscoring worries that discursive formations concerned with self-reflexivity and the improvement of practice are deeply implicated within a wider culture of auditing and accountability, shaped by transnational managerial domains (Strathern 2000; Lazzari 2008). There is much variability amongst the ways in which agencies of these domains address political and ethical issues. However, practices rooted in techno-science 'risk management' grounded in assumptions that 'risk issues' can be separated from 'ethical issues' and 'public concerns' continue to be quite widespread (Strathern (ed.) 2000; Felt and Wynne (eds) 2007).

Some Backgrounds and Implications of Problematical Aspects of 'Risk Management'

A detailed examination of this pattern lies beyond the scope of my aims here. Instead I attempt to summarise something of their backgrounds and manifestations, with attention to implications of trends in 'risk management' for concerns about impacts of 'transnational managerial domains'.

Some of the most problematical aspects of 'risk management' relate to trends, which have predominated in much science and technology policy until quite recently. Here, two bear stressing. One has to do with remarkable changes that have taking place in the scale and the political and economic significance of techno-science, together with beliefs on the part of government and commercial policy authorities that super computer technologies provide a 'sound scientific basis' for adjudicating:

- investments in technological innovation (research and development, R&D),
- 'risk issues' and 'ethical issues' (or 'public issues') (Jasanoff 1995; Wynne 1996, 2006; Nowotny 2000),
- relationships between science and society (Hacking 1983; Galison and Hevly 1992; Galison 1996; Daston 2000; Daston and Galison 2007).

A related trend has been restrictions of the roles of the humanities and social sciences in policies having to do with 'environmental issues' largely restricted to:

- providing information to physical sciences on human activities which perturb natural processes,
- formulating and 'testing' hypotheses about social and economic consequences of these perturbations,
- designing programmes for 'educating' publics into better understanding and appreciating scientific knowledge and prescriptions (cf. Leach et al. 2005; Wynne 1994; Koerner and Wynne 2008).

In many parts of the world, there is now considerable appreciation that sustainable development is not restricted to purely 'environmental issues' in the sense of approaches that envisage the environment as though it were independent of those who live in it (cf. Layton and Ucko (eds) 1999). There is also considerable awareness on the part of those involved in science and technology policy making of the importance to successful 'risk management' of facilitating greater upstream public participation in deliberating the social ends of investments in instrumental employments of technological innovations. The European Union (EU) is an example of a region that has been actively committed to developing fresh approaches to challenges of sustainable development, reframing risk issue in social and ethical terms, and democratising participation in the policy processes. Thus research on the historical background and current state of science and technology policy in the EU can bring light not only to some of the most promising developments, but also to beliefs and practices that form barriers to realising even the best intents (Felt and Wynne (ed.) 2007; Koerner and Wynne 2008). Examples of beliefs that continue to form barriers to democratising science and technology include those that perpetuate presuppositions that:

- innovation occurs in a linear progression from 'pure research' and resolving disagreements amongst experts to application outside of the 'laboratory',
- 'establishing the facts' involves no prior normative commitments,
- innovation-oriented R&D can be considered wholly independently from normative social ends and understandings of risk (Strathern (ed.) 2000; Felt and Wynne 2007).

These are clearly normative presuppositions, reflecting conceptions of social ends of science and technology with powerful cultural heritages (Aho et al. 2005). But such insights are very frequently eclipsed by other beliefs and practices, which legitimate claims that 'ethical issues' and 'public concerns' can be separated from 'risk issues,' including practices that reduce:

- extraordinary diversity of social and physical uncertainties (climate change, nuclear, chemical and biological hazard) to supposedly determinate 'risk measures,'
- trust to monetary 'cost-benefit' models, and of public concerns to the question: 'Is it safe?',
- problematic new versions of 'public deficit of understanding science' models.

The list below illustrates the range of sources of uncertainty at stake with reductions of complex jointly social and ecological uncertainties to risk calculations:

Sources of uncertainty being reduced to 'risk'

Risk: A situation where one presumably knows the causes, consequences, probabilities of possible harmful events and their associated magnitudes (kinds and levels of damage).

Uncertainty: Where one knows the types and scales of possible harms, but not their causes or consequence probabilities. Under conditions of uncertainty, the term 'risk assessment' is actually not applicable.

Indeterminacy: A situation where one is dealing with processes that are not conceivable as predictable outcomes from what are assumed to be the 'same' initial conditions.

Ignorance: A situation where one does not know causes, consequences or even if one is asking the right questions.

Ambiguity: A situation where a variety of divergent—but equally reasonable framing assumptions (about definitions of objects of research, questions, approaches, variables) precludes the imposition of any single scheme of outcomes.

Such reductions simultaneously eclipse issues of emergent complexity (a fundamental feature of the complex systems that concern climate, nuclear, chemical and biological sciences) (Kauffman 1993, 1995; Rheinberger 1997) by ignoring that:

- physical risks are always created and affected in social systems,
- the magnitude of the physical risks is therefore a direct function of the quality of social relations and ecological processes,
- the primary risk, even for the most technically intense activities (indeed perhaps most especially for them) is therefore that of social dependency upon institutions and caters to who may be—and arguably are increasingly—alien, obscure and inaccessible to most people affected by the risks in question (Beck 1992; Lash and Wynne 1992).

Attention to the complexity of 'risk issues' challenges presuppositions that 'public issues' and 'ethical issues' are reducible to the question: 'Is it safe?' Unfortunately, it has and is assumed that computer based 'cost—benefit' analyses provide adequate means of addressing matters of public concerns (for example, Raiffa 1985). Such assumptions lead to:

- defining risk as a context independent entity, which can be quantified into universal units of risk probabilities and magnitudes that supposedly have monetary equivalents,
- treating risk as tradable with benefits according to monetary equivalents,
- marginalising issues that do not fit (a) and (b), including concerns about trustworthiness of authorities (Wynne 1996).

Trustworthiness is not a 'mere' social problem. It can and has caused technologically complex disasters with implications for existential and moral catastrophe—as demonstrated all too clearly by Bhopal. Chernobyl, BSE and numerous other examples. One of the starting points for a recent Report of the Expert Group on Science and Governance to the EU Science, Economy and Society Directorate (*Taking European Knowledge Society Seriously*) (Felt and Wynne (eds) 2007) was the directorate's diagnosis of a growing uneasiness which affects the relations between science and society. The idea that new means to govern science in order to regain public trust is nowadays a central EU policy preoccupation. This preoccupation is marked conflicts between: (a) the value of policy makers' appreciation of problems with relations between science and publics, and (b) policy makers' resistance to changing deeply entrenched modes of envisaging the social ends of science innovation and public participation.

Today it is a mainstream commitment in many parts of the world to cultivate 'two-way public engagement with science.' 'Deficit models' are said to be a thing of the past. But new versions continue to emerge, including versions centring on supposed public deficits of: (a) trust, (b) being able to grasp processes, (c) objectivity about science and value, (d) understanding of benefits, and (e) ability to deal with contingency (Koerner and Wynne 2008). For those who envisage publics in these ways, it is impossible to appreciate that fear of uncertainty is not a prerogative unique to publics. Thus, it is unlikely to be the primary source of skeptical public reactions to risk and innovation policies (Wynne 2006). The public usually takes for granted that things are not as predictable as scientific knowledge experts claim them to be, but is skeptical about their claims to certainty. Many public concerns are motivated by experiences with problematic social behaviours of quite a number of policy institutions. These behaviours include these institutions' modes of representing public perceptions and beliefs and communicative styles towards publics (Felt and Wynne 2007).

Conservation and Conciliation

Several contributions to the present volume highlight the relatively long history of awareness of these sorts of issues in archaeology and anthropology—an awareness that is likely to inform a number of arguments that the World Heritage Convention represents both the best and the worst of contemporary perceptions of and approaches to the tangible and intangible legacy of countless generations of ancestors (Cleere 2006). I conclude this section's introduction with reference to an example of a project that exhibits features shared by especially promising approaches, with attention to theoretical and methodological implications. This is the project pursued under the direction of James Wescoat (2008) at the newly designated World Heritage Site of Champaner-Pavagodh in north-west India (see also, Wescoat 2008).

The idea that conflict can give rise to constructive change is of extraordinary antiquity, and has often been associated with critiques of problematical ways of framing 'public issues' (for instance Vico 1948 [1744]). During the 20th century this association was developed by a number of 'pragmatists', including the philosophers, diplomats and activists, Walter Lippman and John Dewey. Wescoat (2008: 587) has written on the relevance for historical geography of common themes in the work of Dewey and the geographer Gilbert White's 'blend of research and teaching…in the fields of resources and hazards of human environmental relationships…, international diplomacy, public service and activism.' These themes can be summarised as including:

- the historical contingency of environmental precariousness, and the social causes of vulnerability,
- the idea that there are no context independent problems ('We never experience or form judgements about objects and events in isolation, but only in connection with a contextualised whole. This latter is called a situation' [Dewey 1938: 66-67]),
- learning from experience, and problems that can be expected to follow from disregard of the importance of the past for the sustainability of the future,
- the importance to democracy of public participation in policy processes (Dewey 1927; Lippman 1925; Wescoat 2008: 587-601).

For Wescoat, such themes are highly relevant for addressing challenges posed by controversies over for and by whom heritage is managed for a variety of reasons, including that they go against the grain of:

- presuppositions that truth claims can be separated from politics and morality,
- claims about the greater efficacy of context independent rather than context dependent policy making processes,

- the long history of equations of agreement with intelligibility and disagreement with unintelligibility,
- disregard of the rationality and logic of plurality of local public grounds of truth.

Importantly, such themes are not only relevant for critically reflexive approaches. They indicate that the humanities and social sciences can go beyond restricted roles of offering advice on social consequences of policy processes, for instance, by widening public participation in deliberating such questions as:

- Why are so many resources being invested in these heritage sites and monuments?
- Why are some many resources being invested in these technologies?
- Why not others?
- Who needs them?
- Under what conditions will they be enacted?
- Who benefits? Can they be trusted?
- Whose aspirations do they promote, and whose hopes do they stifle?

Wescoat's project illustrates something of the roles 'lived heritage' conservation can play, in that it:

- extends beyond monuments and sites to the rich topographic sense of place,
- focuses on human-environmental relationships,
- encompasses multiple historical layers,
- operates on multiple geographical scales,
- helps to illuminate diverse people—place connections (Wescoat 2008).

One of the project's key features is the attention it draws to the multiple historical layerings that make plurality of contemporary people and place relationships possible. Wescoat (2008) begins his chapter with a passage from Rajamon Gandi's *Revenge and Reconciliation: Understanding South Asian History* (1999: 410):

> A word finally, on Delhi…Can Delhi's accumulated offences be washed away?… Every caring act—of fellowship, considerateness, nursing, apology, forgiveness, greening, or flowering—perhaps heals something of Delhi's torment…and… speaks to all of South Asia.

Wescoat draws no dualist distinctions between ecological and social problems, risk and ethical issues, or expert knowledge and public concerns. This highlights the plurality of conflicting conceptions of the social ends and accountabilities at stake at Champaner-Pavagodh. Whilst he lists the range of these conflicts from the least to the most tractable at the beginning of the chapter, he later lists possible contexts for conciliation in the reverse order:

- conflict amongst professional agencies (or 'experts'),
- conflict resolution amongst community stakeholders,
- harmonisation of heritage conservation and economic interests,
- prevention of intentional destruction,
- places of violence and sanctuary, as heritage.

Wescoat's insight compares with the approaches of Lippman and Dewey to social and environmental problems (see also Marres 2005). For Wescoat, the importance to heritage conservation of upstream participation on the part of publics with diverse cultural backgrounds and social experiences is difficult to overstate. The most immediate departure point for Dewey's *Public and Its Problems* (1927) was Walter Lippman's question in *The Phantom Public* (1925). What is the public? For Lippman, much contemporary citizenship hinges upon public understanding of science, and concerns that if citizens continue to fail to understand science there will be no basis for the foundations of a representative regime—that is a democracy. For Dewey, Lippman's arguments concerning means to address issues posed conflicted. On the one hand, Lippman (1925: 121) that the 'The hardest problems are those which institutions cannot handle. They are the publics' problems.' On the other hand Lippman (1925) suggested that government institutions; (a) seek advice on policy of scientific expertise, and (b) use 'psychological' and 'level of education' data on 'ordinary citizens' to educate and convince them about the benefits of policy decisions being implemented. For Dewey (1927) such contradictions related to failures to appreciate that publics are not irrational or ignorant but 'eclipsed,' and the challenge of the 'public and its problems' is to make it possible for new 'public relationships to emerge.' For Dewey, publics are invisible to policy makers because the latter are unable to place themselves into the frame of questioning in interaction with others. It is impossible for us to genuinely hear what anyone is saying, if we are impose our own projections onto our supposed 'listening' relationships. So, while the correct notes may be played, the music somehow fails to appear, and it does not take specialist musicians to notice the difference (see also, Wynne 2006). Similarly, much of the efficacy of the Champaner-Pavagodh project hinges upon including its own backgrounds and aspirations democratically in discussions of conflict conciliation, and emphasising the diversity of forms of expertise and involvement needed to [re-]design heritage conservation aspirations and accountabilities.

Writing on theoretical and methodological implications of such projects for challenges posed for anthropology and archaeology by changes taking place in dynamics of pedagogical institutions and publics' concerns with increasingly complex social and ecological problems, Darvill 2007 notes that some of today's most immediately relevant research questions have to do with 'quality of life issues'. As is the case for much research carried out in relation to Stonehenge—a site deeply rooted in the complex histories of anthropology and archaeology (e.g., Trigger 1989; Pinksy and Wylie 1989; Schnapp 1993)—the Champaner-Pavgodh illustrates something of the theoretical and methodological significance of arguments that 'World Heritage Sites should be exemplary situations' not only for research, but also for 'creat[ing] and maint[aining] different kinds of knowledge… and contexts for enriching quality of life quality of life' (Darvill 2007: 436).

References

Aho, E., Cornu, J., Georghiou, L. and Subira, A. 2006 *Creating an Innovative Europe*. Europe Commission Report of the Independent Expert Group on R&D and Innovation. Luxembourg: European Commission, EUR 22005.

Beck, U. 1992 *Risk Society: towards a new modernity*. London: Sage Publications.

Carman, J. 2002 *Archaeology and Heritage: an introduction*. Leicester University Press: London.

Cleere, H. 2006 The World Heritage Convention. Management for and by whom?, in R. Layton, S. Shennan and P. Stone (eds) *A Future for Archaeology*, 65-74. London: Routledge.

Darvill, T. 2007. Research Frameworks for World Heritage Sites and the conceptualization of archaeological knowledge. *World Archaeology*, Vol. 39, 3: 436-457.

Daston, L. (ed.) 2000 *Biographies of Scientific Objects*. Chicago: University Chicago Press.

Daston, L. and Galison, P. 2007 *Objectivity*. London: Zone Books.

Dewey, J. 1927 *The Public and Its Problems*. New York: Holt.

Felt, U. and Wynne, B. (eds) 2007 *Taking European Knowledge Society Seriously*. Brussels: European Commission of the Directorate-General for Research, Economy and Society.

Foucault, M. 1970 *The Order of Things*. London: Tavistock.

Friedman, J. 2001 Indigenous struggles and the discreet charm of the bourgeosie, in R. Prazniak and A. Dirlik (eds) *Politics and Place in an Age of Globalization*, 53-70. Lanham, MD: Rowman and Littlefield Publishers.

Galison, P. 1996 Computer simulations and the trading zone, in P. Galison and D.J. Stump (eds), *The Disunity of Science: boundaries, contexts, and power*, 119-157. Stanford: Stanford University Press.

Galison, P. and Hevly, B. (eds) 1992 *Big Science: the growth of large-scale research*. Stanford, Calif.: Stanford University Press.

Gandi, R. 1999. *Revenge and Reconciliation: understanding South Asian history*. NY: Penguin.

Hacking, I. 1983 *Representing and Intervening: introductory topics in the philosophy of science*. Cambridge: Cambridge University Press.

Holtorf, C. 2001 Is the Past a Non-renewable Resource? in J. Bintliff (ed.) *A Companion to Archaeology*, 286-298. Blackwell, Oxford.

Jasanoff, S. 1995 *The Fifth Branch: science advisors as policy makers*. Cambridge: Harvard University Press.

Kauffman, S. 1993 *The Origins of Order: self-organisation and selection in evolution*. Oxford: Oxford University Press.

—— 1995 *At Home in the Universe. The search for laws of self-organisation and complexity*. Oxford: Oxford University Press.

Koerner, S. and Wynne, B. 2008 Wittgenstein and Contextualising Reduction and Elimination in the Social Sciences, F. Stadler et al. (eds) *Wittgenstein and Contemporary Debates in the Humanities and Social Sciences*. Vienna: IWS.

Lash, S. and Wynne, B. 1992. Preface to U. Beck, *Risk Society: towards a new modernity*, i-xx. London: Sage Publications.

Layton, R., Shennan, S. and Stone, P. (eds) 2006 The World Heritage Convention. Management for and by whom?, *A Future for Archaeology*. London: Routledge.

Layton, R. and Ucko, P. (eds) 1999 *The Archaeology and Anthropology of Landscape*. London: Routledge.

Lazzari, M. 2008 Topographies of Value: ethical issues in landscape archaeology, in B. David and J. Thomas (eds) *A Handbook of Landscape Archaeology*, 644-658. Walnut Creek: New Left Coast Press.

Leach, M., Scoones, I. and Wynne, B. (eds) 2005 *Science and Citizens*. London: Zed books.

Lippman, W. 1993 [1925] *The Phanthom Public*. New Brunswick: Transaction Publishers.

Logan, W. 2007 Closing Pandora's Box: Human rights and cultural heritage protection, in H. Silverman and D. Fairchild Ruggels (eds), *Cultural Heritage and Human Rights*, 33-52. London: Springer.

Marres, N. 2005 *No Issue, No Public: democratic deficits after the displacement of politics*. Doctoral Dissertation, University of Amsterdam.

Meskell, L. 2002 Negative Heritage and Past Mastering in Archaeology. *Anthropological Quarterly* 75(3): 557-574.

Nowotny, H. 2000 Transgressive Competence. The narrative of expertise. *European Journal of Social Theory* 3(1): 5-21.

Pinsky, V. and Wylie, A. (eds) 1989 *Critical Traditions in Archaeology*. Cambridge: Cambridge University Press.

Raiffa, H. 1985 Back from Prospect Theory to Utility Theory, in M. Thompsen, A. Wierzbicki and M. Grauer (eds) *Plural Rationalities and Interactive Decisions*. Berlin: Springer.

Rheinberger, H.-J. 1997 *Toward a History of Epistemic Things: synthesizing proteins in the test tube*. Stanford: Stanford University Press.

Schnapp, A. 1993 *The Discovery of the Past*, translated by I. Kinnes and G. Varndell. London: British Museum Press.

Silverman, H. and Fairchild Ruggels, D. (eds) 2007 *Cultural Heritage and Human Rights*. London: Springer.

Smith, L. 2007 Empty Gestures? Heritage and the Politics of Recognition, in H. Silverman and D. Fairchild Ruggels (eds) *Cultural Heritage and Human Rights*, 159-171. London: Springer.

Strathern, M. (ed.) 2000 *Audit Cultures: anthropological studies in accountability, ethics and the academy.* London: Routledge.

Trigger, B. 1989 *A History of Archaeological Thought.* Cambridge: Cambridge University Press.

Vico, G. 1948 [1744] *The New Science of the Common Nature of the Nations of Giambattista Vico*, unabridged third edition translated by T.G. Bergin and M.H. Fisch. London: Cornell University Press.

Wescoat, J.L. 2008 The Indo-Islamic Garden: conflict, conservation, and conciliation in Gujurat, India, in H. Silverman and D. Fairchild Ruggels (eds) *Cultural Heritage and Human Rights*, 53-78. London: Springer.

Wynne, B. 1994 Scientific Knowledge and the Global Environment, in M. Redclift and T. Benton (eds) *Social Theory and the Global Environment*, 169-189. London: Routledge.

Wynne, B. 1996 May the Sheep Safely Graze? A reflexive view of the expert-lay knowledge divide, in S. Lash, B. Szerzynski, and B. Wynne (eds), *Risk, Environment and Modernity. Towards a New Ecology* 44-83. London: Sage.

—— 2006. Public Engagement as a Means of Restoring Public Trust in Science. *Community Genetics* 10: 1-20.

Chapter 8
Coming Full Circle: Public Archaeology as a Liberal Social Programme, Then and Now

John Carman

It can be argued that archaeology in the English-speaking world has ceased to be an elite or specialised subject, and has become one that belongs to all. The rise of 'community', 'collaborative' and 'democratic' archaeologies is now well documented (*World Archaeology* 2002; Field et al. 2002; Moser et al. 2002; McDavid 2002; Faulkner 2000; McDavid 2004) and recent discussions in the UK (e.g., Clark 2006) have focused upon the 'public value' of archaeology as part of 'social inclusion' policies and the creation of a sense of national identity based upon shared values rather than ethnicity. What is interesting to me is that we have been here before: the connection between preservation and promotion of the archaeological heritage and social welfare, rather than being a new invention, goes back to the creation in the late 19th century of what we now call 'heritage' (Carman 2005b). What is new is the concurrent focus upon the economic values of heritage—as material to have a 'price' placed upon it as advocated by accountants (e.g., ASB 2006; Carnegie and Wolnizer 1995; Hone 1997); or as a resource with 'instrumental' and 'institutional' values as well as 'intrinsic' values (DEMOS 2005; Clark 2006a). I want to review some of these developments here in order to link them with other concerns of mine which I hope will help us to view these matters in a new way.

Making Good Citizens Then and Now: The 19th Century versus the 21st Century

In the late 19th century archaeology was at the heart of disputes about the way the future would look (see also Carman 1996: 67-96; Carman 1997; Carman 2005b). In the 'Concluding Remarks' to his book *Prehistoric Times*, first published in 1865 and drawing upon the earlier contents, John Lubbock connected improvements in physical welfare to increases in population and vice versa (Lubbock 1872: 593-594). These in turn were linked to the mental condition of the members of a society, since 'the pleasures of the civilized man [sic] are greater than those of the savage' (Lubbock 1872: 597). All this was then given support by citing the correlation between illiteracy and criminality as evidenced by data from British prisons (Lubbock 1872: 600). Drawing on Darwin, he argued that natural selection works to advance both physical and moral conditions over time and that a knowledge of science makes people more virtuous (Lubbock 1872: 601). Accordingly, he argued that advances in learning lead to a greater understanding of nature, to technological advance and thus to increased happiness. Then, brilliantly reversing the argument, Lubbock led us to the conclusion that scientific advances will necessarily lead to both physical and moral improvement among modern European populations. It follows from all this that 'Utopia turns out…to be the necessary consequence of natural laws' (Lubbock 1872: 603) and the proof of this lies in the archaeological and ethnographic record.

This is in essence a statement of late 19th century Liberal politics with a time dimension added. Lubbock was a Liberal politician and elected Member of Parliament as well as an archaeologist. He was one of a new breed of rising middle-class professionals (Perkin 1989; Carman 1996: 70-89) assisting in the creation of the distinct disciplines of archaeology and anthropology, classifying and defining appropriate behaviour patterns for these new sciences. He was also the prime mover in legislating the preservation of prehistoric monuments and active in promoting the scientific education of the wider public. It was largely through his efforts that the first legislation in Britain to preserve ancient remains was passed, marking the beginning of what today we call 'heritage'. *Prehistoric Times* should accordingly be read as a strategy in contemporary politics whereby a scientific explanation of how human society functions gave support to the Liberal political programme of national public education and welfare (Patton 2007). The same applies to the foundation of such organizations as the National Trust in the 1890s (Gaze 1988) and the growth of public and private museums (Hudson 1987; Hooper-Greenhill 1992) which were dedicated to the improvement of the rural and industrial working classes. In the later 19th century, the discourse of heritage was always about politics and society: it was about making good citizens.

We have now returned to such an idea. 'Essays' by the then UK Minister responsible for cultural matters (Jowell 2004; 2005) sought to outline an approach to valuing cultural resources in terms of their usefulness to creating viable communities and promoting good citizenship. Speeches by both the senior and junior ministers for culture at the conference *Capturing the Public Value of Heritage* in January 2006 developed the theme (Jowell 2006; Lammy 2006). The way Tessa Jowell and David Lammy talked about these issues is at once instructive and redolent of their 19th century forebears. Within the context of developing an idea of 'Britishness', Jowell wished to 'explore the links between public value and heritage and the public benefit that comes from developing a sense of shared identity'; and she declared that she has 'always regarded our historic environment as being a vital part of the public realm—part of what [she] would define as those shared spaces and places that we hold in common and where we meet as equal citizens' (Jowell 2006, 7-8). Lammy explicitly referred to the work of William Morris, John Ruskin and Octavia Hill in promoting 'a radical desire for the British people to enjoy their national heritage [quoting Octavia Hill] to 'make lives noble, homes happy and family life good" (Lammy 2006, 65); and set his own task as 'to revive that radical, empowering conception of heritage…; and to help build a Britain at ease with its present because it understands, values and is able to access its past' (Lammy 2006, 69).

The manner in which these ideas are being operationalised as part of policy-making, however, is rather different from Lubbock's. Apart from the extremely nationalist tone of these statements, the emphasis is placed upon the utilitarian value of the heritage: classed as 'institutional' values (i.e., those returns that accrue to wider society from the activities of the managing or owning institution); and 'instrumental' values (i.e., those that derive from the various uses to which heritage objects can be put) (Hewison and Holden 2006). This moves us beyond previous attempts to value the heritage by borrowing from economics the general idea of use and non-use values (Darvill 1995; see also discussions in Carman 2002, 155-167; Carman 2005b) because the current imperative is towards the identification of *specific* measurable outcomes from investment in heritage objects (Clark 2006b; Jones and Holden 2008). But while the idea of use value retains its force as an analytical category, other types of value are also recognised by economists (e.g., Sinden and Worrell 1979: 411; Eggertsson 1990: 6).

A newer breed of economists have come to recognise that the 'economic approach has not been successful in explaining certain' phenomena because 'modification of social values—that is, changing ideologies—are a major factor in institutional change' (Eggertsson 1990: 72-5). Such thinking in economics focuses upon so-called 'transaction costs', which are the costs associated with maintaining the systems that control, limit and define economic and other relations (Eggertsson 1990: 14). New institutional economists 'are interested in the impact of various structures…on the wealth of nations…Rational individuals [they say] will compete not only to maximise their utility within a given set of rules, but also seek to change the rules and achieve more favourable outcomes than was possible under the old regime' (Eggertsson 1990: 12). The structures here referred to are especially those of property rights, and the rights to claim the benefits that use of resources can bring. This brings us finally to the core but unstated notion which sustains all of the value structures of the economic approach to heritage but which ultimately denies the ubiquity of the heritage: that value only accrues to things that are in some sense, and in some way, *owned.*

Heritage as Property

Let us be clear: all heritage objects are in fact objects to which property relations apply. As if in confirmation of this, the term 'Cultural Property'—although challenged (e.g., by O'Keefe and Prott 1992; Carman 2005a)—remains the preference for describing such things in legal discourse. In practice, all heritage objects are subject to rules about their ownership. In many countries of the world, all archaeological objects (large and small) are, upon discovery, declared to be the property of the State: in others, they may belong either to the finder or to the owner of the land on which they were found; none are declared an 'open-access resource'. Similarly, museums, while frequently claiming to hold objects as 'stewards' or 'custodians' for a wider community, in fact are generally the legal owners of their collections and removal of the objects from their care without proper permissions constitutes theft. Disputes over the proper location of heritage objects—whether the Parthenon (or Elgin) Marbles, Benin Bronzes or other artefacts—dissolve into disputes about 'true' ownership, legal or moral (Greenfield 1989). Larger features of archaeological interest will also be subject to issues of ownership: while ownership of the land on which it stands may remain with the landowner, other rights of ownership—such as those of destruction or use—may pass to another agency (Carman 2005a: 77-9).

A problem that is frequently stated is that the heritage does (or should) not represent any kind of property (see e.g., O'Keefe and Prott 1992; Brown 2004; Rowlands 2004). A way of overcoming this problem that has been proposed is to treat cultural resources as if they are a form of shared intellectual property—akin to a copyright or a patent. The purpose of such regimes is to create a shared sense of common interest and a recognition of benefits accruing to all. Two economists in particular have offered criteria for evaluating the success of such common property regimes:

Table 8.1 Criteria for evaluating common property regimes

Bromley (1991: 3)	Ostrom (1990)
The resource not squandered The resource is enhanced A lack of 'anarchy' among co-owners	The resource is clearly defined Monitoring of projects Agreed sanctions for breach of rules Low-cost dispute resolution

Against these criteria, community archaeology projects reported on in the literature appear to do well (see especially Carman 2005a: 86-90). In all cases examined, the resource—the site and its products from research—are not squandered: local involvement and leadership in the project ensures that the resource is given value; and where the project offers opportunities for local employment, this ensures further a direct economic incentive in preventing damage and looting. The resource is in various ways enhanced: by the process of research, it yields new information for archaeologists and others; outreach and educational programmes—for locals and others—ensure that this information is made widely available, while formal academic publications give the resource intellectual value; where recovered items are retained in a local museum, the nearby site from which they come is given greater prominence and the objects themselves are more closely contextualised.

The lack of 'anarchy' among co-owners is attested by the continuing nature of these projects and their success in maintaining and enhancing the resource, as well as establishing and retaining good will among the local community.

Also in all cases, the resource—a local site or the archaeology of a given area—is clearly defined. In most cases the purpose of the project—whether pure research into the past or to encourage tourism—is made clear. It is frequently the community who make the rules regarding how the project will be conducted, and accordingly they will meet local needs and conditions, while there is also in all cases common agreement as to the rules and how they may be changed. Monitoring of the project is conducted jointly by archaeologists and community alike in both formal and informal ways: the peer-review process evident in formal academic archaeology serves to support such monitoring, as does auditing of financial and other arrangements. Sanctions for rule-breaking or failure to perform will be appropriate for the project and agreed by parties, and will involve external recognition of the arrangements in place. Whether dispute resolution—an important component of total transaction costs—is low-cost or otherwise will depend upon specific arrangements: like transaction costs generally, this is difficult to assess. In terms of archaeologists' time and effort, however, and the way in which this will inevitably impact upon their specifically archaeological performance, it will seem that these costs are quite high.

Such a point may appear to invalidate any argument made in favour of treating such projects as successful common property regimes. If such regimes rely for their success upon low transaction costs, then if these costs are high they do not meet suitable tests. A common factor in all the cases discussed above, however, has been a willingness on the part of those with a specifically non-economic interest to invest in the community project. At the Levi Jordan site in Texas, Carol McDavid (2000: 222) is recognised as a project leader and takes time and trouble to ensure community wishes are respected and disputes addressed. Stephanie Moser and her colleagues in Egypt (Moser et al. 2002) and Judith Field and hers in Australia (Field at al. 2000) were at pains to build significant trust and willingness to collaborate among members of the communities they work with. In Fiji (Crosby 2002), government funding and the provision of expert guidance was a pre-requisite for a project. In other words, the archaeologists involved in the project took on the effort and time of managing the system of collaboration to keep it in place. To the community whose archaeology it is, the costs of managing the system would appear low because they would not have to bear them.

Against Economics

In the case of cultural heritage, the common interest and recognition of benefits to all that lies at the heart of community projects is an example of what has been termed 'cognitive ownership' (Boyd et al. 1996). Such 'cognitive ownership' allows each claimant to a resource full access to it without interference from or interfering with access by others. By giving full reign to such cognitive claims, each can take from the resource without placing any restraint upon similar taking by others. It is not so much of a stretch from here to exercising a voluntary physical restraint on actual use of the resource so as not to deny it to others. This concept of 'keeping while giving' is well established in anthropology (see e.g., Weiner 1992). Choosing to exercise restraint on one's own use of a resource does not serve to deny it completely: but it does create the conditions under which others may have access to it as well. The idea of voluntary restraint on use is not as alien to Western thought as some students of economics would suggest: it is at the core of revolutionary anarchist thinking (Kropotkin 1972: 61-70); and has been applied in the international sphere to protect unoccupied natural environments from appropriation (see e.g., UNOOSA 1967; SCAR 1959). At the core of such treatments of resources lie two ideas: that there is enough of it to go around (see e.g., Kropotkin 1972: 41-50), and recognition that your use of it does not infringe my enjoyment of it, even if they represent different kinds of use altogether.

The archaeological heritage meets the conditions of the first requirement (that there is enough of it to go around) admirably. Heritage is a highly abundant resource: everybody can make a claim on one or more heritages, and some part of any of these heritages may be classed as 'archaeological'. It is also a global phenomenon: the only parts of the world that do not provide archaeological material are those—such as Antarctica, the highest mountain ranges, and the depths of the ocean—that have never been occupied by human beings. Indeed the non-archaeological status of these areas, and others, are currently under challenge: the archaeology of Antarctic exploration (Harrowfield 2004) or mountaineering is a viable field, as is that of Outer Space exploration (Spenneman 2006). As Tunbridge and Ashworth (1996: 8-9) have put it, 'there is an almost infinite variety of possible heritages, each shaped for the requirements of specific consumer groups'. Despite disciplinary rhetoric to the contrary (much of it the standard rhetoric of the archaeological discipline: see e.g., Darvill 1987: 1; McGimsey 1972: 24; Cleere 1984: 127) archaeology is not a 'finite resource' that is diminishing: in fact, it is a resource that is on the increase (Holtorf 2002) by various mechanisms, such as the discovery of new archaeological sites (Darvill and Fulton 1998: 4-7), the recognition of entirely new classes of archaeological material, and the constant deposition of new material to become the archaeology of the future (Carman 1996: 7).

For the second requirement (that your use does not infringe my use), in the case of the archaeological heritage initial notice taken by professionals may serve to awaken interest in a local or wider community such that a range of different values and uses for the object become evident (see e.g., Jones 2004; Boyd et al. 2005). In this case, 'cognitive ownership' can be claimed by all without any loss of the resource. Once this has happened, the conditions are met for voluntary restraints on actual use to be applied without the need for the allocation of specific property rights. Since the value of the object or place has been increased, and new ones ascribed, it becomes an object or place of significance to all its cognitive owners: none has an interest in its depletion or destruction. No agency need appropriate use rights in order to deny them to others; and no rights of access need be controlled by a custodian or steward. Here is the ultimate link between value and property: when the value is a social value held by all, then to conserve the resource no-one need be granted any right of ownership, and there is no need for measurable outcomes.

Conclusion

It follows that to value something in terms of its institutional and instrumental benefits—the kinds of value meaningful to economists—those benefits from use must accrue to someone. In other words, someone must own them or at least the rights to them. The problem is that this is not what current efforts in community archaeology provide, nor what—it seems to me—the 19th century creators of heritage intended; nor what their 21st century successors—at least according to their rhetoric—desire. Instead, by meeting quite well the criteria set by economists for communal property, community archaeology projects provide instances of a non-economic value that exists in the real world. Because they represent communal property they also exist outside the realms of conventional exclusive property relations, one with which economists have difficulty dealing.

This, it seems to me, is the essence of archaeology as a public endeavour: not one with measurable use, institutional and instrumental values and outcomes; but one that has value beyond immediate returns. In the 19th century no measurable outcomes from archaeology were sought: its existence as a source of inspiration was enough. This is the idea we must return to.

References

Accounting Standards Board 2006 *Heritage Assets: can accounting do better?*

—— 2006 Discussion paper. London, Accounting Standards Board.

Boyd, W.E., Cotter, M.M., O'Connor, W. and Sattler, D. 1996 Cognitive ownership of heritage places: social construction and cultural heritage management *Tempus* 6, 123-40.

Boyd, W.E., Cotter, M.M., Gardiner, J. and Taylor, G. 2005 'Rigidity and a changing order…disorder, degeneracy and daemonic repetition': fluidity of cultural values and cultural heritage management in Mathers, C., Darvill, T. and Little, B.J. (eds) *Heritage of Value, Archaeology of Renown: reshaping archaeological assessment and significance*. Gainesville: University Press of Florida, 43-57.

Bromley, D.W. 1991 *Environment and Economy: property rights and public policy.* Oxford: Basil Blackwell.

Brown, M.F. 2004 Heritage as property in Verdery, K. and Humphrey, C. (eds) *Property in Question: value transformations in the global economy*. Wenner-Gren International Symposia Series. Oxford: Berg, 49-68.

Carman, J. 1996 *Valuing Ancient Things: archaeology and law*. London: Leicester University Press.

—— 1997 Archaeology, politics and legislation: the British experience. In G. Mora and M. Diaz-Andreu (eds) *La Cristalizacion del Pasado: genesis y desarollo del marco institucional de la arqueologia en Espana*, Malaga, Universidad de Malaga, 125-32.

—— 2002 *Archaeology and Heritage: an introduction*. London and New York: Continuum.

—— 2005a *Against Cultural Property: archaeology, heritage and ownership*. Duckworth Debates in Archaeology. London: Duckworth.

—— 2005b Good citizens and sound economics: the trajectory of archaeology in Britain from 'heritage' to 'resource' in Mathers, C., Darvill, T. and Little, B.J. (eds) *Heritage of Value, Archaeology of Renown: reshaping archaeological assessment and significance*. Gainesville: University Press of Florida, 43-57.

Carnegie, G.D. and Wolnizer, P.M. 1995 The financial value of cultural, heritage and scientific collections: an accounting fiction. *Australian Accounting Review* 5.1, 31-47.

Clark, K. (ed.) 2006a *Capturing the Public Value of Heritage*, the proceedings of the London conference 25-26 January 2006. London, English Heritage.

—— 2006b From significance to sustainability in Clark, K. (ed.) *Capturing the Public Value of Heritage*, the proceedings of the London conference 25-26 January 2006. London, English Heritage, 61-2.

Cleere, H.F. 1984 World cultural resource management: problems and perspectives in Cleere, H.F. (ed.) *Approaches to the Archaeological Heritage: a comparative study of world cultural resource management systems*. Cambridge: Cambridge University Press, 125-31.

Crosby, A. 2002 Archaeology and *vanua* development in Fiji *World Archaeology* 34.2, 363-78.

Darvill, T. 1987 *Ancient Monuments in the Countryside: an archaeological management review*. London: English Heritage.

—— 1995 Value systems in archaeology in Cooper, M.A., Firth, A., Carman, J. and Wheatley, D. (eds) *Managing Archaeology*. London: Routledge, 40-50.

Darvill, T. and Fulton, A.K. 1998 *The Monuments at Risk Survey of England 1995: summary report*. London: English Heritage.

DEMOS 2005 *Challenge and Change: HLF and Cultural Value. A report to the Heritage Lottery Fund*. London, DEMOS.

Eggertsson, T. 1990 *Economic Behaviour and Institutions*. Cambridge Surveys of Economic Literature. Cambridge: University Press.

Faulkner, N. 2000 Archaeology from below *Public Archaeology* 1.1, 21-33.

Field, J., Barker, J., Barker, R., Coffey, E., Coffey, L., Crawford, E., Darcy, L., Fields, T., Lord, G., Steadman, B. and Colley, S. 2000 'Coming back': Aborigines and archaeologists at Cuddie Springs *Public Archaeology* 1.1, 35-48.

Gaze, J. 1988 *Figures in a Landscape: a history of the National Trust*. London, Barrie and Jenkins/National Trust.

Greenfield, J. 1989 *The Return of Cultural Treasures.* Cambridge: Cambridge University Press.

Harrowfield, D.L. 2004 Archaeology on ice: a review of historical archaeology in Antarctica. *New Zealand Journal of Archaeology* 25, 5-36.

Hewison, R. and Holden, J. 2006 Public value as a framework for analysing the value of heritage: the ideas in Clark, K. (ed.) *Capturing the Public Value of Heritage*, the proceedings of the London conference 25-26 January 2006. London, English Heritage, 14-18.

Holtorf, C. 2002 Is the past a renewable resource? in Layton, B., Stone, P. and Thomas, J. (eds) *The Destruction and Conservation of Cultural Property.* London: Routledge, 286-97.

Hone, P. 1997 The financial value of cultural, heritage and scientific collections: a public management necessity, *Australian Accounting Review* 7.1, 38-43.

Hooper-Greenhill, E. 1992 *Museums and the Shaping of Knowledge.* London, Routledge.

Hudson, K. 1987 *Museums of Influence.* Cambridge, University Press.

Jones, S. 2004 *Early Medieval Sculpture and the Production of Meaning, Value and Place: the case of Hilton of Cadboll.* Edinburgh: Historic Scotland.

Jones, S. and Holden, J. 2008. *Its a Material World: caring for the public realm.* London, DEMOS.

Jowell, T. 2004 *Government and the Value of Culture.* London, DCMS.

—— 2005 *Better Places to Live: government, identity and the value of the historic and built environment.* London, DCMS.

—— 2006 From consultation to conversation: the challenge of *Better Places to Live* in Clark, K. (ed.) *Capturing the Public Value of Heritage*, the proceedings of the London conference 25-26 January 2006. London, English Heritage, 7-13.

Kropotkin, P. 1972 [1892] *The Conquest of Bread.* London: Allen Lane.

Lammy, D. 2006 Community, identity and heritage in Clark, K. (ed.) *Capturing the Public Value of Heritage*, the proceedings of the London conference 25-26 January 2006. London, English Heritage, 65-9.

Lubbock, J. 1872 *Pre-historic Times as illustrated by ancient remains and the manners and customs of modern savages.* 3rd edition. London, Williams and Norgate.

McDavid, C. 2000 Archaeology as cultural critique: pragmatism and the archaeology of a southern United States plantation in Holtorf, C. and Karlsson, H. (eds) *Philosophy and Archaeological Practice: perspectives for the 21st century.* Göteborg: Bricoleur Press, 221-39.

—— 2002 Archaeology that hurts; descendants that matter: a pragmatic approach to collaboration in the public interpretation of African-American archaeology *World Archaeology* 34.2, 303-14.

—— 2004 Towards a more democratic archaeology? The Internet and public archaeological practice in Merriman, N. (ed.) *Public Archaeology.* London, Routledge, 159-87.

McGimsey, C. 1972 *Public Archaeology.* New York: Seminar Books.

Moser, S., Glazier, D., Phillips, J.E., Nasr el Namr. L., Mouier, M.S., Aiesh, R.N., Richardson, S., Conner, A. and Seymour, M. 2002 Transforming archaeology through practice: strategies for collaborative archaeology and the Community archaeology Project at Quesir, Egypt *World Archaeology* 34.2, 220-48.

OKeefe, P.J. and Prott. L.V. 1992 'Cultural heritage' or 'cultural property'? *International Journal of Cultural Property* 1, 307-19.

Ostrom, E. 1990 *Governing the Commons: the evolution of institutions for collective action.* Cambridge: Cambridge University Press.

Patton, M. 2007. *Science, Politics and Business in the Work of Sir John Lubbock: a man of universal mind.* Aldershot: Ashgate.

Perkin, H. 1989 *The Rise of Professional Society: England since 1880.* London, Routledge.

Rowlands, M. 2004 Cultural rights and wrongs: uses of the concept of property in Verdery, K. and Humphrey, C. (eds), *Property in Question: value transformations in the global economy*. Wenner-Gren International Symposia Series. Oxford: Berg, 207-26.

Scientific Committee on Antarctic Research (SCAR) 1959 *The Antarctic Treaty.* Available at: <www.scar.org/treaty/>. Accessed 4 March 2009.

Sinden, J.A. and Worrell, A.C. 1979 *Unpriced Values: decisions without market prices*. New York: John Wiley and Sons.

Spenneman, D. 2006 Out of this World: issues of managing tourism and humanitys heritage on the Moon. *International Journal of Heritage Studies* 12.4, 356-71.

Tunbridge, J. and Asworth, A. 1996 *Dissonant Heritage: the management of the past as a resource in conflict.* Chichester: Wiley.

United Nations Office for Outer Space Affairs (UNOOSA) 1967 *Treaty on the Principles Governing the Activities of States in the Exploration and Use of Outer Space, Including the Moon and Other Celestial Bodies.* Vienna: UNOOSA. Available at: <http://www.oosa.unvienna.org/oosa/SpaceLaw/outerspt.html>. Accessed 4 March 2009.

Weiner, A.B. 1992 *Inalienable Possessions: the paradox of keeping-while-giving.* Berkeley: University of California Press.

World Archaeology 34 February 2002 *Community Archaeology.*

Chapter 9
Manual and Intellectual Labour in Archaeology: Past and Present in Human Resource Management

Nathan Schlanger

Let us begin with a postcard, which might have been titled: 'the man who knew too much'.[1] Under the plane trees in the village square, two men linger. The one in white, relaxing in his folding chair, walking cane suspended to his knee and cigarette to his mouth (it was only in his forties that he learned to smoke, living for a while with Spanish muleteers) lends an ear to the man in black leaning by his side, also smoking and beret-clad (see Figure 9.1). There is also a photographer present, to take and then print the snapshot on postcard paper for the seated man, who annotates its verso and inserts it in one of his autobiographical photo albums, now deposited at the *Musée des antiquités nationales* (recently renamed the *Musée d'archéologie nationale*) in Saint-Germain-en-Laye.

It is when you turn the photograph over to find out what our two cronies are on about that this image of casual complicity turns into an anomaly, good to think with about what archaeology really is. Here is what the caption says:

> Gargas 19. IX. 36
>
> *Le vieil ouvrier, qui a travaillé avec moi à Gargas en 1912, me raconte qu'il a lu dans Dumont d' Urville que des indigènes de certaines îles se coupent les phalanges en signe de deuil, & que cela doit expliquer les mains de Gargas à doigtscoupés.*

---

1 This chapter develops on an earlier presentation at the 2006 EAA Krakow meeting, and a brief text in a publication for Jean-Paul Demoule, first president of INRAP, the French national institute for preventive archaeological research (Schlanger 2007). What follows are for me but preliminary indications of research perspectives initiated in the framework of the EC funded AREA network: 'Archives of European Archaeology', as well as the 'Archives Breuil' ACI project. Reflections on contemporary professional practices emerge within the EC funded ACE network: 'Archaeology in Contemporary Europe. Professional Practices and Public Outreach'. Thanks are due to all partners in research and archive holding institutions in France, Britain and South Africa.

Figure 9.1 'The man who knew too much…'

Source: Archives Breuil, Autobiographical album, *Musée d'archéologie nationale*—MAN.

Gargas 19. IX. 36

> The old labourer who has worked with me in Gargas
> in 1912 tells me that he has read in Dumont d'
> Urville that the indigenes of some isles chop
> their phalanges as a sign of mourning, & this could
> explain the hands of Gargas with fingers chopped.

Cronies? Colleagues? Even if the Abbé Henri Breuil makes nothing of it—for it is indeed him, seated, member of the Institute and Professor at the Collège de France, the world-renowned globetrotting pope of prehistory at the height of his glory—there is in this exchange a flagrant infraction. Facing him is a worker, a labourer, a mere paid hand and nameless at that, intrigued enough by some archaeological observations to go and read through specialised primary sources,[2] prise out an ethnoarchaeological hypothesis with which to interpret upper Palaeolithic symbolic behaviour, and then, having mulled it over for as long as a quarter century, submit it to the leading authority in the land.[3] Not impossible of course, but it takes one so supremely entrenched in his superiority as Breuil not to remark on the anomaly which he himself records. It is certainly rare, if not unique, to encounter documented visual evidence of such transgression. After all, amongst us indigenes of the North Seas, with our globalised occidental cosmology, it is rather a deeply-entrenched separation that prevails, economic, social, moral, between manual and intellectual work, between the means of production and the forces of reasoning, between matter and spirit—indeed a dualism, as clear cut as black and white, whose endorsement and perpetuation appears constitutive of our very modern condition.

Except perhaps—this is in any case the suggestion I submit here for reflection—except in archaeology. Taking a closer historical look at the discipline as it is practiced, it may be argued that demarcations in terms of legitimacy and fields of competencies between workers and scholars have never been so clear-cut or unbridgeable. Over the centuries, archaeological labourers have frequently taken roles other than those of mere executants, while many armchair scholars ensconced in libraries or laboratories have readily endorsed whip-yielding, 'soiled' scientist personae.

2 A harbinger of imperial and industrial modernity, Jules Dumont d'Urville (1790-1842) not only wrote well known accounts of his voyages to the South Seas, but also contributed to the discovery and elevation of the Venus de Milo as an icon of antiquity, and also happened to have met his death in one of the first ever accidents of the railway age.

3 The Pyrenean cave of Gargas had long been explored, but its stencilled hands with missing fingers were only recognised in 1907. By 1910, Breuil raised the possibility of 'mutilated fingers', but with reference to (unspecified) Australian and bushman ethnography (Cartailhac and Breuil 1910: 135). On Breuil more generally see Coye 2006.

To be sure, there have been throughout the making of archaeology constant issues of manpower (and where appropriate, womanpower) to deal with, indeed basic challenges of human resources management, monitoring and control. Boucher de Perthes, precursor as he was to the establishment of high human antiquity 150 years ago, often fell for the fakes proffered by his Picard workers, who conveniently exposed bifaces they had themselves knapped behind the shed and left to soak in buckets of rusty waters to absorb the patina of age and pass off as authentic (see Boucher de Perthes 1847-1864, Aufrère 1940). Others have been far less naive in their dealings: the vigilant Egyptologist W.F. Finders Petrie surveyed the breadth of his Pharaonic tells with a telescope to better monitor, himself unobserved, the comings and goings of the swarming fellahin (Flinders Petrie 1904, passim). Likewise it has not escaped keen tactician Mortimer Wheeler that his five-metres-by-five excavation method served well to single-out and oversee the diggers enclosed within each 'box', be they foremen, navvies, gangs, shovel-hands or basket-boys, be they home-grown working classes or tropical subalterns. Indeed his own chosen example of a model excavation in south India, aptly captioned 'discipline' (see Figure 9.2), depicts a scene redolent with colonial control and hierarchy (see Wheeler 1954, and Chadha 2002, Boast 2002, Eberhardt 2008 and especially Lucas 2001).

Surveillance was then the order of the day, insofar as the ground rules had it that labourers, once recruited, would naturally seek to be 'working less for earning more'. This postulated morality of lucre was readily integrated in the managerial practices of archaeological operations. Innumerable examples can be given, from both normative field manuals and actual site reports: the two cases I give here present each an additional twist.

In 1820s Tunisia, early Dutch archaeologists devised a sophisticated procedure of attribution and reward (Halberstma 2008: 29-31):

> (Borgia) paid the workmen in the excavations 8 till 10 caroubes each day, which amounts to about 45 Dutch cents; The overseers who managed the excavations and paid attention that no objects were stolen by the workmen were paid 80 cents each day; the number of overseers was calculated on the number of workmen; A Mamelouk, or officer of the Bey, needed for keeping order, was paid one guilder and 60 cents each day; The gold and silver objects which were discovered, in whatever form, were estimated and the counter-value of the metal was paid by Count Borgia to the Bey; Copper, bronze and all terracotta objects could by law be kept by Count Borgia. But when statues were discovered, their value had to be estimated by an artist and half of their value was paid to the Bey.

So reliable was the spirit underlying this tariff that Dutch agent Humbert could readily subvert and unleash it to harm an excavating rival at Carthage:

(*a*) Chaos: excavation in the East, 1935

(*b*) Discipline: excavation at Arikamedu, South India, 1945 (See p. 80)

Figure 9.2 'Discipline'. Excavations and Arikamedu, South India, 1945
Source: Wheeler 1954.

> A well placed rumour about a treasure chest hidden beneath a Roman mosaic floor launched a group of Tunisian soldiers to pillage the site and destroy all visible remains, before the eyes of the consul.

Quite different was Gertrude Caton-Thompson's experience when she was called in 1929 to resolve the mystery of Great Zimbabwe, in (then) Northern Rhodesia. An Egyptologist in the Flinders Petrie mould, she brought with her to sub-Saharan Africa moral and managerial expectations which only led to delays and frustrations, as she records in her diaries (Caton-Thompson 1983: 119):

> we had half a day yesterday as the English have taught the natives not to work on Saturday afternoon, curse them (6 April 1929); I have had much difficulty with the workmen, who begun by being an unruly mob without any conception of work of any kind. However they are beginning to get broken in, and about 4 out of 20 promise quite well (19 April 1929); Those wages were 10/- a month plus 10/- for food. I protested but was told it was the Government wage and must not be exceeded. However I retaliated as best I could by stimulating their work and offering small awards for any object found of any archaeological value, as was done in Egypt; and as this included shreds and beads it far more than doubled their wages and their interest. Some became really efficient (19 April 1929).

Thus external constraints ('restrictive' colonial labour laws) and internal properties (poorly developed entrepreneurial spirit) conspired to hamper Caton-Thompson's efforts to confirm the indigenous origins of the famous ruins.

But at the same time, a renewed scrutiny of these issues through historical and archival sources can also lead us to recognise an equally persistent pattern of labourers who, bread-winning apart, have come to invest in their remunerative occupation also at a personal level, deploying industry, zeal and fascination for the task way beyond the call of duty. Already the magnanimous Boucher de Perthes acknowledged that some of his quarry workers displayed such astuteness and perseverance that could entitle them to believe that their contribution was of a nobler, learned kind:

> Their satisfaction knew no bounds when they perceived their names inscribed on the labels (of the pieces they had found). There are some who even refused to take the price of the items, telling me 'I only wish that I be spoken of in the book' (Boucher de Perthes, quoted in Meunier 1875: 37).

What could be more scholarly than this slightly vainglorious and yet eager abnegation? In parenthesis, talking of labels and posterity, here is one contribution which brought labourers immortality (albeit rarely recognised or accredited)—that of nomenclature and terminology. From northern Europe to southern Africa, and alongside more established practices of toponymic or find-spot designation, their vernacular folklore and idiomatic imagination have infiltrated the archaeological discourse: so far as prehistoric artefacts are concerned, innumerable are the mentions of 'amandes', 'ficrons', 'limaces', 'coups-de-poing' and 'langues de chat' strewn in the literature, not to forget the long-lasting 'thunderstones' and the ubiquitous and still very official 'tortoise-cores', rendered in antipodian Afrikaans as 'skilpad', 'perdehoef' and 'hoenderbek' (cf. Brézillion 1968, Schlanger 2005). Beyond issues of employment and livelihood, then, there have always been eager and enterprising workers who stood out from the mass—'trustworthy' is an adjective frequently used—drawing the attention of pioneering Stone Age specialists like John Evans and Joseph Prestwish, or of General Pitt Rivers of 'total' field-recording notoriety. In the Orient, Flinders Petrie had high regards for the experience and skills of some of his workers (in comparison with fresh-faced undergraduates just shipped from England). Following local customs, he made the villagers of Quft into a veritable guild of professional excavators, now in their forth or fifth generation rendering archaeological service for both foreign missions and the Supreme Council of Antiquities. In sub-Saharan Africa, the smaller scale of archaeological operations and the enforced bush intimacy between the boss and the 'boys' led to occasional nominal appreciations. Thus Kenya-born, Kikuyu-speaking, self-acclaimed 'White African' Louis Leakey reported that:

> Juma, our keen native worker, visited a cave at Karatinna near Nyeri during a short holiday, and found in it a coup-de-poing, which he says was dug out of the deposits in the cave by some burrowing animal (Leakey, May 1929, University of Cambridge Museum of Archaeology and Anthropology—UCMAA, 105).

The enterprising Juma reappeared two years later in Leakey's first convoy to Oldoway:

> The personnel of the expedition which set out thus consisted of Professor Reck, Mr Hopewood, Captain Hewlett and myself. The native staff consisted of two lorry drivers, a cook and a general camp boy; Professor Reck's own two boys, Captain Hewlett's two boys, Juma my best bone searcher, a mammal skinner who is preparing the skeletons of modern animals collected for comparative study, and ten of my excavators (Leakey, September 1931, University of Cambridge Museum of Archaeology and Anthropology—UCMAA, 105).

A snapshot can illustrate such 'co-laboration'. Some fifteen years after we saw him attentive to the old labourer reading Dumont d'Urville, Breuil is now weighted-down by age and the African torpor: to record some 'bushman' rock painting, not far from the notorious 'White lady' of the Brandberg, he can only remain seated with his arm in the air, his gesture faithfully mirrored in contact with the rock (see Figure 9.3).

Figure 9.3 'The extended arm'

Source: Breuil Archives, Autobiographical photo-album, *Musée d'archéologie nationale*—MAN.

Never mind that the lad is probably wetting the surface for the colours to be better seen and recorded—he and his reaching arm are an instrument, a tool, a limb, making body with, making happen, extending the will and design of the scholar, a indispensible link in the phenomenological arc of being and knowing in the world. So if the boundaries between manual and mental labour in archaeology appear fairly flexible and contingent, this is also because (unlike so many other economic, technical, scientific and cultural undertakings) there seems to prevail there a perceptible coherence or even compatibility between the production of knowledge and its contents. Already the very act of archaeological exhumation and inscription seems in tune with the equally tangible materiality of the past: could we not easily see (Figure 9.3) an old Master and his apprentice *drawing* the panel? In addition to its compatible scale, this material engagement is moreover readily reified in time and space: without denying the importance of the library or the laboratory, it is during the few weeks of the expedition, on-site in the field, that energies are concentrated and consciousness sharpened, the excavation's confines becoming a scene on which scholars and workers, in a kind of collective effervescence, rub arms and gather dirt together while orienting their common efforts, brain and brawn, towards the remote past. And just as the men break into a rhythm of work, heaping spadefuls of earth onto wheelbarrows, bringing to light remains of dwellings, of inhumations, of daily activities, of technical processes, of prestige items, of symbolic behaviour, their laborious engagement with tangible traces of humanity can spark empathy and lure for adventure, an incitement to imagine and dream, a wish to explain and to understand.

Granted the poetics of history, let us conclude with the politics of the present. Paradoxically enough, the distinctive synthesis of manual and intellectual labour I have highlighted here may be as much a casualty as a hallmark of the modern archaeological condition, that notably ushered in by the widely influential 'European convention on the protection of the archaeological heritage (revised)' (Council of Europe 1992). On the one hand, this 'Malta' convention (see Willems 2007) insists that all archaeological operations are undertaken 'in a scientific manner' and carried out by 'qualified, specially authorised persons' (article 3, passim). The subsequent increase in qualifications, training and specialisations is paralleled by investments in equipments and procedures, making data recovery and recording faster and more efficient, less tedious and hazardous to conduct. Now upheld as the discipline's core identity, this professionalisation perforce leaves less scope for, and sometimes outright excludes, unqualified labour, be they salaried day-jobbers and also plain, eager, motivated volunteers—a sore issue in terms of ultimately sawing the branch on which we sit. On the other hand, though, this incorporation of skills needs to tune in with the laudable objectives of the Malta convention, which are to reconcile the respective requirement of scientific knowledge about the past, archaeological heritage management, and spatial and economic development (an often optimistic euphemism for infrastructure and building works). With developers and their awaiting bulldozers choosing the time and place of archaeological operations (and in the process opening up unprecedented stretches of land for archaeological examination) they are in many countries those who pay and also play the tune ('clearing the grounds') for their contracted archaeologists to abide by. So much so that amidst administrative requirements, time constraints, budgetary limitations, local pressures, political interference, rationalisation audits, pink forms and grey literature, there is a risk in this otherwise valuable apparatus of loosing a sense of purpose, of keeping apart practice and research, economics and academia, profession and passion (Demoule 2005, Bradley 2006). So while a historical vantage point confirms that never before have more people been employed in archaeology, and never have they, individually, better embodied this manual and intellectual synergy, it is also the case, from a different standpoint, that many of today's professionals engaged in preventive or 'developer-funded' archaeology risk ending, collectively, as drabber, low-flying, bureaucratised versions of Indiana Jones, bereft of panache or indeed of any grail to quest.

References

Aufrère, L. 1940 Figures de préhistoriens. I. Boucher de Perthes, *Préhistoire* 7;7-134.

Boucher de Perthes, J. 1847-1864 *Antiquités Celtiques et Antédiluviennes*, Vols I, II, III, Paris, Treuttel.

Bradely, R. 2006 Bridging the two cultures: commercial archaeology and the study of prehistoric Britain, *The Antiquaries Journal*, 86;1-13.

Brézillon, M. 1968 *La Dénomination des objets de pierre taillée: matériaux pour un vocabulaire des préhistoriens de langue française*. Paris, CNRS.

Cartailhac, E. and Breuil, H. 1910 Les Peintures et gravures murales des cavernes pyrénéennes. IV: Gargas, commune d'Aventignan (Hautes-Pyrénées), *L'Anthropologie*, 21;129-48.

Caton-Thompson, G. 1983 *Mixed Memoirs*, Gateshead, the Paradigm Press.

Chadha, A. 2002 Visions of discipline. Sir Mortimer Wheeler and the archaeological method in India (1944-1948), *Journal of Social Archaeology* 2;378-401.

Coye, N. (ed.) 2006 *Sur les chemins de la préhistoire. L'abbé Breuil du Périgord à l'Afrique du Sud*, Paris, Éditions Somogy.

Demoule, J.-P. 2005 *L'archéologie: entre science et passion*, Paris, Gallimard.

Eberhardt, G. 2008 Methodological Reflections on the history of Excavation Techniques, in Schlanger, N. and Nordbladh, J. (eds) *Archives, Ancestors, Practices. Archaeology in the Light of its History*, Oxford/New York, Berghahn Books, pp. 89-96.

Halbertsma, R. 2008 From Distant Shores. Nineteenth-Century Dutch Archaeology in European Perspective in Schlanger, N. and Nordbladh, J. (eds) *Archives, Ancestors, Practices. Archaeology in the Light of its History*, Oxford/New York, Berghahn Books, pp. 21-35.

Lucas, G. 2001 *Critical Approaches to Fieldwork. Contemporary and historical archaeological practice*, London, Routledge.

Meunier, V. 1875 *Les ancêtres d'Adam. Histoire de l'homme fossile*, Paris, Rothschild.

Petrie, W.M.F. 1904 *Methods and Aims in Archaeology*, London, Macmillan.

Schlanger, N. 2005 The history of a special relationship: prehistoric terminology and lithic technology between the French and South African research traditions, in d'Errico, F. and Backwell, L. (eds), *From Tools to Symbols—From Early Hominids to Modern Humans*, Johannesburg, University of the Witwatersrand, pp. 9-37.

—— 2007 Transgressions et vocations: travail manuel et travail intellectuel en archéologie, *Archéopages*—constructions de l'archéologie (hors séries) pp. 95-7.

Wheeler, M. 1954 *Archaeology from the Earth*, Oxford, Clarendon Press.

Willems, W.J.H. 2007 The Work of Making Malta: the Council of Europe's Archaeology and Planning Committee 1988-1996, *European Journal of Archaeology*, 10;57-71.

Chapter 10
'Let's not go to the dogs tonight': Rhetoric as a Strategy of Accountability in Archaeological Outreach

Bo Jensen

Introduction

This chapter focuses on the relationship between experts, expertise and the public. I argue that any discipline needs the public, both to ask new questions and to ensure responsible decisions. To achieve this, I argue that we need many different, often disadvantaged groups to engage with archaeology. Synecdoche may be one way of achieving this, and the dogs of religion may illustrate how synecdoche works.

The Mandarinate and the Crisis of Representation

The heart of the crisis of representation may lie in the recognition that no science is innocent. As Foucault and Haraway have shown, all scientific representation exists in fields of contested power and resistance. Our conclusions are informed by our own values *and* scientifically true at the same time. Multiple different, equally scientific but contradictory findings are possible. This is the nature of the crisis, as I understand it (see Foucault 1991; Haraway 1999; 2004: 247; cf. Turner 1994; discussion in e.g., Thomas 2000: 10).

Habermas seems to have regarded this crisis as a bad thing. He argued for the need to re-establish a clear, shared scientific language and a set of shared scientific values. This would take us back to *Normalwissenshaft*, normal and normative science within a shared paradigm. However, I follow Lyotard in thinking that this would miss the radical potential of the crisis (Lyotard 1984; cf. Thomas 2004: 157). The state of crisis forces science to be explicit about the underlying values behind research and findings, and this, in turn, hinders the creation of mandarinates.

A mandarinate is essentially an administrative expert class, raised above the rest society by its expertise. In imperial China, mandarins were selected through rigorous exams, and formed an administrative elite accountable mainly to itself. In more recent times, similar elites have existed especially in medicine, law and economics. These mandarinates are very much a feature of *Normalwissenschaft*. *Normalwissenschaftliche* scientists and researchers tend to form sealed expert-classes who wield power without being accountable to anyone but themselves. This may create what Haraway describes as an 'invisible culture', where expert opinions are passed off as objective truths, although they remain unavoidably informed by the hidden values of the expert class (Haraway 2004: 233ff).

Obviously, the existence of mandarinates is bad for democratic society. However, the creation of scientific mandarinates is also bad for science. If experts are allowed to advance their theories and views without any challenge, this will not create a robust science. Only through ongoing challenge and debate can we ensure that the science that survives is robust and adapted to society's needs.

Archaeologists have never wielded the sort of social power held by doctors or judges. However, archaeologists, too, share an invisible culture. Most archaeologists share similar backgrounds and interests, and hence ask similar questions of our material and consider similar strategies in our politics. However, this does not mean that these questions are the only interesting ones, or that these claims and strategies always represent the best solutions. If we can engage a non-archaeological public, from very different backgrounds, we may be able to problematise our own set assumptions and make our invisible, shared culture visible.

An example may illustrate this: most archaeologists are seeing, as opposed to Blind. Almost all our documentation relies on vision. Recently, Tilley has analysed Welsh megalithic landscapes based on theories of indivisibility and of monuments quoting the visible shape of other features (Tilley 1994; cf. also Flemming 2005). This is all well and good, but not everyone is seeing. There were presumably blind people in the Neolithic. More importantly, there were other senses than sight in the Neolithic. It is perfectly possible that the monuments Tilley describes were also experienced at night, or that the sound and acoustics of them, the smell and feel, were as important as their appearance. It is possible that the muscular feel of the hard labour involved in building these monuments was as important to many builders as was their eventual appearance (cf. Ingold 2000). An archaeology of landscape that is accountable to the Blind will need to engage these issues, and problematise vision. In turn, this will raise questions valuable to all archaeologists, not just to the Blind.

Archaeology will benefit from new questions. We may need to import these from beyond the discipline itself. The archaeological community is already relatively diverse, and can become even more diverse in the future. However, diversity is limited because the archaeological method is exclusive. Most archaeologists tend to be well-educated adults, for obvious reasons. If we truly want to engage with children's issues, the issues raised by the chronically unemployed, by the handicapped, by religious extremists, or by very small minorities, we need to engage people beyond the discipline.

There is a moral side to this argument, as well as a practical one. Before the crisis, the hegemony of bourgeois *Normalwisseschaft* meant that good science was essentially science approved by the bourgeoisie. It was good because it pleased the powers that were. This is no longer true. Today, good research has to be good *for someone*, and it will only be as good as the goals it serves.

Expertise without Experts

In order to avoid mandarinates, I want to promote expertise without experts, for two reasons. First, no-one is qualified to speculate ahead of the data. No-one can make robust, scientific claims without coherent, transparent arguments and solid data. Contrariwise, anyone who can mobilise these resources can make a scientific argument, no matter what qualifications the researcher has. The merit of the thesis lies in data and logic, not in expert authority. This is a simple, positivist argument. Second, no-one can be right in all contexts. Due to the crisis of representation, multiple equally logical sciences are possible. Consequently, an argument that makes perfect sense within, say, a Marxist feminist logic may not convince Liberals. The argument may be perfectly sound *within* its given frame of reference, yet entirely unconvincing *outside* this frame of reference. This is a relativist argument.

This is not to say that we must do away with *expertise*. Quite the contrary. The extreme relativist approach tried in the early 1990s simply does not work. As a discipline, archaeology has a highly developed methodology and a very good database for making some specific knowledge-claims about the past. However, such claims must be based in this methodology and database, and in an explicitly stated, ethical position, not simply presented as the word from on high by charismatic experts. We need recognised expertises without recognising any expert persons. We need to be able to assess the merit of arguments, both within *their* own logic and within *our* own logic.

The Possibility of an Accountable Archaeology and an Accountable Science

The great benefit of the crisis of representation is an increased accountability on the part of all disciplines, including archaeology. Because our archaeology cannot be innocent, we need to make sure that it reflects values that we can accept. However, we must invite this democratic critique without devaluing the expertise we have so painstakingly built up during the history of archaeological practice. Total relativism is untenable and few archaeologists have ever accepted it (Ucko 1999: 22). We need to engage the public in the negotiation of archaeological values without devaluing our expertise; we need to mobilise that expertise without creating an expert monopoly on discourse on archaeology, let alone on 'archaeological' sites, artefacts and so on.

To clarify this apparent contradiction, I need to stress the difference between information and knowledge. In my rather crude model, information is relatively value-neutral but also largely irrelevant in human terms. It might tell us what we can do, but not what we should do. Knowledge is a more complex phenomenon, a way of structuring information and values into a coherent system of discourse. This is invariably a value-laden process. The knowledge that some site or artefact is important and should be preserved for coming generations is *not objectively true* in the same sense that the date of it is, but it is far more *socially important*. The informed judgement on which sites to protect, which to excavate and which to ignore cannot be made without information, but nor can it be reduced to an objective assessment of information. It must be informed by value-laden strategies of knowledge, shaped by considerations of what we want to know, how badly we want to know it, and which information we can manage without. This knowledge cannot be created by an archaeological mandarinate alone, because archaeologists are not alone in using the sites, artefacts or ideas involved. Archaeologists have a technical expertise on how to know things about the past, but we cannot single-handedly decide what should be known about it, or how to act on that knowledge. To simplify somewhat, we know what sort of questions may be answered, but we need the public to ask all those questions we never think of. Archaeology does not exist in social vacuum (cf. also Haraway 2004: 237f; Thomas 2000: 3; Leone et al. 2000: 461).

The point of all this is that archaeology cannot create knowledge in a vacuum (cf. also Thomas 2004: 40). We can generate information, but to turn information into knowledge, we need values, and *the more different value-sets we can engage, the more complex our knowledge will be*. There are very limited ways of speaking as an archaeologist. There probably always will be: we all share the same broad methods, logic and types of evidence, as indeed we should. Archaeology is a methodology for making specific claims about the world, and must remain limited by this. Engaging with a public with different priorities will allow us to see the limits of our knowledge.

To illustrate this, consider Tanum in Bohuslän in Sweden, a World Heritage Site dominated by Bronze Age rock art. Clearly this site is archaeologically important. However, it is also important in other ways. It is a local recreational area, a spot of scenic beauty, important in local identities, and a major tourist attraction. We can neither argue that the age of the site is more important than the beauty of it, nor that the economic aspect of local tourism is more important than the feeling of belonging or alienation experienced by visitors. Each is important in different ways, and archaeologists cannot claim privileged access to all these aspects. To manage this site for the future, society needs an open, democratic discussion with all interested parties, including those who merely use the site as a shortcut or for walking the dog. Society will not have to privilege all these groups equally, nor to accept all claims as scientific. However, there is a need to recognise that the scientific expert discourse may not always be the single most import one. For Tanum, specifically, archaeologists may argue that this one site is vastly more important than any other comparable site, and that if society is to preserve anything, this should be it. In many other cases, priorities must be different (cf. Wallis 2003: 142ff; Carman 2005).

This is relevant, even for pure research: Archaeologists are reasonably good at dating finds. We may not understand what rock art 'means', nor agree whether it is beautiful, but we can usually date it as accurately as anyone. The method of archaeology is very strong if we want to argue that Tanum is a Bronze Age site, much less so if we want to argue the superiority of figurative over abstract art, or, indeed, the aesthetic beauty of the human form. Archaeology cannot monopolise the decision on which pieces of rock art are the best ones. On this, it may be that an artist may speak with more authority than any archaeologist. In this sense, we can distinguish archaeologists speaking *as archaeologists* from the non-archaeological public. Although archaeologists, too, are human beings who may appreciate scenic beauty, or art, or a place to walk our dogs, we do not usually do this in any *specifically archaeological* way. This said, however, the way we do walk our dogs or appreciate scenic beauty, or art, may colour our archaeologies, as well.

Who is the Public, and to Whom are we Accountable? Subalterns in Academic Discourse

Archaeology has historically been manned and financed by members of the bourgeoisie. Hence, the discipline has also traditionally been accountable to this class. Bourgeois archaeologists working for bourgeois employers and publishing for a bourgeois audience have created a predictably bourgeois archaeology. However, different classes exist, notably the proletariat and the subalterns. The latter, especially, have been widely ignored.

The Italian Marxist philosopher Gramsci defined a 'subaltern' as someone excluded form discourse, as the subalterns' specific position is not recognised as legitimate by either those in power or their recognised opposition. Later research, especially by Haraway and Spivak has emphasised the subaltern position of multiply marked people—those both coloured *and* women or belonging to sexual minorities (Haraway 1991: 129ff). Gramsci argued that the Marxist intellectual elite should try to *speak for* these people, and represent their objective needs from its privileged position. In the face of post-modern critique, the claim of privileged Marxist access to the real, objective needs of any group cannot be maintained, however, nor the Marxist claim that the oppressed have privileged access to un-ideological truth. Hence, Gramsci's suggestion appears patronising (cf. Elam 1997: 183, Foucault quoted in Gates 2000, 296). Moreover, Haraway emphasises the problems inherent in trusting any group to *speak for* anyone else. While vicarious representation may be a practical, temporary solution, the long time goal must be to allow people to speak for themselves (Haraway 1991: 122 and 155; 2004: 87; cf. Latour 1999: 42). We cannot trust any group to represent others correctly, unless we resort to a facile essentialism. We cannot trust ourselves to correctly represent others. This is crucial. Every experience of difference is unique. Although we may make educated guesses about how various differences might affect experience, we need primary information. We cannot speak for the subaltern, because we cannot know the interests of the Other as well as the Other does.

Instead of *speaking for*, Haraway suggest the manoeuvre of *speaking with an interest in* (Haraway 2004: 87ff). Certainly, Marxist intellectuals can and should take an interest in improving the conditions of the wretched of this earth (Turner

1994, xxx). This can be done without reducing subaltern experience to a materialist consequence of economic conditions, however.

An accountable archaeology must be accountable to everyone. The more diverse the groups we can engage, the better our archaeology. For my purposes, then, *'the public' consists of every interested party, and we need to engage every potential interest.* Our collective knowledge is better, the more different 'we' are: Haraway borrows Trinh's phrase, 'in/appropriate(d) others' to describe the irreducible difference crucial to this project (Haraway 2004: 53). To Haraway, the very fact that these other views are inappropriate within traditional science also means that they are not appropriated by it. Diversity, *as* diversity, can challenge the hegemonic mandarinate. A science that limits input to those considered appropriate, and so not really other, cannot do this.

The Challenge of Engaging the Public

So, an accountable archaeology needs a diverse public. As far as possible, we need people to speak for themselves, or rather, to speak as having a personal, subjective interest. There is no room for disinterested objectivity in an accountable archaeology. Nor is engaging the constituted lobbies and special interest groups good enough. These only represent politically and economically strong groups. We need to engage *every* potentially interested party. Because knowledge is created through the combination of information and values, the values informing people's interest are actually more important than the background of the people themselves. So, engaging so many women, members of minorities etc. is not enough, if these people do not speak specifically from their interest as such. So, while feminist standpoints have challenged established views of archaeology in the past, simply being a woman does not automatically invest anyone with feminist values. A few critical feminists can do a lot more for us than a 50 percent representation for women.

This project of representativity is not a matter of majority rule. It does not hinge on the largest groups deciding everything. Rather, it must seek to include the full variation of human experience, including small minorities and subaltern groups. We do not just need to engage such groups for economic or democratic reasons. We need to engage them because they exist, and because they can provide a different perspective. In this regard, it does not matter how common or rare that perspective is, how much or how little the group in question contribute to financing archaeology. The fact that a group is underrepresented or historically wronged is less important than the fact that it has a different frame of reference (contra Shanks and Hodder 1998: 70).

Engaging the public is difficult. Most people, most of the time, do not care too much about archaeology, although they may care about sites, artefacts and so on that archaeologists also happen to care about. It is possible to live a very full, happy life without knowing much about the distant past or material culture. It is possible to do so while believing in completely un-scientific narratives on the past, as do many fundamentalist Christians. It is possible to care about, say, an archaeological site in a completely un-archaeological way. Indeed, the people who care most about archaeology tend to be those most similar to archaeologists: well-educated, modern, agnostic, white, heterosexual Westerners (but see Wallis 2003). However, we need to engage a much wider public in the creation of archaeological knowledge. The public may not need archaeology, but an accountable archaeology needs the public. *We need them more than they need us.*

The Need for Conscious Rhetorical Strategies

To engage this public, I argue that we need a conscious rhetorical strategy, specifically a conscious use of synecdoche. We need to be very conscious of *what* we say and of *how* we say it (see also Joyce 2002). The current practice of publishing archaeology has not been sufficiently successful. Far too many archaeological texts appear dull and irrelevant, and are read only by a small minority. We can do better. We have to.

Part of this relies simply on learning to write. There should be no room for poor argument, incomprehensible language, jargon or unreadable text in any discipline. The basic rules of grammar and logic are not *that* difficult, and any archaeologist ought to learn them. Indeed, our collective knowledge is only as good as our rhetoric discourse. Rhetoric will never make us right, but clear argumentation will make falsification much easier, and hence make it much easier to recognise when we are wrong. If we accept claims merely on the personal, charismatic authority of experts, our archaeology will become incoherent and contradictory. We need a clear, logical rhetoric to create a shared archaeological knowledge. This is not to say that archaeology should speak with one voice. It should not. However, knowing *why* someone makes a given claim about the past will let us accept or reject that claim on the merit of its arguments. We may not all accept the same claims for the same reasons, but we would all benefit from having specific reasons for accepting or rejecting claims.

The uses of advanced rhetorical moves (synecdoche, metaphor, simile, irony) are far harder, but worthwhile. Specifically, I want to focus on synecdoche.

Synecdoche

Synecdoche consists of letting a part stand for the whole. We do this all the time. The Bronze Age was not just an age of *bronze*, but also one of agriculture, a specific social development and so on. Likewise, 'funnel necked beakers' may stand for a whole Neolithic culture, and 'insular' for a complex of mediaeval art styles. Synecdoche, then, is the art of picking an emblem, rather than describing the whole (cf. also Bal 1994: 106ff).

We need this. The general public does not have the decades most of us devote to studying archaeology. They need a digest of archaeology, which highlights the problems and potential of archaeology without bogging things down with long, irrelevant, technical discussions. While any practising archaeologist needs a working knowledge of radiocarbon, sedimentation and so on, not every visitor at Tanum Rock Art Centre may care for that much technicality. Many will be satisfied with knowing roughly how old the site is, without knowing how it was dated. Indeed, as long as the data is solid and uncontested, the discussion of why this is so may not be all that interesting. Conversely, if the public can only give limited time to archaeology, we need to ensure that this time is devoted where the public will make the largest impact. We need the public's input more in those fields where they are likely to disagree with us than in those where they are likely to agree. We need to cut to the chase. So, Carbon-14 dates are robust enough by now that few amateurs or indeed archaeologists have anything constructive to contribute. Rock art interpretation is not. We have far more to gain by engaging the public in discussion of the latter than of the former.

Synecdoche, then, may serve as a rhetoric short-cut that allows us to get to the interesting bits. However, to do this, we need *good* synecdoche, and like good metaphors they obey no constant rules. This is art, not science (cf. Ricoeur 1977).

Between Gods and Dogs: *Leitmotif*, Synecdoche and Rhetoric

I shall examine in some detail one example of useful synecdoche, the relationship between people, dogs and religion. I argue that using this sort of rhetorical strategy may help us to build an accountable, democratic archaeology. If we can engage the public in the discussion of archaeological dogs, we can keep the discipline from going to the dogs, as it were.

My engagement with religious dogs arose in 2006, when I taught a course called 'The archaeology of religion—methods, possibilities and problems' in Copenhagen. The study of religion is heavily burdened with 'expert tradition': from the angels and chosen prophets to the approved, orthodox interpreters and enlightened masters everyone who speaks on religion risks being invested with numious authority (cf. also Sontag 2001: 254f).

This synecdoche arose more or less by accident, largely through the need to sum up structuralism and the linguistic turn in a very brief introduction: as an example, I took Saussure's differentiation of the *signs* (words) 'dog' and 'chien' from the *signified*, the actual animal, illustrated as a French bulldog that walks past my office window every morning. Saussure's point is, of course, that neither word looks or sounds anything like the animal. From Saussure and structuralism, dogs resurfaced in a discussion of Haraway and her 'Cyborgs to companion species' (Haraway 2004: 297ff; cf. Ibid. 126ff), and throughout the course these different dogs were reintroduced as synecdoche for the respective theoretical approaches.

Dogs occurred elsewhere, however: in burials, as burial goods, sacrifices, or primary subjects of burial; as sacrifices in other contexts, whole or butchered; in religious art, as gods, companions of gods and opponents of gods; and in Christianity where they turn up in cameos in Old and New Testaments alike, and even occur as Early Mediaeval saints, or their companions.

This use of dogs as a *leitmotif* served several purposes. First, it created *mnemonic continuity*: each new dog (hopefully) reminded students of all the other dogs they had seen. Each new dog in the material record reminded them of the theories discussed earlier. Hopefully, they will be able to recall the main features of each lecture simply by recalling the problems discussed relative to dogs in that lecture. These included questions such as 'What is burial?'; 'What is the relationship between destruction and sacrifice?'; 'How can we recognise myths in art?'; 'How did mediaeval Christianity differ from modern versions?'; and 'How did the Church interact with secular society?'. By making specific dogs emblems of all these discussions, I created a mnemonic continuum, where every dog should remind students of all the other dogs and hence of the problems they illustrate. So, any grave should remind them of the dog-burials at Ertebølle, and these dogs in turn of the French bulldog and Saussure, and the companion species and Haraway.

Second, dogs provided a certain *light relief* while highlighting my *personal subjectivity*. No student on this course believed that dogs were truly central to religion, so by picking a consistent, idiosyncratic emphasis on dogs, I also emphasised that I neither presented all the material, nor even always the most important material. Clearly, I could have taught a similar course without dogs, and clearly it would have been different. By focusing on dogs, I hope to have shown my students that they needed to make their own selection of the material. Indeed, my mentor, one of our professors, is famous for her love of dogs. This specific synecdoche, then, emphasised the role of our personal and institutional bonds in structuring discourse. In as far as the whole course could be seen as a tribute of sorts, it could never be mistaken for objectivity. By emphasising dogs, I reminded students that much of the record is not dogs (fundamentally a *Verfremdung*, if you will).

Third, dogs are suspended in *complex emotional webs* today. No-one can claim indifference to them—we tend to like, love, pity, fear, admire or hate specific dogs and sometimes even dogs in general. However, like people, dogs are always specific. Few people truly fear a Pekingese as much as a Rottweiler, or love a diseased, rabid mutt as much as a sleek Labrador puppy. Any simple statement that all dogs are exclusively one or the other (food/workers/pets; dangerous/beautiful/loyal) will fly in the face of personal experience. Because of this, I cannot seduce people into believing just anything about dogs. Unlike gods, who may be very obscure and exotic, dogs are part of our subjectivity. So, for instance, Hosler's semiotic analysis of the esoteric meaning of West Mexican metallurgy can only be evaluated from a thorough understanding of metallurgical analysis and a close reading of native and colonial sources with attendant problems of source criticism—something few, if any, of my students were equipped for (Hosler 2001). A discussion of the meaning of Mixtec ('Aztec') dog-sacrifice and the cult of Xolotl ('dog') is far easier to initiate, even if it must remain tentative, pending similar close analysis.

So, dogs worked as a synecdoche in this specific context. Partly, this depended on the audience, but I believe a few traits may be generalised. First, there is *plenty of evidence*. I could find dogs all across the world, from the Palaeolithic to today. Indeed, the chief reason they do occur in Christian literature is that Christians, too, lived with dogs. They are not central to the creed, but they are central enough to practical, human experience that they had to occur somewhere. In Biblical times, as today, the dogs under the table would eat any scraps thrown to them. Second, *our modern relationship with dogs is extremely complex*, meaning that they cannot be reduced to simple symbols. Third, dogs are heavily *invested with emotion in contemporary society*. People *care* about dogs. Because of this, dogs provide a shortcut to engagement. My students may not care too much about the esoteric aspects of the Mixtec cult of Xipe Totec, or Xolotl, but they do care about dogs. So do most other people. More generally, there is such a thing as common human experience. In the post-modern world, gods no longer constitute an unproblematic part of the shared frame of reference, if indeed they ever did. Religion has been revealed as inherently ideological. The polyvalent experience of dogs seems more robust, partly because it is not shaped by any monopolised discourse—everyone is free to experience dogs on her own—and partly because it remains marginal to any larger ideological project.

In sum, I argue that this particular synecdoche allows us to highlight both the breath and the limits of archaeological knowledge. By describing, e.g., dog sacrifice, we can present both our information, our methods, and emphasise their limits: we may reconstruct the size and age of the dogs, date them, discuss nutrition and so on, but the crucial question of *why* they were sacrifices, and what people felt at the time, relies entirely on analogy. For this, archaeology provides little or no data, but we still need to ask the question. In effect, the dogs may provide us with starting-point for showing what archaeology can and cannot do, and how far we depend on our experience as people, rather than as experts, to get to the really interesting questions.

Brief Conclusion

We need to engage the public. We need to preserve the value of our expert knowledge without casting ourselves as an expert class. A conscious, rhetoric strategy, including synecdoche, may serve this purpose. However, there are no hard and fast rules to this. It is art, more that science, and relies on trial and error to see if it works. When it works, however, it may keep archaeology from going to the dogs.

References

Bal, M. 1994 Telling objects: a narrative perspective on collecting in Elsner, J. and Cardinal, R. (eds) *The Cultures of collecting*, Cambridge, MA: Harvard University Press.

Carman, J. 2005 *Against cultural property archaeology, heritage and ownership, Duckwoth debates in archaeology*, London: Duckworth.

Elam, D. 1992 *Romancing the postmodern*, Blackwell: London and New York.

—— 1997 Feminism and the postmodern: theory's romance, in Belsey, C. and Moore, J. (eds) *The feminist reader*, Malden, MA: Blackwell.

Foucault, M. 1991 (1975) *Discipline and punish: the birth of the prison*, translated from French Surveillier et puinier, by Alan Sheridan, Hammondsworth: Penguin.

Gates, M. 2000 Power bodies and difference in Thomas, J. (ed.) *Interpretive archaeology—a reader*, London and New York: Leicester University Press.

Haraway, D.J. 1991 *Simians, cyborgs and women: the reinvention of nature*, London: Free Association Books.

—— 2004 *The Haraway reader*, London and New York: Routledge.

Hodder, I. 1999 *The archaeological process*, Oxford and Malden, MA: Blackwell.

Hosler, D. 1998 Sound, color and meaning in the metallurgy of ancient West Mexico in Whitley, D. (ed.) *Reader in archaeological theory—post-processual and cognitive approaches*, London: Routledge.

Ingold, T. 2000 (1993) The materiality of the landscape, in Thomas, J. (ed.) *Interpretive archaeology: a reader*, Leicester University Press, London and New York. First published in *World archaeology 25*, 1993, London: Taylor and Francis Group.

Joyce, R.A. with Preucel, R.W., Lopparo, J., Gryer, C. and Joyce, M. 2002 *The languages of archaeology—dialogue, narrative and writing*, Oxford and Malden, MA: Blackwell.

Latour, B. 1999 (1993) *Aramis or the love of technology*, translated form French, *Aramis ou l'amour de technologie* by Porter, C., Cambridge, MA and London: Harvard University Press.

Leone, M.P., Potter, P.B. and Shackel, D.A. 2000 Toward a critical archaeology in Thomas, J. (ed.) *Interpretive archaeology—a reader*, London and New York: Leicester University Press.

Lyotard, J.-F. 1999 (1984) The post-modern condition—a report on knowledge, *Theory of History and Literature 10*, Manchester: Manchester University Press.

Riceour, P. 2003 (1975) *The rule of metaphor—the creation of meaning in language*, translated from French, *Le metaphor vive*, by Czerny, R., London and New York: Routledge.

Shanks, M. and Hodder, I. 1995 Processual, postprocessual and interpretive archaeologies in Hodder, I., Alexandri, A., Buchli, V., Carman, J., Last, J. and Lucas, G. (eds) *Interpreting archaeology—finding meaning in the past*, London: Routledge.

Sontag, S. 2001 (1961) Piety without context, originally published in *The second coming* 1961, quoted here from Sontag, S. *Against interpretation*, London: Vintage Books, Random House.
Spivak, G. 1988 Can the Subaltern speak? in C. Nelson, L. Grossberg (eds) *Marxism and the interpretation of culture*, London: Macmillan, 271-313.
Thomas, J. 2000 Introduction in Thomas, J. (ed.) *Interpretive archaeology—a reader*, London and New York: Leicester University Press.
—— 2004 *Archaeology and modernity*, London and New York: Routledge.
Tilley, C. 1994 *A phenomenology of landscape—places, paths and monuments*, *Explorations in anthropology*, Oxford: Berg Publications.
Turner, B.S. 1994 *Orientalism, postmodernism and globalism*, London: Routhledge.
Ucko, J.P. 1995 Introduction in Ucko, J.P. (ed.) *Theory in archaeology: a world perspective*, London: Routledge.
Wallis, R.J. 2003 *Shamans/neo-shamans: ecstasy, alternative archaeologies and contemporary Pagans*, London: Routledge.

Chapter 11
Rethinking the Political Implications of Statics and Dynamics at the Site of Castanheiro do Vento, Portugal

Ana Vale

The aim of this chapter is to address two main issues: how do we (as archaeologists) see the past? And also, how do we represent the past? In asking these questions, my principal concern is to explore the relationship between archaeologists and archaeological sites; the engagement from which past discourse emerges. In order to explore those questions I will take as examples the research that has been carried out in the Iberian Peninsula on the 'walled enclosures' dated from the third millennium BC and my own research on the pre-historic site of Castanheiro do Vento (also a walled enclosure dated from the third millennium), located in the North of Portugal.[1] I'll argue that the traditional interpretative regime that sees these sites as fortified settlements represents the past in frozen images that become stereotypes and, in this way, appear as a symbol of domination and domestication of the past. I also think that those images of the fortified settlements offer a familiar past where we desire to recognize ourselves, the nostalgic other that we imagine we once were (the familiar past is fiction too and has no basis on any reality). This sense of familiarity is achieved by the representation, for example of houses and families, of a gendered division of labour, (women weaving, men transforming mineral into metal), by talking about children, adults and old people. This speech erases all the doubts that the archaeologist experienced during their research. It appears as a clean statement where the past reveals itself as a self-evident thing which carries the familiarity that archaeologists desire in order to feel comfortable with their work.

However the research carried out at the pre-historic site of Castanheiro do Vento (see for example, Jorge, V.O. et al. 2006) made me wonder about my own relationship with the site and with the concept of the past in archaeology. In this way I would like to discuss here the concept of strangeness in order to question a traditional approach, raised within modern science, to open the dialogue, and to problematize the unfamiliarity of the past that seems to resist classification and

1 This site has been studied since 1998. The project was initiated by Vítor Oliveira Jorge and João Muralha Cardoso. Later several researchers joined the team: Leonor Sousa Pereira, Gonçalo Leite Velho, Bárbara Carvalho, Sérgio Gomes and I.

objectification. However, to talk about strangeness is not to talk about the 'exotic other', the completely different and mysterious, that the archaeologists seek to understand (or captivate). We should bear in mind Hodder's (1999) questions when he asks: 'In producing the past as 'other' are we just producing inverse images of ourselves? Are we simply engaged in a play of difference, of relevance only to our contemporary selves?' (156). This line plays also with its opposite, i.e., the past lived by people 'just like us', that we reduce to 'the same as us' (Thomas 2004: 238). As Thomas has stressed 'The problem is one of letting the difference of the past reveal itself *as* itself, rather than allowing it to dissipate into a set of mere images which can be absorbed by the more general economy of signs that dominates contemporary existence' (ibid.). In this way, I suggest that archaeological enquiry should open itself to doubt, to the strangeness of archaeological sites, escaping the fixity of certainty that regulates the politics of knowledge in modern science. Defamiliarizing a discourse or a narrative, allows us to look again and in another way at archaeological sites.

Strangeness is what does not fit in a typology. It is connected with doubt and uncertainty. It is a limited situation where our expectations fall apart. Strangeness is the unfamiliar. It is an uneasy feeling, a struggle between measure and the unmeasured. As Charles Guignon points out, after Heidegger, things have measure and limits and so they are determined and focused, but yet they remain indeterminate, not mastered, concealed, confusing (2001: 42). Strangeness is a concept that remained in silence in the modern science. After Adorno and Horkheimer 'the regression of the masses today is their inability to hear the unheard-of with their own ears, to touch the unapprehended with their own hands' (1992 [1944]: 36). Through the 'Dialectic of Enlightenment', the authors argued that modern science led to the disenchantment of the world, by its objectification, calculation and classification, transforming it in a mathematized world. They questioned the 'impoverishment of thought and of experience', as well as the domination of nature, through science, that is in the enlightenment, definitely separate from myth and from poetry. It is a world dwelt by machines and reason, where 'cognition is restricted to its repetition', while the task of cognition should consist in the determinate negation of each immediacy (1979: 27). So could this strangeness reappear in our work, if we recover the dynamic of enchantment?

In order to go further with this question, I would like to take as an example an article about Los Millares. In this chapter the authors try to 'analyse the metric and geometric features of the construction of Los Millares' (2007: 894), translating walls, entrances and circular structures into numbers and graphics. These numbers and graphics are used in the end to validate the general interpretation of the site as a fortified settlement. However the authors note that the second wall has less important defensive characteristics and the enclosure walls exceeded practical defensive requirements. These two annotations were not emphasized by the authors but it seems to me that these two observations could lead us to question the plausibility of the interpretation, could lead to the doubt, and the unfamiliar because they do not fit in the model. The authors propose a mathematized interpretation of

Los Millares in the mathematized world of nowadays. They even question if these kinds of sites could indicate the emergence of mathematical thinking. In this way the construction of Los Millares is translated into numbers: numbers of work per day, volumes and measures.

A feeling of strangeness arises from the struggle between the visible and the invisible, the presence and the absence. It is a refusal to look for the original, the authentic, a refusal to contemplate but a desire to engage, to mesh with the materials and to recover their enchantment. The feeling of strangeness appears in the moment of encounter with an archaeological site, and from my relationship with it, like an 'auratic' experience, after Walter Benjamin (Benjamin, W. 1999 [1970] a and b). Benjamin describes the aura as a 'strange weave of space and time' (Didi-Huberman 2005, 12 (SW2: 518)), as a 'unique phenomenon of a distance, however close it may be' (1970: 216). Although his concerns were connected with a different sort of engagement that which photographs or films allow with the public, I would like to follow Didi Huberman's approach to the concept of aura. This author proposes that if previously aura was imposed in religious cult images, now it is to be found in artists' studios (2005: 5). And he adds 'Benjaminian supposition of the aura and of the 'origin' understood as a reminiscent present where the past is neither to be rejected nor to be reborn, but quite simply to be brought back as an anachronism' (2005: 7). In this way, the auratic experience could be related to the encounter with what seems an experience of an anachronic time, both too distanced and to close at the same time, the auratic experience is the liminal experience of an encounter.

Strangeness is an encounter with the gap between past and present, materialized in what we call the archaeological record. However this gap is not waiting to be discovered, but appears as a creation. A gap that is discrepant, after Mieke Bal, and I quote 'Discrepancy, to my mind, is a brilliant word to indicate the gap between past and present, as well as to suggest the two—or more!—sides of the gap, without prejudging the kind of cuts, joints, and erasures needed to make that discrepancy something we you can look at and learn from'. (2002: 60). If we follow Derrida, what we have from the past is 'disjointed remains' that are adjusted, I will say, by the archaeologists, in order to establish a coherent narrative. 'On Derrida's account, history is not linear, developmental, logical or coherent. Due to the fact that it contains within itself gaps and secrets, ghosts and holes, it can never tell us who we are' (Dooley and Kavanah 2007: 4).

As I started to say in the beginning of this chapter, the main questions that rise from this approach are: how do we understand the past, how do we represent it and how do we translate it? Is the interpretation of the past a continuous process, made with creative images and multiple approaches and stories, or do we pretend to present a static interpretation, established in well defined images where the archaeologist is rendered invisible? Probably I am addressing the problem in two extreme ways, however, I would like to argue that the research about the Iberian Peninsula's so called 'fortified settlements' has highlighted the second approach to the past, rendering it in a familiar dimension, interpreting the gap between past and present as a void that could be filled up with moments of building and moments of occupation. I suggest that the majority of past narratives denote a belief that archaeology can provide sequential stories that in the end allow us to reconstruct the missing gap, as a bridge, from the origins to the present, searching for the authentic meaning of each object, context, site or landscape.

Seeking for authenticity, through the manipulation of a real thing, from where we can extract a real meaning, seems to be one of the base problems in Iberian research on the fortified settlements. This way of understanding things is built on the premise that meaning is encapsulated in the object, and that we can reveal the authenticity of an argument by the analysis of material things. The material things are studied, organized into different types, and then each of these is attached to a function. Their presence or absence will determine the function/explanation of that context, without considering for example the associated pattern or even the constructed devices that constitute that context. This analysis take for granted the non-intervention of the structures (in this context, most of the time, in stone and clay), because they are interpreted as quick constructions, they provide the scenario for the set of activities that the archaeologist has inventoried through the study of the material culture. This line of thought, makes it possible to extract from the material things, like loom weights and arrow heads, an authentic meaning, that is also a static meaning, and what changes the meaning of the site is the absence or presence of specific objects in specific contexts.

This approach seems to implicitly carry a desire for an authentic past, represented by stereotyped images, by fixed images, that can be static because they have lost the temporality of past actions. In a fixed image 'Time is caught in a loop by constant repetition of the same action' (Biesenbach, K. 2001/2: 20, quoted by Ross, C. 2006: xvi). In the images suggested by the explanation of the 'fortified settlements', past actions seem to be frozen in stereotypes that can represent several generations. But a problem arises when the image becomes reality and not interpretation. It turns into a frozen reality, a frozen activity, that in the end constitutes a powerful barrier to dialogue by blocking the discourse and creating condensations of meaning that tend to resist critical examination. However, as Dovey argued, 'The fundamental paradox emerges out of our very attempts to find and recreate a lost authenticity, a lost world of meaning' (Dovey 1986: 47). Following the author,, authenticity is not a quality of material things but is generated by our engagement with the world. Authenticity 'is only found and generated in the dwelling practices of everyday life'(1986: 44) (through a Heideggerian perspective).

It seems to me that looking for authenticity is also a search for the original, the past as it really happened, and an attempt to find our origins in an original past. In the case of the research on the 'fortified settlements', there is an implicit desire to explain the origins of, for example, metallurgy, stratified societies, social inequalities, private property, proto-urbanism…in this way the fortified settlements interpretative model emphasized the beginning of a new social order, based on the construction of a new kind of site that represents the emergence of elites that had power and wealth, who controlled the exchange and storage of agricultural goods, and who were in charge of ritual practices…These fortified settlements would reflect the beginning of social inequalities, translated in the distribution and dimensions of domestic units, like in Los Millares, where the elite would live inside a small precinct, called a 'citadel' (Molina and Cámara 2005). Also Carlos Tavares da Silva and Joaquina Soares (1976/77: 266), when interpreting the fortified settlements located in the south of Portugal, argued that these sites reflected the social division of labour, and Michael Kunst (2000) has considered the beginnings of war, understood as a group of warriors organized in armies, in the copper age. However, as Vítor Oliveira Jorge (2008) pointed out, we are always astonished with the radical strangeness and arbitrariness of the world, which makes impossible the writing of a coherent narrative about our own origins. The interpretations presented previously, have been held by archaeologists for several years, and with no reflexive revision of those works, with rare exceptions like those who conduct their research on the pre-historic sites of Castelo Velho and Castanheiro do Vento (Northern Portugal). The traditional approach to the 'fortified settlements' is not connected with doubt, and the past appears as something that can be grasped by the archaeologist, which apparently closes down the discourse to strangeness, and because of its extreme familiarity with the way modernity sees the world, it becomes anachronistic.

Castanheiro do Vento has been interpreted as a maze of intermeshed paths and practices, with multiple meanings and connections with different times and memories. The images that we have drawn don't have well defined lines, since Castanheiro do Vento is understood as a set of paths and relations, and not as a group of things, as a group of containers of hidden meanings that the archaeologist is able to recover through the right use of archaeological tools. The hill of Castanheiro do Vento could be apprehended in different ways, depending on which place you see it from and depending on the where you are in. And 'The maze is more a world of immersion than one of vision' (2006: xx), as Ross argued. In this way, Castanheiro do Vento could not be seen as just a place to see or to be seen, and we cannot just ask what it would have looked like in the interior of its walls. 'Merleau-Ponty (1968) argues that our entire western tradition has been dominated by what we call the 'philosophy of vision', which is guided by the desire to grasp reality all at once and at distance, that is, to attain knowledge as a unified and objective totality' (Presnell and Deetz 1996: 306). Is it this 'philosophy of vision' that gives us the framework to question the archaeological sites as potential containers of well defined past images? However if we agree that Castanheiro do Vento is a place of immersion, and if all the practices are connected with different times, memories and hopes, it is impossible to raise a single picture. In other words, once I'm thinking of Castanheiro do Vento as a labyrinth, I question the 'philosophy of vision' in which archaeology arose as a discipline. Thereby, my analytical categories don't allow it a classification…and so feelings of strangeness arise. It is the impossibility of objectification and classification that I translate as the strangeness of my relationship with the site and with the 'archaeological past'. It is the encounter with the discrepant gap, with the disjointed points. It is in the end, my own encounter with doubt.

For Derrida, the past is translated in traces, cinders and remains, that haunt, disturb and inspire us (Dooley and Kavanah 2007: 8). They can never bring the past to the present, but, for the same reason, continue to impassion historical investigation. However those remains are not neutral or treated as simple objects that can reveal a happened past, but they could appear as ghosts, the absent presences of the past, that haunt us, that make it strange. And, following Derrida, it is only by the work of mourning that we can address the past, not to bring it to the present, but as an interminable work that can promise a future, which 'would determine a future'. 'If memory testifies to the fact that we can never fully recollect the past, then mourning affirms that we are never finished with the past: that the task of comprehending the past always lies ahead of us' (Dooley and Kavanah 2007: 8). It is a never-ending process of questioning, which keeps open the possibility of understanding things in other ways.

Freud (2001 [1919]) explored the concept of the 'uncanny', to translate the feeling experienced when something is 'secretly familiar, which undergone repression and then returned from it' (245). Uncanny is what is not new, however uncomfortable. What is frightening most of the time is because of its extreme familiarity, and so it is not just the unfamiliar. It could be reached, for example, by the repetition of the same thing, not expected, as a strange coincidence. Following this line, could it be argued that, the familiarity of the plans of the fortified settlements raise uncanny feelings? It is my opinion that their similarities awaken in the viewer a strange feeling of familiarity and in this way the familiar images about the past are uncanny images, as they can be frightening and uncomfortable. It is common to agree that the known and the unknown are two different spheres of our lives, one familiar and the other dangerous, hidden and most of the times secret. However what if the barriers of this binary system of thought fall apart? Following Freud's steps, I would say that the feeling of the uncanny arises. It is the uncanny feeling towards the familiar images about the past, proposed by other archaeologists, as well as the strangeness that I felt and feel within the relationship and encounter with Castanheiro do Vento that gave me the questions that I have tried to address in this chapter. It is the encounter with the endless proximity and distance of the Past. The questioning of these issues should make us adopt a disquieting attitude towards our own work, towards the discourses that we weave, and to feel discomfort and unease towards established narratives.

I started this chapter by asking questions about seeing the past and representing the past, but maybe now at the end of the chapter, I can reformulate the question: how do we see again what we represent? In always heeding the warning signs of static images of the past that are frozen as a stereotype, and in wanting to represent these sites differently through dynamics, I take on a struggle between essentialist concepts of the 'person' (man, woman, or child) and deconstructive workings of identity. Perhaps I could phrase this another way as a tension that exists in my work, where my attempts to make images are generated from a tension between wanting to make a space for children, women and men in prehistory, but wanting to further complicate these categories of person by looking at the way in which identities constantly come into being through a myriad of connections and disconnections between things and people that go somewhere else. Rather than seeing this as a failure to work a pure theory of representation, perhaps this struggle, or tension in the ways in which we work, is a maturing process in our thinking image.

Acknowledgements

This chapter has evolved from a presentation originally submitted at the Theoretical Archaeological Group (Southampton 2008) in the session 'Desires from the past. What do archaeological images want?', organized by Vítor Oliveira Jorge, and with the title: 'Representing the past: strangeness vs. familiarity. The case of the so-called 'fortified settlements' of Iberia. However this is the first attempt to address this questions and it is integrated in a research programme that has been funded Portuguese Foundation for Science and Technology. I'm grateful to Vítor Oliveira Jorge, Julian Thomas, Stephanie Koerner, Lesley McFadyen, Sérgio Gomes and Irene Garcia Rovira with whom I discussed this chapter.

References

Adorno, T.W. and Horkheimer, M. 1992 [1944] *Dialectic of Enlightenment*. Translated by John Cumming, London, New York: Verso.

Benjamin, W. 1999a [1970] On Some Motifs in Baudelaire. In *Illuminations*, Edited and with an Introduction by Hannah Arendt, Translated by Harry Zorn, London: Pimlico, 152-96.

—— 1999b [1970] The Work of Art in the Age of Mechanical Reproduction. In *Illuminations*, Edited and with an Introduction by Hannah Arendt, Translated by Harry Zorn, London: Pimlico, 211-44.

Didi-Huberman, G. 2005 The Supposition of the Aura: the Now, the Then, and Modernity. In A. Benjamin (ed.), *Walter Benjamin and History*, London and New York: Continuum, 3-18.

Dooley, M. and Kavanah, L. 2007 *The Philosophy of Derrida*. Stocksfield. Acumen Publishing Limited.

Dovey, K. 1985 The quest for authenticity and the replication of environmental meaning. In D. Seamon and R. Mugeraver (eds), *Dwelling, Place and Environment*, New York and Oxford: Columbia University Press, 33-49.

Freud, S. 2001 [1919] The 'Uncanny'. In *The Standard Edition of the Complete Psychological Works of Sigmund Freud, Volume XVII (1917-1919) An Infantile Neurosis and Other Works*. Translated from German under the general editorship of James Strachey, in collaboration with Anna Freud, assisted by Alix Strachey and Alan Tyson. London: Vintage.

Guignon, C. 2001 Being as Appearing: Retrieving the Greek Experience of Phusis. In R. Polt and G. Fried (eds) *A Companion to Heidegger's. Introduction to Metaphysics*. New Haven and London: Yale University Press, 34-56.

Hodder, I.R. 1999 *The Archaeological Process: An Introduction*. Oxford. Blackwell.

Jorge, V.O. 2008 *O jogo da presença e da ausência*. Available at: <http://trans-ferir.blogspot.com/2008/04/o-jogo-da-presena-e-da-ausncia.html>. Accessed 20 November 2008.

Jorge, V.O. et al. 2006 Copper Age 'Monumentalized Hills' of Iberia: The Shift from Positivistic Ideas to Interpretative Ones. New Perspectives on Old Techniques of Transforming Place and Space as Result of a Research Experience in the NE of Portugal, In Jorge, V.O. et al. (eds), *Approaching 'Prehistoric and Protohistoric Architectures' of Europe from a 'Dwelling Perspective'*, JIA, vol.8, special issue, Porto, ADECAP, 203-264.

Kunst, M. 2000 A Guerra no Calcolítico na Península Ibéria, *Revista (era) Arqueologia*, No 2, 128-42.

Molina, F. and Cámara, J.A. 2005 *Guia del yacimiento arqueológico Los Millares*. Junta de Andalucía: Consejería de Cultura.

Presnell, M. and Deetz, S. 1996 Interpersonal Icons: Remembered Images and the Closure of Discourse from a Lacanian Perspective. In H. Mokros (ed.) *Interaction and Identity: Information and Behavior*. New Brunswick, NJ: Transaction Press, 297-315.

Ross, C. 2006 *The Aesthetics of Disengagement: Contemporary Art and Depression*. University of Minnesota Press.

Silva, C.T. and Soares, J. 1976/77 Contribuição para o Conhecimento dos Povoados Calcolíticos do Baixo Alentejo e Algarve, *Setúbal Arqueológica*, 2-3, 179-272.

Thomas, J. 2004 *Archaeology and Modernity*. London and New York: Routledge.

Chapter 12
Integrate Plurality of Landscapes and Public Involvements into Maltese Environmental Policy[1]

Steven Vella and Marlene Borg

Introduction

Our thesis is that it is important to use visual media to link Environmental Impact Assessments (EIA) and public involvement in planning processes. Using the full capabilities of available analytical technologies, such as Public Participant GIS (PPGIS), would instigate collaboration and interdisciplinarity between consultants while actively involving stakeholders and facilitating transfer of information understood by all parties. Interfacing PPGIS and context model databases used with conventional applications such as CAD to develop interactive 3D virtual reality models, combined with other visualization techniques, would improve public involvement and the process as a whole.

Recent governance models suggest that direct public participation could assist mediation and reduce conflict if stakeholder concerns were integrated into the environmental decision making process.[2] Since environmental assessments and Social Impact Assessment (SIA) in particular have traditionally been associated with public participation, we suggest that integrating these tools would enhance environmental governance. We also suggest that public involvement should be 'early, substantiative and continuous' (Shepherd and Bowler 1997: 733). This may not always be the case as disaster and risk management diplomacy experience has

1 This article draws upon the shared reflexive experiences of a visual anthropologist and an archaeologist who respectively work as a Social Impact and a Cultural Heritage Consultant within the framework of Environmental Impact Assessments in Malta. This is based upon research and experience in the field, a paper we presented at the 14th Annual International Conference of the European Association of Archaeologists and a second paper given by Vella at the ASA conference in Bristol, in April 2009 through a Wenner Gren Fund grant. Our aim was to evaluate public participation in the local EIA process and the use, or lack of it, of visual media.

2 Extrapolated from discussions during the conference 'Environmental Conflict Mediation and Social Impact Assessment: approaches for Enhanced Environmental Governance?' Helsinki, 14-15 February 2008.

shown that grassroots involvement can also be counterproductive at high-level strategic policy making (Kelman 2008), since the bottom line of the planning process is a political power game, that while noted in academic research on planning, it is rarely if ever examined in actual projects.

Since EIA forms part of that political process it stands to reason that there are several discrepancies between what there is on paper on what EIA is and the guidelines on how it should be performed and how it actually fits in the process 'in the real word' of planning and consultancy.

EIA, SIA, CHA and Public Involvement within the Planning Process

EIA is a process that seeks to blend administration, planning, analysis and public involvement in assessment prior to taking decisions related to physical, biological, social, cultural and economic components (Barrow 1997). EIA, then, is a procedure that systematically assesses the likely significant environmental effects of certain types of projects before they can be given development consent by the competent planning authorities. This helps to ensure that decision makers and the public alike, properly understand the predicted effects, and the scope for reducing them.[3]

Wood (1995: 88-9) explains that many times EIA takes place too late in the decision making process to influence crucial choices under the jurisdictions of different authorities. Secondly, EIA is dealt with in different ways in different countries. For example, the EU directive avoided, or played down the role of socio-economic considerations and gave more importance to the physical components. Following the 1998 signature by the EU of the Aarhus Convention on public participation in environmental matters, the directive was amended in 2003 and according to official guidelines, stakeholder participation, in theory now enjoys a central role with timely access to the process for all interested parties. A number of member states though have not yet made the necessary changes to their EIA legislations, or have not taken provisions to enforce more involvement of public participation. Bowler and Shepherd argue that public participation should not be included just because it is required by law but because it is democratic and meets citizens' needs and ensures that the best decision for the common good is taken (Shepherd and Bowler 1997: 157).

The Social Impact Assessment (SIA) is an often-peripheralized component of EIA, traditionally associated with public involvement. In theory, it evaluates alternative sites, techniques and technologies in terms of their social impact, and proposes the changes and management solutions that will lead to the enhancement of positive effects and a reduction of the adverse ones. It is generally believed that affected communities should be involved in most, if not all stages of impact

3 Environmental Impact Assessment: A Guide to Procedures, DETR and The National Assembly for Wales, Thomas Telford Ltd 2000, Downloaded from the Internet: http://www.communities.gov.uk/documents/planningandbuilding/pdf/157989.pdf.

assessment. This is because they are in a better position to say how they will be affected, and what their priorities are. These can then be matched by scientific positions on the issues (see for example Bews 2004; Burdge 2003; 1996; Burdge and Vanclay 1995; Goldman 2000; Okpoko 1998; Summerville et al. 2006; Vanclay 1999; 2002). In essence, as Okpoko points out, 'a compromise must be struck between the subjectivity of value judgements and the objectivity of scientific approach…' (Okpoko 1998: 35).

Cultural Heritage Assessments (CHA) officially deal with the assessment of impacts of a project on cultural heritage assets. For example in Malta these include military, archaeological, historical, architectural, landscape context, rural and vernacular features, including rubble walls (Environmental Impact Assessment Regulations 2007, LN 114 of 2007). It therefore covers a plethora of features. Inevitably it is one of the elements that create discussions and debates among the public during the EIA process, involving many stakeholders with different and differing opinions.

EIA is influenced by an intricate set of relationships delineated by socio-economic politics. The major players, depending on the stakes of the project, perceive the process and the affected landscape differently. The highly politicized nature governing the EIA process, substantial economic considerations, political and power affiliations together with social relations, all affect the outcome of a development application (see for example Bews 2004; Goldman 2000). Unfortunately, the EIA turns into a balancing act of push and pull factors making it very difficult for policy makers to integrate and take advantage of the full potential of EIA, resulting in unhappy trade-offs (Okpoko 1998; Sadler in Morrison-Saunders and Bailey 2000). Furthermore, policy makers are only interested in research that produces models that can be applied in more than one setting (Summerville et al. 2006).

This is no different in Malta where the planning apparatus is relatively in its infancy, having been set up in the early nineties (MEPA 2005). Public involvement suffers since it is limited to submissions in writing by stakeholders, first during a 21-day period when the Terms of Reference or TOR are being formulated, and later during the public hearing, after the EIA has been submitted (Sub-regulation (3) of Regulation 10 and Regulation 24 respectively of LN114 of 2007). Most of the experience decision makers have of public consultation is of public hearings with very unsatisfied NGOs and affected members of the community on one side, feeling that they have been silenced, ignored, or short changed by the system, and a well-rehearsed crowd who are in favour of the project, cheering or booing depending on who happens to be making his or her point.

It comes to no surprise, then, that the public criticizes the structure of the EIA and its various components, including SIA and CHA. Interviewees argue that the studies do not really look for solutions and are useless, because decisions have already been taken, contracts signed and the Impact Assessments are only there to fulfil legal obligations by the proponents. There is little or no mitigation with the communities that are impacted by the project, no monitoring of social impacts or enforcement of recommendations resulting from the EIA. In Malta, due to budgetary and political constraints, SIA does not follow through most part of the project process, and stakeholder participation is kept to a strict minimum. Sometimes, it is even omitted from the TOR of the EIA, resulting in an EIA without the legally required SIA component. These are only a selection of the questions that interviewees pose and reflect the citizens' attitudes towards development policy and management and a complete distrust in the system (also see Moran 2006: 121).[4]

The case-by-case impact assessments rarely, if ever, look at the overall picture, the macro-level impacts of projects and the Area of Influence (A of I) of a project is usually as small as the budget and time constraints allow. Whilst international standards of best practice advocate comparative studies that should also inform on the efficacy of policies that govern development schemes and environmental change within the country, such studies are not included within a study, since such considerations do not fall within the terms of reference (TOR) outlined by the governing body, even though many times such considerations do impact people's lives even at project level.

Policy makers have yet to realize that instigating SIA in the early conceptual stages of a proposed development scheme and making provisions to enforce public participation during the SIA study and for SIA to follow through the various stages of the project, would often be advantageous for all stakeholders, minimizing the costs and enhancing the benefits for communities, developers and local governments alike.[5]

Consultancy and Academia

The academic community criticizes SIA as not being proper anthropology, with an ongoing debate on the roles of anthropology and anthropologists in the environmental and bureaucratic sectors (Milton 1996; Okpoko 1998; Sillitoe 2007). In the increasingly cross, inter and multi-disciplinary academic and consultancy spheres, anthropological fieldwork methods have to be adapted to the kind of work the anthropologist is doing, with an urgent deed to adapt and negotiate between consultancy experience and academic endeavour (Strathern 2004). Here one should refer to what Golsinga and Frank call the "multi-sited ethnography' of the shadows that emerge when the personal fails to fit within the framework of the professionalism' (Goslinda and Frank in McLean and Leibing (eds) 2007: xii).

4 Whilst the above is derived from direct involvement with stakeholders during SIA in Malta, similar questions are found in the literature (for example, Goldman, in a private communication with Stuart Kirsch, made similar summations (Goldman 2000, 14).

5 See for example, Barrow 2002; Commission of the European Communities 2002; Coxs and Miers 1995; Interorganizational Committee on Guidelines and Principles for Social Impact Assessment 1994; Lane 1997; Sadler and Fuller 2002; Summerville et al. 2006.

Closer to home, a PhD candidate who conducted research on one of the projects for which Vella performed the SIA, rightly criticized the report for failing to analyse the power and political dynamics (Morell 2008). The official, legitimate and rather simplistic counter-argument is that such an analysis is not part of the remit or TOR of the SIA consultacy and therefore not desired within such a report by the competent authorities and the consequently, the EIA coordinator. This is especially so when large projects that may have significant impacts on the socio-physical environment are already sensitive on their own account. To factor in those very sensitive issues into an official and public document, can open the proverbial pandora's box, fuelling pressure groups and NGOs with their own agendas who are already on the war path. Whatever is considered out of context, irrelevant for the report—such as direct quotes from informants—are deleted or commented upon, on submission of the first draft to the EIA coordinator. The argument is that the report is NOT an academic article and is only based on anthropological fieldwork. It is as though the project was removed from the socio-political landscape, a stand-alone, when of course we all know that it is not.

Consultants rarely know who the fellow team members are. For instance, the two authors of this article have never collaborated on the EIAs in which they were both consultants. This obviously limits the flow of information, let alone an inter-disciplinary and collaborative approach towards the research to be conducted or the propensity for pro-active collaborative learning (Moran 2006; Wolsink 2004) between the various consultants. This dovetails with Marcus and Okely's debate on holistic anthropology (Marcus and Okely 2008), where Okely points out 'whether cross disciplinary collaboration is translated into 'team work' at the empirical outset or post fieldwork, when acquired knowledge bequeaths a continuing lateral questioning of all things taken for granted' (ibid; also see Boucher 1995), here in relation to the EIA process.

We superimpose this on the need to overcome the often-rehearsed distinction between 'hard' technocratic biophysical scientists and 'soft' social scientists in the process of EIA, without which process SIAs (and CHAs) lose much of their efficacy (see for example Bews 2004; Ervin 2000; Lane 1997; Summerville et al. 2006). This is also reflected in the TOR. While other sections of the EIA have detailed TORs, the SIA has only a paragraph to its name, leaving it to the coordinator to interpret them and for the SIA practitioner to justify his/her methods. Some consultants are considered to gather their data in a more 'scientific' manner than others since they use scientific methods, such as mathematical analysis of data, sampling techniques and chemical analysis in laboratories. This is perceived as more realistic and impartial than opinion-based data like those collected by CHA and SIA. This is especially so in attributing value to cultural features and in predicting impact significance in the case of CHA.

Currently, it is the developer who pays for and ultimately chooses the consultants, so that the consultants have to keep as low a budget as possible to be competitive, while providing the most accurate, impartial and detailed information as possible. Sometimes budgets are not adequate to meet the demands of the TOR and the consultant has to balance between the professional ethic and livelihood!

Those who use the landscape to be affected by the proposed project, from farmers, residents, entrepreneurs to ramblers, find it difficult to distinguish between the developer and the consultants who are mistakenly considered as one and the same thing. They think that consultants do their utmost to justify the development and to twist reality, for example, the presence or absence of cultural features, in favour or against the development, depending on their pre-set opinions. On the other hand, consultants occasionally find that users obstruct their studies when limiting accessibility by enclosing fields, blocking doorways or leaving unattended guard dogs on their property. In this scenario, the discrepancies mentioned above may be somewhat overcome by involving the public by the use of technology, such as PPGIS. This might bring the stakeholders closer and even create a working relationship.

Mapping of data on GIS platforms, may provide data that can be used by various consultants and stakeholders for analytical purposes. In Malta, GIS is only used at a very late stage to present data visually to the Planning Authorities, rather than as a means of generating data and communicating it to the various consultants, stakeholders or the public, as part of the ongoing process. After all this lies at the heart of GIS and more importantly PPGIS.

A Hands-on Experiment

A simple exercise the authors carried out for the purposes of this chapter proved that we could justify the appeal for interdisciplinarity and collaboration among EIA Consultants locally, and better use of available GIS technology. Data had been collected separately from a fertile agricultural area in the northern part of Malta. The area consisted of fields enclosed by a network of rubble walls and was rich in vernacular features including corbelled huts some of which are still used by farmers. The CHA provided a catalogue of features, including rubble walls, and their state of preservation that were recorded from a field survey and archival research, while the SIA provided the occupation of the famers and the level of importance they attributed to the land they worked, based on a number of interviews that were carried out. This data was plotted using a GIS software.

It was evident that part-time farmers are more tied to their land than full-timers, possibly because it is a way of spending their free time rather than a concern to generate income for subsistence. This was mostly corroborated by the data collected from the CHA that showed that features and rubble walls in fields belonging to part-time farmers were in a better state of preservation.

In other cases, farmers who said that land was important to them, hardly maintained the rubble walls bordering their fields as shown in Figure 12.1. Was this land important, now that a development was threatening the land they tilled or was there another reason for not maintaining such an important feature on one's field? If this data had been available to the SIA practitioner beforehand,

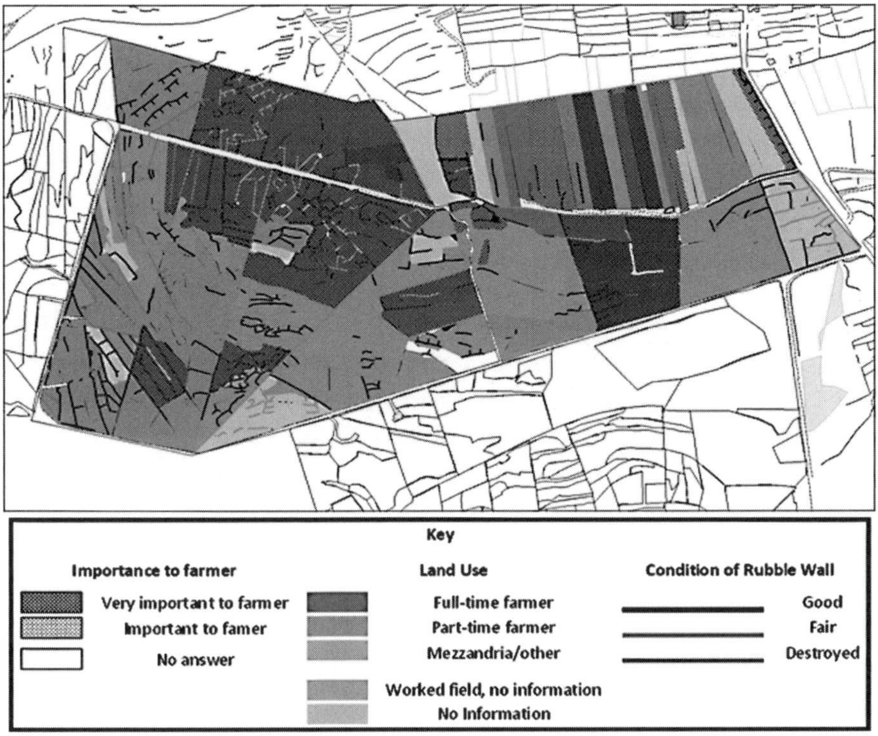

Figure 12.1 Diagram showing the importance to farmers as compared to the condition of the rubble walls around their fields

discrepancies from the original results by the individual studies, that became evident when plotted on GIS, would have been addressed and the results of the SIA amended within the Environmental Impact Statement (EIS).

Moreover, Vella, as the SIA consultant, also obtained a cultural heritage inventory collated by the farmers themselves, who had also formed a group mostly made up of the farmers and local residents to stand up against the proposed development scheme. Obviously this inventory differed from that of the archaeologists in that the definition of cultural heritage is subjective and naturally some information was not available. This was only available after the CH study had been concluded. Therefore, this simple exercise also shows that collection of data can be corroborated and improved upon, if insider knowledge is included during fieldwork. The photo superimposed upon the GIS diagram contains the farmers' inventory where other features had been recorded (Figure 12.2). If PPGIS was used during this assessment more information would have been collected and the cultural heritage inventory would have been more complete.

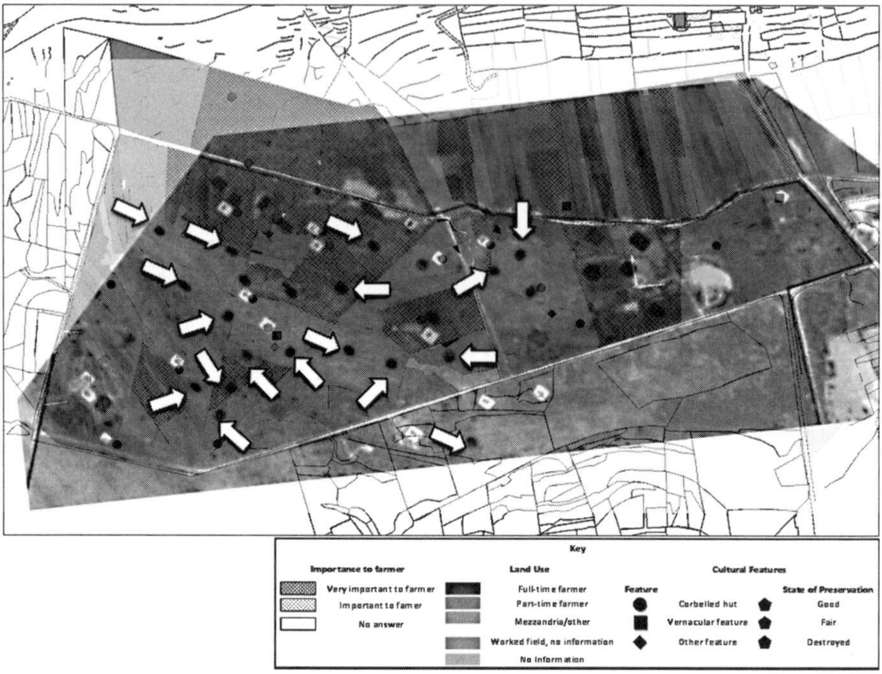

Figure 12.2 A GIS diagram showing various layers that include land use, importance to farmers and cultural features

Note: Superimposed is an aerial photograph with cultural features supplied by the farmers. The arrows indicate features that had not been identified during the CHA study and therefore did not feature in the GIS layer.

The integration of local knowledge with data collected in the field by the specialist would lead to a situation which Brown describes as 'an opportunity to move away from a focus on objects and sites as ends in themselves, toward assessing/managing the material record and intangible heritage values in their historical, social and broader landscape contexts' (Brown 2008, 1).

Visual Representations and Environmental Assessments

What about visual representations during fieldwork? SIA practitioners depend on oral explanations of the proposed project during informal interviews, using maps, architectural drawings, artistic impressions and sometimes, printed 3D models, as the only visual aides to describe the proposed project. Architects and marketing companies prepare such visualizations but only for promotional purposes to attract investors, or to make meetings with planning officials more comprehensive. It is only at the discretion of the EIA coordinator that printouts are given to the SIA consultant to use during fieldwork. These would include photomontage-views of the proposed scheme. Scaled down models may also be commissioned, but these are hardly ever available to the public or stakeholders. All these visual representations are divorced from the immediate physical landscape where the project is to be built, and from other projects taking place at the same time. The public, especially users of the area surrounding the development, have to try and imagine where one's house, place of business etc. is in relation to the model.

Visual orientation as Collier (1967: 9) puts it, can then be difficult and alienating, especially when one is not familiar with the visual medium presented, such as maps or technical drawings. Instead of being the proverbial 'can opener' or 'golden key' (ibid.: 12), the visual representations, even though they are requested by those being interviewed, become a divider between the confused interviewee and the fieldworker. The fieldworker then, has to assume the role of expert mediator.

Miller (in Ball, Cappani and Watt 2008: 81) argues that information has no intrinsic value and only becomes knowledge once there is understanding based on a multitude of contexts, such as culture, educational background, gender, personal history, politics, expertise and so forth. When it comes to the use of and interpreting photographs, audio-visual material and ethnographic films in an anthropological context, others make similar arguments (Pink 2001; Collier 1967; Ruby 2000; MacDougall 1998; Okely 2001). In landscape planning, by mapping these out comprehensively, the values attributed to a particular landscape can start to be understood by the various stakeholders, becoming a round table for mitigation.

Benson (in Bishop and Lange 2005: 184-92) notes that the results of visualizations in Visual Impact Assessments and Zone of Visual Impact are debatable, and there is little research done to quantify these inaccuracies. In other words, since as Kyungjin An (2005: 190) points out, the framework and guidelines to safeguard against technical and ethical misuse of visualization technologies in planning and consultation are scant, visualizations are still mostly only used as a marketing tool.

Benson also points out that however accurate and realistic the visualization, the key issue is, the perception of visual effects, because at the heart of EIA lies significance, which is most subjective and intractable (ibid.: 190). As Appleyard (1979) puts it, technical planning and environmental decisions are not only value based, but also identity based. We argue that this is where integration with SIA should come in, rather than SIA and VIA being unconnected components of the EIA only to be collated together by the coordinator after the studies have been performed.

Visualizations and Public Participation

It has already been established that stakeholders actively use a variety of visualizations in participation processes (Miller et al. 2005: 175-183). Oppermann and Tiedke (in Warren-Kretzschmar and Tiedtke 2005: 160) point out that the choice of visualization methods is dependent on the audience, the size of the site, resources, as well as the planning objectives and participatory situation, i.e., Internet, workshop, town meeting and so forth.

Effective consultation programmes tend to use multiple techniques together over time, providing a variety of opportunities in which to participate, review and improve the public involvement process (Canter 1996 in Shepherd and Bowler 1997: 735). In conjunction with other tools, such as soft system analysis, debates and round tables, visualizations can positively assist with public consultation and participation within the planning process. Current research has shown that audiences also desire temporal landscape processes of visualizations—in other words, flexible visualization techniques that can interactively visualize new ideas or edit the existing visualization 'on the fly' during meetings. The Macaulay Institute's Virtual Landscape Theatre or VLT, for example, can move components around and change parameters in real time, even weather conditions, to show participants a number of suggestions that come up during meetings.

While visualization tools supporting community participation are becoming more widely adopted; and acceptability and understanding of the Virtual Reality world is increasing, there still remains a gap between exploiting GIS in public participation and the need for a GIS expert as an interface between the public and the data. In fact Tiedke and Blaschke (2005: 77) caution that although GIS development and the technological advances that are taking place should be taken into consideration, the workflow should not be technology driven. One should not simply be guided by what the technology can do, but should assess whether these increases in capabilities can enhance the usefulness of visualizations (see also Ervin 2001; Appleton and Lovett 2003).

We must also keep in mind the following:[6]

- Which characteristics are significant to support citizen participation in the planning process?
- Which methods are best suited for different landscape planning tasks?
- Which methods are best suited for communities in particular socio-cultural settings? I.e., what works in Germany or Scotland may not necessarily be successful in a setting such as Malta.
- How can visualization be successfully employed in citizen participation activities, both online and offline?
- Which organizational aspects are important? Who is going to shoulder the financial and technical responsibilities?

6 Based on Tiedtke and Warren-Kretzschmar, in Buhmann et al. 2005: 157.

This last one is particularly important when keeping in mind the problems already present within the process, some of which have been described above.

Conclusion

There is consensus for social investigation in efforts to address these issues and re-thinking the traditional instruments to assess environmental impacts (see for example Devuyust et al. 2001; Milton 2004; Moran 2006; Okpoko 1998; Weston 2003). Unfortunately, initiating such change is usually 'fraught with problems', as Goldman (2000) puts it when talking about the passageway from SIA recommendations to project implementation. It is useless recommending, for example, a particular method of public consultation such as the Virtual Landscape Theatre (VLT), or that a post-SIA forum process is set up with all the stakeholders to discuss, dismiss and prioritize the SIA findings, if such a solution is not adequate within the national socio-political and cultural frameworks. It is difficult at best to get consultants to share their findings and work inter-disciplinarily, when there is no official framework to implement such practice, depending entirely on the good will of the EIA coordinator. In Malta, for example, even scoping before an EIA commences is not mandatory by law. Therefore there is the need for the introduction of a tool that would necessitate individual consultants to share their findings, for the tool to work effectively.

Our contention is that introducing visualizations proactively, especially PGIS and Interactive Virtual Reality would induce a number of necessary changes to the way EIA is conducted. As we have seen, the technology is there and is being used in one way or another, but not for public participation or comprehensive transfer of information. Besides, current research and development is striving towards interfacing the technologies that are already being used, utilizing open source programming protocols to facilitate such interfacing and minimally affecting budgets. What we are saying here is that visual representations are not the be-all and end-all. Visual representation and PGIS can become tools that involve all the stakeholders within the planning process, from EIA consultants to local experts (such as farmers, as we have seen in our example), which can be used in conjunction with other participatory tools to make this process more democratic and inclusive. Of course, to continue with the tradition of other authors stressing the need for more research, even here, it is important to understand that such a hypothesis needs to be validated by extensive interdisciplinary research and case studies.

The first step would be a willingness to change and to collaborate on the sharing of data between all stakeholders including the competent authorities. Then it would be a question of setting up guidelines and a policy framework that would streamline visualizations within Impact Assessments (An 2005: 193) together with an ongoing research evaluating the efficacy of the process and amending the methods adopted accordingly.

References

Albrow, M. 1997 Travelling Beyond Local Cultures: Socioscapes in a global city in Eade, J. (ed.) *Living The Global City: Globalisation as Local Process*, Routledge, 1997.

Alder, J. 1993 Environmental Impact Assessment—The Inadequacies of English Law, *Journal of Environmental Law* Vol. 5 No. 2, 203-220.

An, K. 2005 Implementation of Real-time Landscape Visualisation for Planning Process, in, *Trends in Real-Time Landscape visualization and Participation*—Proceedings at Anhalt University of Applied Sciences 2005, Buhmann/Paar/bishop/Lange (eds) 2005 Wichmann, Germany, 185-94.

Appleton, K. and Lovett, A. 2003 GIS-based Visualization of Rural Landscapes: defining 'sufficient' realism for environmental decision-making. In *Landscape and Urban Planning* 65, 117-31.

Appleyard, D. 1979 The Environment as a Social Symbol: Within a theory of environmental action and perception, *Journal of the American Planning Association* 45 (2):143-45.

Ball, J. and Cappani, N. 2008 Virtual Reality for Mutual Understanding in Landscape Planning. *International Journal of Social Sciences*, 2 (2), 78-88.

Barrow, C.J. 1997 *Environmental and Social Impact Assessment: an introduction*. London, UK, Arnold, Hodder Headline.

—— 2002 Evaluating the Social Impacts of Environmental Change and the Environmental Impact of Social Change: an introductory review of social impact assessment. *Environmental Studies* 59 (2), 185-95.

Becker, H. 1997 *Social Impact Assessment: Method and experience in Europe, North America and the Developing World*. London: UCL Press.

Bender, B. 1993 *Landscape: Politics and Perspectives*. Oxford: BERG.

Bews, N. 2004 *Social Impact Assessments, Theory and Practice Juxtaposed—Experience from a South African rapid rail project*. Seminar, RAU, Deaprtment of Sociology, Johannesburg.

Boucher, G.W. 1995 *The Necessity of Including the Researcher in One's Research*. C.M. Roland Tormey, (ed.), Department of Sociology, Trinity College Dublin and TCD Sociology Postgraduates Group.

Brown, S. 2008 Integrating cultural landscape approaches in cultural heritage impact assessment, *International Association for Impact Assessment conference*, Perth, Western Australia, 4-9 May.

Burdge, R.J. 2003, The Practice of Social Impact Assessment—background. *Impact Assessment and Project Appraisal*, 84-8.

Burdge, R.J. and Vanclay, F. 1995 Social Impact Assessment. In F. Vanclay, D. Bronstein, F. Vanclay and D. Bronstein (eds), *Environmental and Social Impact Assessment* (pp. 31-65). Chichester: John Wiley and Sons.

Burdge, R.J. and Vanclay, F. 1996 Social Impact Assessment: a contribution to the state of the art series. *Impact Assessment*, *14*, 59-86.

Cash, D. and Clark, W. 2001 *From Science to Policy: Assessing the Assessment Process*. Faculty Research Working Papers Series; START/CIRA Chapter draft; John F. Kennedy and Harvard University. Downloaded from http://papers.ssrn.com/abstract=295570.

Cock, J. 1994 Sociology as if survival mattered, *SA Sociological Review 6* (2), 14-31.

Collier, J. 1967 *Visual Anthropology: Photography as a Research Method*, Holt, Rinehart and Winston, USA.

Commission of the European Communities 2002 *Communication from the Commission on Impact Assessment.* COM(2002) 276 Final; 5 June.

Council for British Archaeology 2006 *DCLG Consultation on Amended Circular on Environmental Impact Assessment Eia: A Guide To Good Practice And Procedures*, Comments from the Council for British Archaeology, September.

Coxs, G. and Miers, S. 1995 *Social Impact Assessment for Local Government: a handbook for Councillors, Town Planners and Social Planners.* NSW: NSW Government Office on Social Policy.

Dale, A.P. and Crisp, R. 2001 Institutionalising social assessment in Queensland: The Social Impact Assessment Unit 1993-1996. In A.P. Dale, N. Taylor and M. Lane, *Social Assessment in Natural Resource Management Institutions.* Queensland: CSIRO Publishing.

Dale, A.P., Chapman, P. and McDonald. 1997 Social Impact Assessment in Queensland: why practice lags behind legislative opportunity. *Impact Assessment* 15, 159-79.

Department of the Environment, Heritage and Local Environment, Republic of Ireland 2007 *Review of Archaeological Policy and Practice in Ireland: Identifying the issues.*

Devuyust, D., Hens, L. and Lannoy, W.D. (eds) 2001 *How Green is the City? Sustainability and the Management of Urban Environments.* Columbia University Press.

Dupagne, A. and Teller, J. 2001 The application of EIA/SEA procedures to the urban cultural heritage active conservation, Integrating cultural heritage into the living city. Available at: <http://www.lema.ulg.ac.be/downloads/Teller-22.pdf>.

Edwards, E. 2005 *An investigation into the quality of coverage of cultural heritage impacts in UK Environmental Impact Assessment*, Thesis presented in part-fulfillment of the degree of Master of Science in accordance with the regulations of the University of East Anglia.

Environmental Impact Assessment 2006 A Guide to Procedures, DETR and The National Assembly for Wales, Thomas Telford Ltd, 2000, Available at: <http://www.communities.gov.uk/documents/planningandbuilding/pdf/157989.pdf>.

Ervin, A. 2000 *Applied Anthropology: Tools and Perspectives for Contemporary Practice.* Boston, MA: Allyn and Bacon.

Ervin, S.M. 2001 Digital Landscape Modelling and Visualization: a research agenda, *Landscape and Urban Planning* 54, 49-62.

Faludi, A. 1998 Planning by Minimum Consensus: Austrian 'Co-operative Federalism' as a Model for Europe? *European Planning Studies*, 6 (5), 485-504.

Ginsburg, F. 1999 The Parallax Effect: The Impact of Indigenous Media on Ethnographic Film, in J.M. Gaines and M. Renov (eds) *Collecting Visible Evidence*, University of Minnesota Press, 156-75.

Goldman, L. (ed.) 2000 *Social Impact Analysis: An Applied Anthropology Manual.* Oxford, New York, UK, USA: BERG.

Goslinda, G. and Frank, G. 2007 Forward: In the Shadows: Anthropological Encounters with Modernity; in McLean, A. and Leibing, A. (eds) *The Shadow Side of Fieldwork: Exploring the Blurred Borders between Ethnography and Life*. Blackwell Publishing.

Institute of Environmental Assessment, 1996, *HSE Manual.* Lincolnshire, UK.

Interorganizational Committee on Guidelines and Principles for Social Impact Assessment 1994 *Guidelines and principles for social impact assessment.*

Kelman, I. 2008 *Learning from Disaster Management Diplomacy.* Paper presented during the International conference on Climate Change Diplomacy; 7 February 2008; Malta.

Lane, M. 1997 Social Impact Assessment: Strategies for improving practice. *Australian Planner, 34* (2), 100-102.

Laws of Malta, Cultural Heritage Act, 2002.

Laws of Malta, Environmental Impact Assessment Regulations, 2007, LN 114 of 2007.

MacLeod, G. 1999 Place, Politics and 'Scale Dependence' Exploring the Structuration of Euro-regionalism. *European Urban and Regional Studies*, 6 (3), 231-53.

Madanipour, Hull, and Healey (eds) 2001 *The Governance of Place—space and planning processes.* Aldershot: Ashgate.

Malta Environment and Planning Authority, url: http://www.mepa.org.mt.

Marcus, G.E. and Okely, J. 2008 Debate Section: 'How short can fieldwork be?' in *Social Anthropology*: With permission from the authors.

Massey, D. 2006 Landscape as a Provocation: Reflections on Moving Mountains, *Journal of Material Culture* 11; 33. Sage Publications.

Mifsud Bonnici, U. 2008 *An Introduction to Cultural Heritage Law*. Malta: Midsea Books.

Miller, F.J. 2002 'I = 0 (Information Has No Intrinsic Meaning)', *Information Research* Vol. 8 (1), paper no. 140. Available at http://InformationR.net/ir/8-1/paper140.html.

Milton, K. 1996 *Environmentalism and Cultural Theory: Exploring the Role of Anthropology in Environmental Discourse.* London: Routledge.

Moran, E. 2006 *People and Nature: An Introduction to Human Ecological Relations.* Oxford: Blackwell Publishing.

Morell, M. 2008 *Paisatges d'urbanització i ciutadania. Etnografia i anàlisi política de la valorització d'entorns urbans en la ciutat neoliberal: El cas de Tigné*

(Sliema, Malta). Unpublished Advanced Studies Diploma final dissertation. Ciutat de Mallorca: Universitat de les Illes Balears.

Morrison-Saunders, A. and Bailey, J. 2000 Transparency in Environmental Impact. Assessment decision-making: recent developments in Western Australia, *Impact Assessment and Project Appraisal* 18 (4), 260-70.

NOAA Technical Memo 1994 Guidelines and Principles for Social Impact Assessment Prepared by The Interorganizational Committee on Guidelines for Social Impact Assessment.

Okely, J. 2001 Visualism and Landscape: Looking and Seeing in Normandy, *Ethnos* 66:1, 99-120.

Okpoko, P.U. 1998 The Necessity for Anthropological Forum on Environmental Impact Assessment, *Research Review (NS)* 14 (2), 33-41.

Provincial Government of the Western Cape: Department of Environmental Affairs and Development Planning 2005 *Guidelines for involving Heritage Specialists in EIA Processes*, June.

Pink, S. 2001 *Doing Visual Ethnography*. SAGE Publications.

Ruby, J. 2000 *Picturing Culture: Explorations of Film and Anthropology*. Chicago University Press.

Ryall, Á. 2007 EIA and Public Participation: Determining the Limits of Member State Discretion, *Journal of Environmental Law* Vol. 19 No2, 247-57.

Sadler, B. and Fuller, K. 2002 *Topic 13: Social impact assessment*. Retrieved 25 January 2008 from http://www.iaia.org/Non_Members/EIA/ManualContents/Intro_manual.pdf.

Shepherd, A. and Bowler, C. 1997 Beyond the Requirements: Improving Public Participation in EIA, *Journal of Environmental Planning and management* 40(6), 725-38.

Sillitoe, P. 2007 Anthropologists Only Need Apply: Challenges of Applied Anthropology. *Journal of the Royal Anthropological Institute, 13*.

Simpson, I.A., Parsisson, D., Hanley, N. and Bullock, C.H. 1997 Envisioning Future Landscapes in the Environmentally Sensitive Areas of Scotland. *Transactions of the Insitute of British Geographers* 22 (3), 307-320.

Strathern, M. 2004 Commons and Borderlands: working Papers on Interdisciplinarity, Accountability and the Flow of Knowledge. Oxford: Sean Kingston Publishing.

Summerville, J. et al. 2006 The implementation of social impact assessment in local government. In C. Hopkinson and C. Hall (ed.) *Proceedings Social Change in the 21st Century Conference* (pp. 1-10). Carseldine QUT, Brisbane: QUT.

Taylor, C. et al. 1990 Social assessment: Theory, Process and Techniques. *Studies in Resource Management* (7).

Tilley, C. 2006 Introduction: Identity, Place, Landscape and Heritage. *Journal of Material Culture*, 11 (7).

Tiede, D. and Blaschke, T. 2005 A Two-way workflow for Integrating CAD, 3D Visualization and Spatial Analysis in a GIS Environment, in *Trends in Real-Time Landscape visualization and Participation*—Proceedings at Anhalt University of Applied Sciences 2005, Buhmann/Paar/bishop/Lange (eds), 2005, Wichmann, Germany.

Torgerson, D. 1999 *The Promise of Green Politics: Environmentalism and the Public Sphere.* Durham, NC: Duke University Press.

Vanclay, F. 1999 Social Impact Assessment. In J. Petts, *Handbook of Environmental Impact Assessment* (Vol. 1, pp. 301-326). Oxford: Blackwell Science.

—— 2002 Impact assessment and the tripple bottom line: competing pathways to sustainability? In C.E. Cheney (Ed.), *Sustainability and Social Science: round table proceedings.* The Institute for Sustainable Futures, Sydney; CSIRO Mineral, Melboune.

—— 2004 The Triple Bottom Line and Impact Assessment: How do TBL, EIA, SIA, SEA and EMS relate to each other? *Journal of Environmental Assessment Policy and Management (JEAPM)*, 6 (3), 265-68.

Warren-Kretzschmar, B. and Tiedtke, S. 2005 What Role Does Visualization Play in Communication with Citizens?—A field Study from the Interactive Landscape Plan, in Buhmann, Paar, Bishop and Lange (eds) *Trends in Real-Time Landscape Visualization and Participation—Proceedings at Anhalt University of Applied Sciences 2005*, Wichmann, Germany.

Westerlind, A. 1999 *Cultural Heritage and Strategic Environmental Assessment: A Double Callenge, Proceedings from the 3rd Nordic EIA/SEA Conference*, 22-23 November 1999.

Weston, J. 2003 *Is there a future for EIA?* Response to Benson, *Impact Assessment and Projects Appraisal* 21(4), 278-80.

Wolsink, M. 2004 Institutional capacity for collaborative learning in wind power implementation. In S. Thor (ed.), *Acceptability in implementation of wind turbines in social landscapes* (pp. 55-64). Stockholm: International Energy Agency/FOI.

Chapter 13
The Academy and the Public

Don Henson

> Archaeology's ultimate goal—if it is to have any meaning or justification—must be to convey its findings not only to students and colleagues, but above all to the public which generally footed the bill for the work and paid the salaries.
>
> Bahn 1996: 83

> …not only is archaeology fun and highly educational and intellectually satisfying to the individual, it is also crucial to the survival of man on this planet…
>
> Rahtz 1985: 3

Archaeology in Britain is a young profession but a long-standing discipline. Its origins in the United Kingdom lie in antiquarian activity in the 16th and 17th centuries. Very few antiquarians or archaeologists were employed as such. The first Keeper of Archaeology in a museum in Britain was employed in 1683 at the Ashmolean Museum in Oxford; a position held by the celebrated Welsh antiquarian Edward Lhuyd from 1690 to 1709. The British Museum was not created until the middle of the 18th century (1753), at which time antiquarians began to organise themselves in national societies. The Society of Antiquaries was formed in 1707, to be followed by equivalent bodies in Scotland and Wales, as well as a British Archaeological Association, and a Royal Archaeological Institute by 1844. The 19th century saw the formation of the county archaeology societies, 32 being founded between 1836 and 1878. Archaeological activity and membership of these societies in the 18th and 19th centuries was largely the preserve of the intellectual or gentleman amateur, as exemplified by William Stukeley, Sir Richard Colt Hoare and Thomas Bateman (Daniel 1975). Archaeology was overwhelmingly male, and upper or upper middle class. The first university to appoint an archaeologist was Cambridge with the Disney Professor of Archaeology in 1851. Male domination continued well into the 20th century. Dorothy Garrod became Disney Professor of Archaeology in 1938, causing much controversy at a time when Cambridge University would not allow women to graduate with degrees. Archaeology became a field profession for the first time with the appointment of General Augustus Pitt-Rivers as Inspector of Ancient Monuments in 1882, while the Royal Commissions on Ancient and Historical Monuments were formed in 1908. Along with the Ordnance Survey and the Ministry of Works, these were the main employers for archaeologists well into the 20th century. As late as 1961, Dame Kathleen Kenyon could enumerate only 189 professional archaeologists in Britain outside museums

(Kenyon 1961). In the voluntary sector, the period after World War II saw the foundation of over 150 local archaeological societies, at a more local level than the county (and possibly with a wider social base?). Until the mid-1970s, most archaeology was carried out either by universities or by the amateur societies.

This began to change from the 1960s, with the creation of the first local authority archaeology services, e.g., the Oxfordshire Sites and Monuments Record, and the earliest field archaeology units, e.g., Leicester Museum Field Archaeology or Colchester Archaeological Trust. By 2002, a major study of archaeological employment (Aitchison and Edwards 2003) found an estimated 5,710 archaeologists, 41 percent of whom worked for the field units (Table 13.1).

Archaeological investigation of the past became increasingly dominated by the professional units, especially after the introduction of Planning Policy Guidance Note 16 (PPG16) in England in 1990 (followed by similar guidance in Wales and Scotland by 1993), which gave archaeological investigation a role within the planning process. This development of professional field archaeology has been separate from the continuation of a thriving voluntary sector. The early county societies of the 19th century proceeded without professional support, but the increasing sophistication of archaeological methodology has meant that this was not a viable model for public participation in the 20th century. W.G. Hoskins began the first archaeology and local history extra-mural courses at Leicester University in 1934 (Speight 2003), and Maurice Barley began the tradition of summer field schools engaging part-time students in real research projects at Lincoln in 1949 (Speight 2002). The university based adult education sector became the main means whereby amateur archaeologists could gain up-to-date knowledge and skills from their professional colleagues. The main expansion in local archaeological societies happened in the 1960s and 1970s, often growing out of university extra-mural classes. The number of extra-mural archaeology courses increased from just under 200 in 1961 to nearly 700 in 1980 (Brown 1986). By 1999, the Council for British Archaeology could count 1,327 separate courses in its database of adult education courses covering the physical heritage of the past (wider than just archaeology). There were very few parts of the country where people did not have access to a course within relatively easy travelling distance

Table 13.1 Employed archaeologists, 2002

Employment sector	Number	%
Government	880	15.4
Local government	1,250	21.9
Universities	890	15.6
Units/Trusts	2,360	41.3
Other	330	5.8
TOTAL	5,710	

Source: Aitchison and Edwards 2003.

(and in the Scottish Highlands, Aberdeen University pioneered remote teaching techniques to reach remote rural areas). Teaching on these courses was by a mix of university academic staff, postgraduate students learning valuable teaching skills and using their own research to bring the latest ideas and findings to an adult audience, and some long-standing part-time tutors who themselves had voluntary archaeology experience and skills. The latest academic archaeology could therefore be fed into the voluntary sector and the growing divide between volunteer-based and professional field investigation was ameliorated. University based research of local areas could thus appeal to communities based in the area itself.

Recent developments in public archaeology have continued the trend towards more localised engagement between people and their historic environment. The creation of the Local Heritage Initiative (LHI) in 2000, curiously administered by the Countryside Agency, saw an explosion of applications for funding by local heritage groups. The LHI funded more local projects than the parallel funding stream of the Heritage Lottery Fund (HLF), providing some £23,000,000 to over 1,400 local communities. Using a narrow search for archaeology projects on the LHI database (www.lhi.org.uk/index.html) yielded 233 archaeological projects being funded over the six years of the initiative. A similar search of the HLF database (www.hlf.org.uk/english/grantsdatabase/) produced 172 projects funded between 1994 and 2006 (although a wider definition of archaeology would yield many more). Some of the groups funded under these schemes were existing local societies, but many were new and had an intensely local focus. For example, the Friends of Judy Woods, south of Bradford in West Yorkshire, devoted themselves with LHI funding to the archaeology and ecology of a 40-hectare area of woodland. The Brain Valley Archaeological Society in Essex gained LHI support for an excavation in Cressing churchyard. The Cambridge Archaeology Field Group, on the other hand, had a wider area of interest with a landscape archaeology project over several parishes in South Cambridgeshire. Alongside the funding of local groups, a national initiative, funded by the HLF and by government, was helping to support work by individual finders of archaeological evidence. This was the Portable Antiquities Scheme, begun as a pilot in 1997 and extended to cover the whole of England Wales in 2003. Perhaps the most significant aspect of the scheme has been its reach into social groups that often feel alienated from mainstream heritage activity. Many people from these sectors of society make their way into metal detecting clubs, a section of the public traditionally viewed with great suspicion—in some cases even outright hostility—by mainstream archaeologists. Yet, metal detecting as a pastime appeals greatly to a wide section of the public. The PAS has recorded that 47 percent of finders who use the PAS come from social grades C2, D and E (skilled and unskilled manual workers, those on pensions, benefits and the unemployed, forming 49 percent of the population), compared with the 29 percent of museum visitors who come from these sections of society (MLA 2004). An earlier national project, run by the Council for British Archaeology, was the Defence of Britain Project from 1995 to 2002, where 600 volunteers recorded and mapped more than 20,000 anti-invasion defences of the

20th century in the United Kingdom. In both heritage groups and the work of the national schemes, we see an engagement with heritage that is meaningful and personal, that gives a great deal of emotional satisfaction to individuals.

Of course, professional archaeologists also derive personal satisfaction from what they do. However, this is within the context of an academic training that emphasises the intellectual exploration of the past, and a methodology at once highly technical and scientific. Archaeology as professionally practised, and by many non-professional groups, is an intensely rational pursuit. A difference between the rational and the emotional approaches to the world has a long pedigree. Perhaps, the ancient Greek Stoics would recognise and welcome what they would find in most archaeological sites and laboratories. The emancipation of learning from an ecclesiastical context during the Renaissance and enlightenment has born fruit in the tradition of western higher education where the study of the human world proceeds according to reason and science. Such a context for research distrusts the emotional and the spiritual, just as the Stoics turned their back on the gods and understood the world as far they could by applying reason to its manifestations (Hamlyn 1987: 81). The emotional appeal and subjective experience of local people engaging with heritage would be far more familiar to a different school of Greek philosophy, the ancient Sceptics, who turned away from a search for 'truth' to a enjoyment of the world as it was, rejecting belief and being content with appearances (Hamlyn 1987: 86). Personal experience was all. It is no wonder that personal exploration of heritage, where enthusiasm and subjective experience reign supreme is mistrusted by the academically trained establishment in archaeology. Outside the boundaries of academia, there are many people whose engagement with the past is not grounded in theory or devoted to answering the big questions about the past, but who revel in a love of artefacts and a direct, subjective contact with the past. One large group of such people are those who use metal detectors as a hobby. Complaints by archaeologists about the activity of metal detectorists have usually been couched in terms of the potential destruction of archaeological sites and loss of information through poor field techniques. An early reaction by the archaeological establishment was the creation of a campaign, STOP or Stop Taking Our Past, in 1980 to oppose the activities of metal detectorists (Thomas 2006). This opposition was sparked off by rogue detectorists committing trespass on land and damaging archaeological sites. While the authors of the campaign were acting to preserve the past from destruction, the use of 'Our' in the title of the campaign could be seen to be somewhat ambiguous. Was it a past belonging to all or the past as belonging to the archaeologists that was being defended? A statement by the Council for British Archaeology (CBA) on metal detecting of 1979 equated metal detecting with treasure hunting and contrary to the national interest, while also positioning the methods and disciplines of archaeology as the safeguard for that interest (Dobinson and Denison 1995: 66). We can detect in the complaints against metal detectorists a more fundamental distrust of the 'outsiders' who approach the past from a different intellectual and emotional standpoint to ourselves. As well as the opposition between the rational and the emotional approaches to the past, the open warfare that existed between the archaeological profession and the metal detectorists in the late 1970s and early 1980s reflected a socio-economic divide (what used to be called 'class'), with a largely university educated middle class defending the past from the rampages of an unskilled and archaeologically uneducated working class (Gregory 1986: 26). More recently, archaeological attitudes have changed and a working relationship has been found between the two camps, beginning with a CBA report in 1995 (Dobinson and Denison 1995), and latterly through the Portable Antiquities Scheme. Fences have been mended, although attitudes on both sides are still liable to be misunderstood (Addyman and Brodie 2002, Cleere 1984).

There are many 'camps' within archaeology. Professional and amateur are merely two possible poles within the discipline. Archaeology may be an overwhelmingly university originated discipline when looked at from the professional pole, in that most of its professional practitioners have at least one (and often more than one) university degree, but the view from the opposite end of the discipline is of a vast number of amateur practitioners and supporters without archaeology degrees. From this viewpoint, the academic training in archaeology inducts its recipients into a special world of the initiated and removes them from everyday life and language. Academic study is reported in third personal neutrality, striving to be objective and contributing to a search for 'truth'. This model of learning has been attacked in recent years by many from within academia, to which a label of post-modernism has often been applied. The dominant mode of academic archaeology from the late 1960s was the so-called 'New Archaeology', which valued a scientific approach to evidence and the search for laws of human behaviour as though archaeology were a social science (Binford and Binford 1968, Clarke 1968, Trigger 1989). The coming through of a newer generation of scholars to challenge the establishment brought with it the importation of post-modern ideas into archaeology, although it could be argued that these developments were a natural growth from within the discipline (Johnson 1999). This so-called 'post-processual' archaeology of the 1990s was one of contested meanings and an attempt to provide a more empathetic understanding of the subjective experiences of people in the past (Hodder 1986, Shanks and Tilley 1987). If history teaches us one thing of value, it is that nothing is new. There is little that arises that has not been thought of, done or experienced before. Our current post-modern affliction is but a resurgence of Neo-Platonic ideals in a modern scholarly guise, wherein the search for 'truth' is futile since the underlying reality of the world is beyond our intellectual grasp. Better then that we should engage in a mystical intuition and reach our own personal connection with whatever reality reaches our emotional core. These ideals have a long pedigree in archaeology. There have been those within archaeology who would by no means be considered 'post-modern' who have argued against a dry academicism, and for an engagement with more popular modes of understanding, e.g., Gregory (1983) who argued that the success of metal-detecting was an indication of the failure of archaeology to appeal to the public at large. Likewise, Jacquetta Hawkes argued cogently against the then emerging 'New Archaeology', writing the following about current archaeological discussions in 1968:

> [They] have seemed to me so esoteric, so overburdened with unhelpful jargon, so grossly inflated in relation to the significance of the matters involved, that they might emanate from a secret society, an introverted group of specialists enjoying their often squalid intellectual spells and rituals at the expense of an outside world to which they will contribute nothing that is enjoyable, generally interesting or of historical importance (Hawkes 1968).

Hawkes's own archaeology was more humanistic than scientific, as embodied in her poetic and imaginative book *A Land* in 1951; an approach which other academics dismissed as lightweight, too interdisciplinary and too subjective (Finn 2001). Her appeal against esoteric studies and unnecessary jargon has gone unheeded. Yet it could be argued that Hawkes was more in tune with the public appetite for the past than most archaeologists. There are many who seek a vision of the past that satisfies their emotional needs rather better than the dry, rational and evidential past presented by the establishment. The post-modern attack on the western intellectual tradition has opened up space for those who would offer a version of the past based on narrative, on mystery, on grand claims and inflated ideas, or even of those for whom the past is a quarry of ideas, motifs and situations to be exploited as creative fiction. The phenomenal success of Dan Brown's book *The Da Vinci Code* published in 2003, and devoured by a public willing to embrace its grand conspiracy theories, shows that a 'history' that is sensational, mysterious and controversial can appeal to a wide readership, many of whom believed that the novel was really representing a deeply hidden truth. This is not a new phenomenon. The books of Erich von Däniken since the 1960s provided similar satisfaction for an eager and credulous public, e.g., *Chariots of the Gods* in 1968. More recently, others have sought to provide sensationalist theories which strain the bounds of rationality, e.g., Robert Bauval *The Orion Mystery* in 1994, Graham Hancock *Fingerprints of the Gods* in 1996.

The world has come along way since the 18th century when learned societies and individuals could form a small elite studying the past, divorced from any popular audience. The vast bulk of the population then saw the past as a simply ruins to be avoided, worked around, removed or re-used. The antiquarian William Stukeley (FRS, FSA and so member of two of the most respected learned societies of the day) would rail against the damage being done to the heritage of the past by local people who would cart off the stone of Avebury for building materials (Stukeley 1743). Instead of this simplistic split between an educated intellectual vanguard and a largely uncaring and ignorant public, the modern world has a fractured, multi-faceted heritage appealing in different ways to a wide range of disparate, educated and interested audiences. The approximately 6,000 professional archaeologists are only a small proportion of the heritage audience. It is hard to come up with an accurate estimate of the number of voluntary archaeologists. There are perhaps 200 local archaeological societies. A study in 1995 estimated that there were c.250 metal detecting clubs and perhaps c.20,000 metal detector users (Dobinson and Denison 1995). We can add to these, the newer local heritage groups that have grown up over the last decade as a result of heritage Lottery Fund or Local Heritage initiative support. Casting the net wider than traditional archaeology, we might include the approximately 850 Civic Societies, representing 250,000 people, 260 Building Preservation Trusts and c.400,000 volunteers looking after or explaining heritage (English Heritage 2007). These are people with an active interest in the past, seeking to engage with it intellectually and emotionally as a part of their lives. There are also those who like to visit heritage for pleasure, as an occasional leisure activity. Some 56 million visits were made to historic environment attractions in 2005, and a survey in 2006 found that 69 percent of adults had visited a historic environment site during the last 12 months (English Heritage 2006a). Many people prefer a more passive engagement with the past and receive their archaeology delivered to the comfort of their armchair through the television. The phenomenal success of *Time Team* on television has led to an explosion in programmes and series on archaeology and heritage. Viewing figures of three million are not unexpected, and even higher figures can be generated by popular topics such as ancient Egypt. Nearly 10 million people watched *Pompeii: The Last Day* broadcast by the BBC in 2003. Research by Angela Piccini for the Council for British Archaeology (Piccini and Kulik 2006) has revealed that 96 percent of adults watched at least one television programme about heritage during the year, and that 20 percent of adults watched at least 99 programmes a year. For those who prefer a romanticised or mysticised past, they can read historical novels or blockbuster best-sellers, such as the popular series begun by Jean Auel in *The Clan of the Cave Bear* in 1980. There will also be those, impossible to quantify, who follow their own path of research divorced from any rational regard for evidence, and who belabour heritage agencies with their often bizarre theories based on complex mathematical theories, visitors from outer space or myths and legends of figures like King Arthur (seeking the Holy Grail at Stonehenge seems to be popular). For some, archaeology means enjoying the exploits of the scantily clad and buxom Lara Croft in high sensory impact computer games. Lara's popularity was a major feature of the gaming world since 1996, spawning spin-off films such as *Tomb Raider* in 2001. The professional field archaeologist or the university-based academic is no longer the guardian of an archaeological 'truth' working away at his his/her research for the benefit of us all. They have become just another sectional interest that has to share the heritage space with others, some of whom they would rather avoid.

Not all archaeologists have sought to maintain the boundaries between the discipline and the public however. The main arena for reaching out to the adult public has traditionally been the university part-time education sector. This has undergone several changes in name, including extra-mural education, continuing education and lifelong learning, reflecting changes in the nature and funding of part-time higher education. As noted earlier, this has been the mainstay of professional engagement with the amateur sector for over 70 years. While the figures for 1999 quoted earlier showed a healthy and vibrant adult education scene, the picture has become less rosy over the last seven years. There was a fall of 26 percent in the number of universities offering continuing education courses in archaeology by 2006, a 20 percent fall in the number of courses, and a massive 49 percent fall in the number of locations where courses were held. Some large areas of the country now no longer have access to adult part-time university courses in archaeology. Changes in funding have seen money diverted towards providing qualifications like certificate and diplomas, where the idea of studying for the sheer fun of it, or for purely personal development no longer has validity. More recently, funding has been further diverted towards courses that relate directly to employment. While this is worthy in itself, it does rather beg the question whether we live our lives only for work, or whether work is something we do to enable us to live our lives more meaningfully. The basis of the government's approach has been a narrowly utilitarian vision of education. The greatest happiness of the greatest number has become transmuted in the 21st century to the economic well-being of the nation (which of course thereby guarantees the well-being of all its citizens!). Individual educative goals and a humanistic vision of education as liberating must therefore give way to the goal of providing employable skills, so that all can joyfully take their place as a cog in the great economic machine. It may be doubted that Utilitarianism's founder, Jeremy Bentham, would have appreciated such a view of happiness that reduced individuals to servants of society as a whole. (Bentham was a strong believer that education should be open to all, and his ideas underlay the founding of University College London to challenge the restricted entry to Oxford and Cambridge Universities.) Someone who inspired the young Bentham was scientist, philosopher and educationalist Joseph Priestley, who had this to say about the study of the past, and which still stands as one of best humanistic justifications for history and archaeology:

> History enabled students to understand change and cause and effect, to improve their judgement and understanding, to lose their prejudices, and to learn from the past how to improve affairs in the future and appreciate the wide variety of human nature (Priestley 1803: 25).

A humanistic vision of lifelong learning and a place for archaeology and history in that vision is facing a two pronged attack. On the one hand a neo-Platonic, post-modernism has sought to undermine the very basis for archaeology and history, and on the other a neo-Epicurean, narrow materialism refuses to admit a social utility for the past. For post-modernist historians like Jenkins (1991), the past is unknowable. If this is so, then the study of the past is impossible and archaeology and history become the telling of tales, indistinguishable from fiction. The utilitarian perspective of government is perhaps exemplified in fact that the Minister for Higher Education, Bill Rammell, could openly welcome a decline in applications to study at university subjects like history (a 10 percent drop in that year) in favour of more 'vocational' subjects:

> Students are dropping subjects such as philosophy and history in favour of courses that will be more useful to their careers, the higher education minister, Bill Rammell, said today. The minister said the trend—which also hit classics and fine art—was 'no bad thing' (*The Guardian*, 15 February 2006).

We are led to believe that the past is both impossible to know, and has no relevance within the present. Most archaeologists and historians would beg to differ. The influential Institute for Historical Research held a conference in 2007 called *Why history matters* which addressed some of the issues about the role of history in society. The archaeological and heritage communities have sought to establish their social relevance through various studies and reports. English Heritage published a wide-ranging account of the social value of the historic environment (English Heritage 2000). The Archaeology Forum issued a leaflet on the benefits to society of archaeology (The Archaeology Forum 2005). The wider benefits of heritage in general were the subject of a publication by Heritage Link (2004). The National Trust has done work to establish a methodology for measuring the value of the past to the present (The National Trust and Accenture 2006). Most recently, a major conference, *Capturing the public value of heritage*, was organised by the Heritage Lottery Fund, English Heritage, the National Trust and DCMS (the government's Department for Culture, Media and Sport), with the proceedings published in 2006 (English Heritage 2006b). Archaeology has thus sought an instrumentalist justification by demonstrating its economic and social value. Archaeology is concerned with the recovery and management of the nation's heritage, a prime economic asset. The role that heritage can play in raising local pride, providing jobs and stimulating inward investment is well established. The way in which archaeology can act to bring people to together and provide an outlet for community activity is being acknowledged more and more. However, we also need to admit that archaeology is more than an instrument of socio-economic improvement, but that it also has enormous emotive power. It enables people of today to extend a hand across the ages and almost touch people in the past. We can literally touch what they touched, and so have an empathetic contact that is very powerful. It is the individually inspiring, humanistic quality of archaeology that is its real strength, and why people of all ages and walks of life wish to engage with it. Archaeology thus becomes a subjective and emotional experience. Yet, we are trained to be objective, dispassionate and rational in our approach to the past. This is what differentiates modern archaeologists from romantic wish-fulfilling pedlars of alternative realities. It is also this that separates academic and professional archaeologists from the public they serve. Hawkes's criticism of archaeologists still stands today as a cautionary tale (see above, Hawkes 1968).

What of the future? The work being done to establish the utility of the past can count on solid support among the people. A MORI survey in 2004 (English Heritage 2004), showed that 62 percent of people had visited a heritage site during the previous 12 months, that 76 percent of people felt that heritage was directly relevant to them, that 82 percent thought heritage was fun, that 92 percent wanted to keep the historic environment during redevelopment, and that 94 percent wanted children to be taught about heritage. These are astonishingly high figures and show just how public support could be garnered by the archaeological community. Conferences like *Your place or mine?* in 2006 or *History and the public* in 2007 brought together a range of people from different areas of the heritage world, all with a passion for reaching out to the public. There are increasing numbers of community archaeologists, outreach officer and others actively engaged in bridging the divide between the academic or professional discipline and the public. We are also seeing the transfer of knowledge and skills from the professional to the voluntary sector. For example, Heritage Lottery and Local Heritage Initiative funding has enabled local heritage groups to be trained in the use of modern, digital survey techniques and in the use of geophysics equipment. The transfer need not be one way. Other initiatives provide volunteer input into areas that would normally be the preserve of the professional sector, but which could never be adequately covered by hard pressed local authority services or field units. The Leicestershire and Rutland Archaeological Network has built on over 30 years of community archaeology in Leicestershire to set up a network of parish heritage wardens and archaeological wardens. These are volunteers, now in over half the parishes in the two counties, who can help to investigate, monitor and conserve the heritage of their areas based on sound and detailed local knowledge. This local dimension is too easy to forget with our concerns for top-down designation of heritage through scheduling and listing according to national standards for what should be significant in heritage. Yet, it is the local heritage that is most meaningful for people. With the growth of community archaeology, we are beginning to see the democratisation of archaeology, where concern for heritage will be led by people and their concern for what they deem to be their heritage. The Council for British Archaeology welcomes this, and can help to provide a space for community archaeology to flourish through its newly launched Community Archaeology Forum (http://www.britarch.ac.uk/communityarchaeology). This is intended as a space where community groups can upload and share the results of their work, make contact with each other, have access to guidance and help, and discuss issues of common concern. Groups can also create their own webpages using templates available on the site.

The intellectual concerns of academia, and the rigid separation of disciplines such that archaeology and heritage studies largely inhabit separate worlds in higher education, effectively undermine the ability of archaeology to reach out to potential allies among the public. However, in spite of this, there are signs that times are changing, and that universities have a role to play in this change. Universities are taking a lead in tackling some issues of public involvement, e.g., how to ensure access to fieldwork for students with disabilities (through a project led by Reading University, http://www.hca.heacademy.ac.uk/access-archaeology/inclusive_accessible/). Although there has been a decline in university-based continuing education, there are still universities that are prepared to engage in part-time provision of archaeology and history for the public. Sussex University even has a Professor of Archaeology based in its Centre of Continuing Education, while other hotspots of activity include Birkbeck College London, Bristol, Cambridge, East Anglia, Lancaster, Nottingham, Oxford and Reading. Other universities engage in particular outreach projects to reach new audiences. Exeter University ran a community landscape project in partnership with Devon County Council, Devon Archaeological Society and local archaeology groups from 2001 to 2004, which engaged local volunteers in academic landscape archaeology research. Bradford University has a community engagement strategy, partly funded through its Higher Education Innovation Fund, whereby local schools can be brought into the university to work with archaeologists on the local historic environment. Especially noteworthy has been the way the university has been able to broker relationships between schools in different parts of the local area, e.g., an inner city school with high ethnic diversity, and a less diverse rural school to work together and, in the process, learn about each other. Sheffield University is engaging in a partnership with local museums, libraries and archives, and the Workers Educational Association in a South Yorkshire Community Archaeology Project which will transfer skills and knowledge to the local voluntary sector in Barnsley, Doncaster, Rotherham and Sheffield. Some archaeologists in universities like Cambridge and University College London are using the widening participation agenda (through the Aim Higher initiative to raise the number of young people entering higher education to 50 percent) in order to reach out to teenagers. Bringing a wider range of young people from all socio-economic backgrounds is vital for the long term health of the discipline. Often the work done by universities depends on highly motivated and committed individuals like Sean Hawken at Exeter, John McIlwaine at Bradford, Sarah Dhanjal at University College London or Carenza Lewis at Cambridge. It has yet to become an embedded and natural part of higher education practice across all university archaeology departments. Nevertheless, the signs are encouraging that universities are more and more engaging with the world outside the boundaries of traditional academia, and this can only be a force for good in the future.

What we are seeing is the interaction of two communities that have different ways of engaging with the past. It is perhaps overly simplistic to label thee as the rational and the emotional, yet these terms do signify important aspects of each community. Thinkers and writers over thousands of years since ancient Greece have debated and opposed the rational and the emotional, and these different ways of experiencing life have often been placed in opposition to each other. In modern times, we have seen the two worlds of science and art erected into opposing modes of experience (Snow 1959), where the old Renaissance ideal that the whole of human experience and knowledge can be appreciated and understood by one mind are no longer held. Likewise, post-modern anxieties about whether we can truly know the world we live in are nothing new, and go back thousands of years. Archaeology can transcend these debates and polarities. It is that rarest of subjects; one that crosses the boundaries and enlarges the mind. It has this power because it is an activity, a set of processes. The two cultures of science and art are fundamentally worlds of explanation and expression. One seeks to understand by investigation, the other seeks to explore though expression. Since Snow put forward the idea of the two cultures, a third culture has come into being (Kelly 1998), that of technology. This is not technology in the old sense but of modern, enabling and empowering digital and computing technology, where creative power is placed in everyone's hands. Investigation and expression have been joined by experiencing as a new way of exploring our world. Devotees of the new culture wish to actively engage in doing, in making, and in pushing at the boundaries of knowledge and skills. They wish to experience the creative process and make the world according to their own inner lights. Taking part in process is a key aspect of this new world. It is in this third culture that archaeology most naturally belongs, and offers a way of experiencing the past through its techniques. The desire to take part, to experience the past has long been visible, but not often regarded in a serious light. It is this that underlies the growth of local archaeological societies and heritage groups. It also underlies the activities of the re-enactment societies that recreate the past as living history (and which are somehow less regarded than academically based experimental archaeology). They offer more than just entertainment and are remarkably popular (with nearly 500 such societies listed on one website (http://www.re-enact.com)).

We must rediscover the reasons why we do archaeology. We all have our personal motives, but at the end of the day, the past has a relationship to the present that transcends narrow concerns of intellectual demarcation. So long as we respect the lives of the past and give them meaning through an honest recovery and investigation of the evidence they have left behind, then we will have an archaeology that is bigger than those who practice it; an archaeology that involves everyone, that can be practised by everyone and which serves everyone. The academy is increasingly engaging with its wider audience and becoming part of that wide, diverse and fascinating (and occasionally exasperating) family which we call archaeology, although there is still a long way to go, and many attitudes that need to be changed. We really can learn from the past, and to put this into practice, we can consider the words of Sir Francis Bacon in the 17th century as they might apply to academic or professional archaeology:

> the greatest error of all the rest is the mistaking or misplacing of the last or farthest end of knowledge: for men have entered into a desire of learning and knowledge, sometimes upon a natural curiosity, inquisitive appetite; sometimes to entertain their minds with variety and delight, sometimes for ornament and reputation; and sometimes to enable them to victory of wit and contradiction; and most times for lucre and profession; and seldom sincerely to give a true account of their gift of reason, to the benefit and use of men (Bacon 1605).

References

Addyman, P.V. and N. Brodie 2002 Metal detecting in Britain: catastrophe or compromise?, in N. Brodie and K.W. Tubb (eds) *Illicit Antiquities: The theft of culture and the extinction of archaeology*: 179-84, London: Routledge.

Aitchison, K. and R. Edwards 2003 *Archaeology Labour Market Intelligence: profiling the profession 2002/03*, Bradford: Cultural Heritage National Training Organisation.

Bacon, Sir F. (Viscount St Albans) 1605 *The Advancement of Learning* (edited) G.W. Kitchin 1861, reprinted by Dent and Sons Ltd., London.

Bahn, P. 1996 *Archaeology: a very short introduction*, Oxford: Oxford University Press.

Binford, L.R. and S.R. Binford 1968 *New Perspectives in Archaeology*, Chicago, Aldine Publishing.

Brown, A.E. 1986 Archaeology in adult education: some facts, CBA Education Bulletin 1: 12-20.

Clarke, D. 1968 *Analytical Archaeology*, London: Methuen.

Cleere, H. 1984 Great Britain, in H. Cleere (ed.) *Approaches to the Archaeological Heritage*: 54-62, Cambridge: Cambridge University Press.

Daniel, G. 1975 *150 years of archaeology*, London: Duckworth.

Dobinson, C. and S. Denison 1995 *Metal Detecting and Archaeology in England*, London and York: English Heritage and the Council for British Archaeology.

English Heritage 2000 *Power of Place*, London: English Heritage.

—— 2004 *Making Heritage Count?* London: English Heritage.

—— 2006a *Heritage Counts: the state of England's historic environment 2006*, London: English Heritage.

—— 2006b *Capturing the Public Value of Heritage*, London: English Heritage.

—— 2007 *Valuing Our Heritage: the case for future investment in the historic environment*, London: English Heritage.

Finn, C. 2001 A rare bird, *Archaeology* 54: 38-43.

Gregory, T. 1983 The impact of metal detecting on archaeology and the public, *Archaeological Review from Cambridge* 2, 1: 5.

Gregory, T. 1986 Whose fault is treasure-hunting? in C. Dobinson and R. Gilchrist (eds) *Archaeology Politics and the Public*: 25-7, York: York University Archaeological Publications.

Heritage Link 2004 *The Heritage Dynamo*, London, Heritage Link.

Hodder, I. 1986 *Reading the Past: current approaches to interpretation in archaeology*, Cambridge: Cambridge University Press.

Jenkins, K. 1991 *Re-thinking History*, London: Routledge.

Johnson, M. 1999 *Archaeological Theory: an introduction*, Oxford, Blackwell.

Kelly, K. 1998 The third culture, *Science* 279: 992-93.

Kenyon, K.M. 1961 (2nd edn) *Beginning in Archaeology*, London, Phoenix House.

MLA (Museums, Libraries and Archives Council) 2004 *Visitors to Museums and Galleries*, London: MLA.

Piccini, A. and K. Kulik 2007 Archaeology viewers count, *British Archaeology* 94: 56-7.

Priestley, J. 1803 *Lectures on History and General Policy.*

Rahtz, P. 1985 *Invitation to Archaeology*, Oxford: Blackwell.

Shanks, M. and C. Tilley 1987 *Re-Constructing Archaeology: theory and practice*, London: Routledge.

Snow, C.P. 1959 *The Two Cultures and the Scientific Revolution*, Cambridge: Cambridge University Press.

Speight, S. 2002 Digging for history: archaeological fieldwork and the adult student 1943-1975, *Studies in the Education of Adults* 34, 1: 68-85.

—— 2003 Localising history 1940-1965: the extra-mural contribution, *Journal of Educational Administration and History*, 35, 1: 51-64.

Stukeley, W. 1743 *Abury: a temple of the British druids,* London.

The Archaeology Forum 2005 *Archaeology Enriches Us All.*

The National Trust and Accenture 2006 *Demonstrating the Public Value of Heritage.*

Thomas, S. 2006 Wanborough revisited: the rights and wrongs of treasure trove law in England and Wales, paper presented at *Cultural landscapes in the 21st century* conference in 2005 Available at: <http://www.ncl.ac.uk/unescolandscapes/files/THOMASSuzie.pdf>.

Trigger, B.G. 1989 *A History of Archaeological Thought*, Cambridge: Cambridge University Press.

Chapter 14
Exotic Objects, Uncanny Encounters: Anthropological Knowledge, Tourism and Challenges of Representing a World Heritage Site

Marta De Magalhães

Google 'Tourism Salvador Bahia' and your first hit is likely to be a link to the city government's tourism webpage.[1] Click on the link and you will be taken to an introductory page, where a series of photographs of the city will follow, in slideshow, to the rhythm of a traditional samba. 'Jump intro' to enter the site properly and you will be directed to the site's home page, where a text, prefaced by the injunction 'Know Salvador,' begins:

> Salvador is undoubtedly one of the most beautiful cities in the world. Due to its very beauty and to a series of unique features, it has become a primary destination for international tourism. Famous for its history, for the legacy left by people from other continents, for its religious syncretism and for its hospitable people, the capital of the state of Bahia has staged and has been the object of several studies, conducted by professionals from different fields (http://emtursa.ba.gov.br).

The opening lines provide a most intriguing motto. Unsurprisingly, the city is first praised for its beauty and uniqueness. The rather general praise gives way to a list of more specific, singular traits, leading onto the interest the latter have sparked off among scholars. The city is said to have both 'staged' and 'been' the object of studies conducted by 'professionals from different fields'. The idea of a stage evokes the city as a spectacle. Its association to scholarly work, to the 'staging' of studies, so to speak, might be said to call forth the image of a laboratory, of

1 The webpage is an official endeavour by EMTURSA, the public company responsible for all matters pertaining to the tourist sector, in the city of Salvador. In 1986, the city government, perceiving an urgent need for an independent organ dedicated to the sector alone, founded the company as a mixed-capital venture. Despite its status, the government holds 99 percent of EMTURSA's shares and is effectively responsible for its public administration.

carefully controlled experimentation taking place under the watchful gaze of an interested audience. Whose gaze this might be does not here seem to be of interest. Neither is the question of what, in the city, has been deemed an object of study explicitly addressed. In keeping, the text remains vague about the identity of the many professionals that have 'conducted' the all important studies. Yet the list of traits that preface the observation—'history, a legacy of people from other continents, religious syncretism and its population'— provides a hint.

One the same website, to the left side, the prospective visitor will find a box with a series of links to the various aspects of the city that make her the most memorable of destinations: history, architecture, geography and weather, cuisine, museums, squares, libraries, handicrafts, curiosities and, significantly here, a link rather generally entitled 'culture'. A tourist map pointing towards the city's main attractions completes the box. The section on 'culture' picks up and expands upon the themes introduced in the opening page. It begins:

> Culinary, art, cultural and religious manifestations, architecture and memory. When put together, these elements constitute the legacy of a people: their culture. However, the concept of culture in Salvador goes beyond any definition, comprising an endless universe of riches. And the greatest rich of all is the city's population.

At first sight, there is nothing remarkable per se about the website. Like so many other websites dedicated to the promotion of tourism where tourism has become a prime source of revenue,[2] it is primarily designed to entice the consumer: landmarks are enumerated, maps delineate routes to the must-sees, glossy photographs illustrate the short texts. Within the context of a highly competitive global tourist industry, it is no surprise that Salvador's institutional actors should be so invested in producing an image of the city as a place endowed with an inimitable character. As Harvey (1990) put it, 'the active production of places with special qualities becomes an important stake in spatial competition between localities, cities, regions, and nations' (p. 295). Yet, since neither the abundance of architectural landmarks nor her tropical nature are unique to Salvador, one is still left with a question as to what exactly constitutes the city as special.

The answer, anchored on the idea of Salvador's cultural distinctiveness, is circular—'cultural and religious manifestations,' 'memory' and, as the final line has it, the city's population itself, her 'greatest rich of all.' The reading does,

---

2 According to the Brazilian Institute for Statistics and Geography (IBGE), the entity responsible for collecting, organising and publishing statistical data on the country and its population, in 2003, the economic revenue generated by the tourist sector amounted to 2.23 percent of Brazil's overall GDP. With regard to the state of Bahia, in 2004, the (federal) Institute for Research in Applied Economics (IPEA), estimated that the service sector, where the tourist industry leads the ranks, was responsible for the generation of 40.8 percent of the region's GDP (IBGE 2003; IBEA 2004; Guimarães 2009: p. 11).

indeed, feel like a succession of recursive steps, with each paragraph repeating the first, in increasingly baroque fashion. Nowhere is this more evident than under the rubric of 'culture,' where the text begins by listing the very categories that make up the box of links, among which one had found 'culture' to begin. It then skilfully casts the list aside, proceeding to praise the city's population, the real bearer of 'culture', in what feels like an echo of the general description provided by way of introduction, in the website's homepage. There is, however, one crucial difference between the two texts. For whereas the general description explicitly invokes the scholarly work carried out on the city and its population to strengthen her claim to distinctiveness, the section on 'culture' tells us that her greatest asset may lie in her ability to elude and exceed all manner of descriptive or analytical work—as the text has it, 'the concept of culture, in Salvador, goes beyond any definition'.

There is an inherent tension between the two positions. Nonetheless, such tension is not accidental. The qualities that make the city special cannot be thought independently of the effort consciously made by the various institutional actors responsible for her promotion within the context of the global tourist industry to substantiate her claim to a unique heritage, both material and immaterial. Such effort rests on mobilising a vast network of actors and resources deemed proof of the claim. Paraphrasing Corsín Jiménez (2005), it requires that the project of endowing the city with a distinctive identity, through a variety of networks and partnerships, become 'cultural work' in its own right (p. 14). This vast network is mobilised, on the one hand, by way of a series of real partnerships between different sectors, which involve extensive negotiation, agreement and investment, but the 'cultural work' requires more than these real partnerships, which is where the emphatic statement on scholarly studies of the city—especially on her population—comes in. As previously noted, the scholarly studies deemed proof of the city's claim to her ability to offer something unparalleled by similar locations show up, in the text, without further elaboration on their object or authorship. This is partly to do with the fact that, although the object of such studies may very well have been the city and its population, scholarly work is not usually carried out with the purpose of fostering the tourist appeal of any given place. Yet the significance this work takes, as far as Salvador's tourist sector is concerned, is undeniable. Wittingly or unwittingly, scholars here become important actors, whose work can be called upon as expertise. Their presence, their written work and the continuous interest the city seems to arise among them is mobilised as definite proof of the city's claim to her very own set of 'special qualities,' her inalienable property.[3] Simultaneously, it becomes important to reiterate Salvador's ability to elude every attempt to close off inquiry, renewing her appeal to younger generations of scholars. The renewed appeal constitutes proof of her ability to provide the tourist

3 See Ferry (2006) for a discussion of the analogy between history and memory, on the one hand, and alienable and inalienable property, on the other, as elaborated by the inhabitants of Guanajuato in their descriptions of the value of heritage.

with the unique experience she should undoubtedly be looking for. In other words, the tension is not accidental, but endlessly productive.

Now, the website tells us that a horde of professional scholars have worked on the city, but what is evident, given the tourist sector's strategic emphasis on her cultural distinctiveness, is that some academic work is more suited to the task than other. Unsurprisingly, 'cultural and religious manifestations' along that most intense source of interest, Salvador's population, have led many an anthropologist, Brazilian and foreign alike, to her shores. To everyone invested in promoting the city as a prime tourist destination, this comes as both good news and good publicity. The tourist sector has, indeed, been quick to proclaim Salvador's appeal to anthropologists. That the tourist sector counts a fair share of people trained in the social sciences in its ranks is of significance here. For the status accorded to anthropological work by tourist officials is a product of knowledge imparted as part of training. Anthropologists are invoked as people that are both interested in and expertly trained to appreciate cultural distinctiveness. And this view, fiercely promoted, spills over to the public domain, where it becomes common knowledge, an incontestable fact. Much as one might be tempted to dismiss tourist officials' appropriation of scholarly discourse on the city as somewhat misguided, the problem might be better approached as one where, like in Riles' challenging ethnography of Fijan bureaucrats and activists preparing for and participating in the United Nations Fourth World Conference on Women in 1995, tourist officials' 'analytical categories, devices and practices [...] [approximate] anthropological analysis, [so that] in the systemic nature of the artifact, sociological analysis is [...] part of the indigenous exegesis' (Riles 2001: p. 16). And herein lies the problem: that in this context, the anthropologist's analysis is always in danger of becoming 'a mere replication of indigenous representations' (Ibid.).

The problem, as I am describing it, of producing ethnographic work on an object that deploys itself through the anthropologist's own categories is of no small importance. However, anthropologists of Salvador have tended to ignore it, proceeding as if the indigenous appropriation of anthropological categories for purposes somewhat distinct to those of the professional researcher were a contingent matter. This is puzzling on all counts, not least the fact that to carry out fieldwork in Salvador is to be, at every juncture, reminded of one's position as an expert on culture and made presciently accountable for one's analysis. My own excursions across the ethnographic landscape of Salvador often ended with a heightened sense of puzzlement about the patent absence of reflection on this question within the many tomes written about the city. This article is, to a large extent, an attempt to critically examine that heightened sense of puzzlement. My response to the problem is, first, to recognise that the familiarity of the institutional discourses and practices in question is only partial. This requires that we look closely at the themes that animate actually existing anthropological work on Salvador, at the questions that anthropologists have confronted and at the problems that have informed ethnographic description. To say that these are only partially familiar practices is not to dismiss their purchase as objects of inquiry in their own right. For I would argue that the parallels between these knowledge practices, which are common to late modern institutional knowledge more broadly speaking, and social science are what renders the study of their relation both problematic and necessary to begin with. And here, anthropologist's reluctance to critically engage with 'local interest' in their writings poses as much of a trap as the danger of replication. My aim, then, is to show both that the kinds of knowledge in question are, in fact, different and to so by putting them into practice vis-à-vis one another (Riles 2001).

A few words of caution are in order before I proceed. Although I have thus far talked about 'anthropological work' writ large, my focus here is on work done exclusively by foreign anthropologists.[4] In addition, to put two different kinds of knowledge into practice vis-à-vis one another necessarily entails some degree of replication. This may seem partly like a failure to elucidate difference and, yet, this aesthetic failure is productive. For I have borrowed the idiom in question consciously and deploy it so as to make the paradox of describing that which is seemingly too explicit to warrant analysis or too subtle to analyse explicitly visible. Taking my cue from Strathern (1988), one might say that the necessity of the exercise stems from the way it is (already known) from the start. In other words, where my interlocutors, directly and indirectly, resort to the idiom of anthropology to describe themselves, whilst I may be able to render anthropology's knowledge practices explicit, my ethnographic description, if it is to have analytical purchase, must also do what it says—that is, it must demonstrate the relation between the knowledge practices in question. With caution then, let me now turn to the making of the city as that object that, as the website has it, has become irresistibly attractive to professional scholars.

The Making of an Object

One of the most noteworthy things about the depiction of the city presented by the two website excerpts discussed here (http://www.embratur.ba.gov.br) is that, for all their emphasis on the city's unique character, neither of them contains even a brief mention to Salvador's status as a UNESCO World Heritage Site. The rest of the website, including links to the city's history and architecture, is equally silent on the matter. For anyone acquainted with Salvador this is somewhat startling. That Salvador's historic centre should be a recognised UNESCO World Heritage Site is a matter of central importance for her tourist sector. Achieving World Heritage Status has, after all, had a marked impact on its development, both in terms of capital investment and official policy.

4 This is so for a series of reasons. On a practical level, there is the fact that Salvador has, indeed, been extensively researched by foreign anthropologists, myself included. On the other hand, an examination of local scholarly practice would both take up more space than I have to contend with and yield a set of distinct, albeit interrelated, questions.

By UNESCO's rules, a site must meet a minimum of two (out of ten) criteria to qualify for world heritage. In the case of Salvador, the advisory body (ICOMOS) recommended it on the grounds that:

> criterion iv) it is an eminent example of Renaissance urban structuring [...] The density of monuments makes it the colonial city par excellence [...] and criterion vi) it was one of the major points of convergence of European, African and American cultures of the C16 to C18 [...]

The list of UNESCO criteria rests upon the idea that, to become world heritage, a city must prove a claim to possessing items of 'outstanding universal value.' In Hill's words (2007), 'outstanding universal value is a fuzzy concept based on an eclectic assortment of values, including masterpieces of human genius, important interchanges of human values, unique testimony to cultural tradition, outstanding examples of buildings and landscapes, and tautologically, even ideas that 'possess outstanding universal value' (UNESCO 2005: 19)' (p. 65). When, in 1985, Salvador saw her bid for recognition approved, there could be no question as to the legitimacy of her nomination, though the eclecticism that attends to UNESCO's own definition is such that one is left with a question as to whether it would be possible, in principle, to rule anything out.

Salvador's history—her legacy, as her tourist officials have it—does seem to have endowed it with a peculiarly rich assemblage of heritage objects. Capital of colonial Brazil until 1763, the city grew around a royal administrative district with privileged links to the metropolis and around her port, which was, for sometime, a hub of imperial trade across four continents. With the first slave market in the New World, founded in 1558, Salvador received more slaves than any other place and is today said to be 'the world's blackest city' outside Africa, often suggestively called 'the black Rome of the Americas' (Agier 2000). As slaves poured into her port, the city became indelibly associated with her African population's cultural repertoire: *Candomblé*, an umbrella term for a series of religious and ritual practices of West African origin, as well as *Capoeira*, a martial art rooted in plantation society, have become ubiquitous elements of her landscape. Yet, her singularity is not exhausted by the traces of her African heritage. Salvador is much more than Brazil's most African city, we are told—she is the nation's cradle, the foundational city par excellence. Her inhabitants proudly proclaim that all things Brazilian started there, most significantly samba—and in fact, a form of Bahian samba known as *Samba de Roda do Reconcâvo*, became one of Brazil's first items on UNESCO's Intangible Cultural Heritage list.[5] Alternatively and with some degree of irony, we

5 According to UNESCO's 2003 Convention for the Safeguarding of Intangible Cultural Heritage, intangible heritage can be defined as 'the practices, representations, expressions, knowledge, skills—as well as the instruments, objects, artefacts and cultural spaces associated therewith—that communities, groups and, in some cases, individuals recognize as part of their cultural heritage. This intangible cultural heritage, transmitted

are also told that all things Brazilian endure only there, like her street *Carnaval* (Carnival), the largest of its kind anywhere and one of the yearly high points with regard to tourist revenue.

This assemblage of heritage objects is remarkable, in more than one way. In the words of one of my interlocutors, there is no denying that every one of those objects is 'truly a part of Salvador.' Nonetheless, what is noteworthy about it is its relation to the emergence of an increasingly more sophisticated discourse on the city's tourist vocation. Such discourse did not emerge overnight. From the 1930s onward, a slow consensus begun to crystallise around the idea that Salvador might have just about every attribute of a tourist bestseller (Sá 2006). By the mid-1950s, Salvador's Tourist Board (*Diretoria Municipal de Turismo*) had designed and approved Brazil's first official policy document on tourist development, the so-called Director's plan (Guerreiro 2005). At the same time, Salvador's institute for tourist research was put in place. Its main tasks consisted of collecting information about Salvador's history and culture, promoting contemporary research of interest to the tourist sector and, whenever called upon, advising on tourist policy (Queiróz 2002). With the 1964 military coup, the tourist sector suffered a reversal of fortune, as the military dismantled the political structures that had fostered its development. For the next twenty years, the relation between the political and economic agents that had been central to the early development of the sector and private investment remained incipient. However, by the 1980s, new winds swept across the nation and, as the democratic process begun in earnest, the relation between institutional political power and Salvador's tourist industry regained its momentum. With Salvador's nomination to UNESCO's World Heritage list, the early efforts to unearth the city's distinct cultural patrimony had finally paid off. By 1986, her new status saw every social partner press for a new strategy, one that required strong collaboration and, most importantly, a set of institutional agencies devoted to the relation between 'culture' and tourism. It was then that the new government Secretariat for Culture and Tourism (*Secretaria da Cultura e Turismo*), under which EMTURSA operates, came into being and it was also then that a renewed interest in scholarly work on the city and her people surged forth.

It would seem then that those invested upon promoting the city as a tourist destination were, from early days, keen on exploring scholarly work about her heritage and cultural life to legitimise their claims about her potential allure. In a sense, these were early days both for the tourist industry and for anthropology. Indeed, in the early 20th century, in Brazil as elsewhere, anthropology was only beginning to emerge as a professional discipline in its own right. Salvador played a specific role in this story, for it was there that what we may describe as proto-ethnographic studies of Brazilians African descent were first carried out. By the

from generation to generation, is constantly recreated by communities and groups in response to their environment, their interaction with nature and their history, and provides them with a sense of identity and continuity, thus promoting respect for cultural diversity and human creativity' (http://www.unesco.org/culture/ich/index.php?pg=00002).

1930s, Salvador had begun to attract more than its fair share of ethnographic attention beyond Brazilian shores, especially in North America. Ethnographic work on Salvador was, from its inception, marked by a fascination with her black population. This was partly to do with an interest in what Herskovits (1943) called 'New World Africanisms'[6] and partly to do with the emergence of a discourse claiming that racial mixing (*mestiçagem*), in Brazil, had made room for its peculiarly harmonious model of race relations.[7]

Ruth Landes' *City of Women* (1947) constitutes a prominent example of the genre that was to develop. Landes came to Salvador in the late 1930s intent on studying its black population. In her own words, she had settled on heading to Salvador upon hearing a rumour that 'race relations were exceptionally harmonious there' (p.1). However, as she made her acquaintance with the city, she quickly lost interest in the question of race per se, concluding rather naively that Brazilian people of African descent 'were oppressed by political and economic tyrannies, [...] not by racial ones' (Landes 1947: p. 248). What caught her attention instead would become a staple of studies of Salvador, namely *Candomblé*, which she described as a source of intense freedom (Ibid.). Her Salvador was a city populated by proud descendants of Africa, particularly her female population, richly seeped in a number of remarkable traditions. Like Landes, Pierre Verger, a Belgian anthropologist turned iconic adopted son of the city, would come to Salvador in 1946 and remain there until his death, in 1996. Verger had come across a description of Salvador in the work of a young Brazilian novelist and become interested in the ubiquitous allusions to African ritual traditions that cut across the text. As he arrived, he became enthralled by Salvador's great *Candomblé* ceremonial houses (*terreiros*) and by the city's historical links with the Gulf of Guinea, where he would later conduct fieldwork. In Salvador, Verger would become a public figure, the chronicler of her outstanding African heritage par excellence and a strong asset as far as the tourist sector is concerned. Verger's photographic essays on the city are today widely distributed, found as likely in academic bookshops as in Salvador's airport newsagents.

Recent ethnographers of Salvador have, in a sense, come a long way. Ideas about 'racial harmony' have become practically unfathomable. For, in the early 1970s, the proud black population of Landes' rather poetic ethnography had begun to mobilise—a plural black movement (*Movimento Negro*) was on the rise, determined to play a central role in Salvador's public life. The birth of Salvador's black movement, along transformations within the discipline itself, would change

6 The interest in 'New World Africanisms' became especially prominent in the US around the beginning of World War II. Describing her colleagues' initial reactions to her interest in Brazil, Landes (1947) put it thus: 'We wanted to know how that interracial situation differed from our own in the United States. It was a sociological project that excited the imaginations of few people. Not until a year later did the crash of war make the Negro people and their problem a part of the day's news' (1).

7 For a solid discussion of Latin American racial ideologies, see Wade (1999).

the nature of anthropological work on the city, which now begun to take an interest in the movement itself. However, the significance anthropological work has for the tourist industry would remain unabated. After all, Bahia's black movement too is fundamentally concerned with heritage and cultural practices. In many cases, local historians and anthropologists figure among its most vocal members. Furthermore, many of the better known groups within the movement are directly involved in the institutional efforts to promote Salvador's African heritage on a global scale. In other words, in recent years, anthropologists have clearly sought to avoid the more essentialising aspects of earlier scholarship on the city, turning their gaze to emergent political formations that seemed to contest that scholarship. And contest that scholarship these formations have, indeed, but they have also simultaneously re-ignited some of the old questions and thus, upon closer inspection, what becomes noteworthy is the extent to which recent ethnographic work has remained tied to the themes that animated earlier scholarship.

Candomblé, in its relationship with both the black movement and broader transnational activist networks (cf. Johnson 2002 and Matory 2005), is still widely studied. *Capoeira* (cf. Lewis 1992 and Lewis 1995) as well as all manner of popular festivals, including *Carnaval* (Carnival), that most important source of revenue, when the city is said to awake and display its true colour, have also become popular themes (cf. Agier 2000 and Ribard 1999). There is no doubt that recent ethnographies of Salvador belong to a very different body of scholarship. The concerns of anthropologists have changed considerably—attention to relations of power, to ethics and to ethnographic modes of representation have all been taken on board. And, in fairness, avoidance of themes that are unquestionably central to Salvador's public life—such as her historical entanglement with slavery, the rise of her vigourous black movements, indeed, the city's ongoing refashioning as 'The Black Rome of the Americas'—is not an option. What remains, nonetheless, surprising is that so much of this work should be marked by a reluctance to critically address tourism, the tourism industry's appropriation of anthropological work as 'symbolic capital' (Bourdieu 1984) and the impact that the dissemination of ethnography turned symbolic capital actually has on the city's public life.[8] In other words, if institutional power, in the form of a highly competitive tourist sector, has both done a good job of appropriating anthropological work and become centrally involved in disposing of the cultural artifacts that are the object of anthropological inquiry, would it not follow that to study Salvador, at present, requires that we take tourism to be part of our object? In the next section, I will turn to my own uncanny encounter with this question and attempt a solution by pursuing a different mode of inquiry.

---

8 In a recent article, John Collins (2008) has explored the relation between systems of care, UNESCO's world heritage and knowledge, in Salvador's historic centre. Though not concerned with anthropological knowledge per se, his work takes some steps toward addressing the question of knowledge and discussing the effects of policies that would turn inhabitants of Salvador's historic centre into heritage in need of tutelage.

Intangible Scales

According to the Oxford Dictionary and Thesaurus, 'exotic' can be defined as an adjective to denote something 'introduced from or originating in a foreign (esp. tropical) country,' as something 'attractively or remarkably strange or unusual' or, in its substantive form, as 'an exotic person or thing.' It is not difficult to see how 'exotic,' in its various formulaic definitions, can be said to apply to representations of the city by both tourist authorities and (many) anthropologists alike. There is, of course, the perennial appeal to a legacy of people from other continents and it would not be wholly unfair to say that many an ethnographic description of Salvador's cultural landscape portrays it as 'attractively unusual,' if not outright 'strange.' Yet my reference to the term 'exotic,' in the subtitle, also meant to evoke the archaic meaning of the word—not just 'originating in a foreign country,' but 'foreign to,' which the legacy alluded to above certainly is not.

The word 'exotic' is loaded with associations that most anthropologists would rather avoid. Until now, I too have purposefully avoided it but, at this juncture, it is worth asking whether the questions that have prompted this excursion may not be inextricably linked to the problem of exoticism. Consider, for example, the methodological problems that anthropologists encounter as a consequence of Salvador's present status as world heritage, a city overflowing with artifacts of 'exceptional universal value' and a major destination within a transnational network of heritage objects vying for all manner of resources within a global tourist market that is tied to the successes and failures of their claim to exceptional value. Despite anthropologists' efforts to study Salvador's assemblage of heritage objects with reference to studies of transnationalism, the literature they have produced has not just revolved around their predecessor's themes, but tended to yield familiar analytical paradigms. Notions of identity, 'community' or 'tradition' seem to creep into anthropological attempts to render the novelty of the field. Yet these notions are all but novel, much as the field of transnationalism is not new per se. I would argue that what is new here is 'rather the ethnographic encounter with knowledge practices already familiar to, and already in use by, the anthropologist at precisely the moment at which he or she seeks insight through fresh ethnographic observation' (Riles 2001: 5). This sense of the exotic, then, points to a somewhat different question, namely whether the significance of the anthropologist, from the viewpoint of the city's inhabitants as well as her tourist industry, may not lie in her exotic ability to conjure and legitimise Salvador's entitlement to a powerfully profitable form of exoticism. The problem of exoticism remains central to anthropological practice. And it does so not just because anthropologists of Salvador continue to exoticise through notions of identity, community or tradition—understood as 'methodological devices, in an effort to render the familiar strange so that it might be aprehended as ethnography' (Riles 2001: 5)—but also because that impulse to exoticise is essential to the actors that present the ethnographer with familiarity, in that it allows them to claim back their strangeness.

Let me now pause briefly to describe one of my own vivid encounters with seemingly familiar knowledge practices. Shortly after I had first got to Salvador, I was invited by a high-ranking official in the government's department for culture and tourism to attend a symposium on *Baianidade* in Salvador's newly built convention complex.[9] I was still unfamiliar with much of the city, so I headed to the complex earlier. Among the first to arrive, as I waited, I decided to sit on a wall by the main entrance, writing up some notes. Almost immediately, a short, slim man of mixed ancestry approached me: 'are you an anthropologist?' Startled by the interpellation, I blurted out a not-too-confident 'yes, I suppose so.' 'Ah', he replied, 'you might be interested in my organisation then.' He proceeded to explain that he was one of the founders of a small NGO, based in Salvador's Itapagipe Peninsula, working on the valorisation of Itapagipe's distinct cultural heritage. He thought Itapagipeshould interest me, he said, for the origins of its black population, which differed from the hegemonic association of Salvador with specific West African groups and was, instead, to be located in what is now Angola. I did not interrupt him, neither did he try to know if my work fitted the bill. As the symposium was about to begin in earnest, he seemed to feel a sense of urgency, telling me: 'I will leave you my card…perhaps you could write something about us or put in a good word back home.'

I reflected upon this encounter for a long time. Something in it bothered me. It was not that he thought about the possibilities afforded by ethnographic publicity. Rather, I was apprehensive about both his interpellation and his commentary on what *should* interest me. Most of all, it bothered me that his pitch, which was not off the mark, presumed a certain type of knowledge about something that then became simultaneously familiar and uncomfortably strange, that is, a certain familiarity with anthropological knowledge practices. What might have otherwise been a casual encounter thus became 'uncanny.'

The term 'uncanny' is often, though not exclusively, associated with the work of Freud (1919)—the 'unheimlich'. I use it here to index something that is, at once, familiar and uncomfortably strange, which is consonant with Freud's discussion of the 'unheimlich.' With some caution, I want to draw further on two specific interrelated aspects of the Feudian 'unheimlich' to reflect upon the knowledge practices in question here. First, Freud talks about the 'uncanny's' capacity to cause a certain cognitive dissonance in terms of it sharing a meaning with its antonym ('heimlich'), which is the sense of something that must be concealed and kept hidden (Freud 2003: 132). What is 'heimlich' (homely) becomes increasingly

9 I cannot, at present, address the term *Baianidade* in depth. The word *Baianidade* comes from Bahia, the name of the state of which Salvador is the capital. The word Bahia is metonymically used to describe Salvador and its hinterland, in Salvador itself. Inhabitants of Salvador call themselves *Baianos*. Bahian social scientist Osmundo Araújo Pinho has described *Baianidade* as an 'ideological discourse' about the city's identity 'that is sustained by a symbolic arsenal and pragmatically deployed by the city's dominant political and cultural agents' (Pinho 1998: 4).

more ambivalent, to the point where it is possible to say 'the uncanny is, in some ways, a species of the familiar' (Freud 2003: 134). Second, Freud approaches the 'uncanny' in association with the concept of 'the double' (Freud 2003: 141-43). With these two aspects of Freud's discussion of the uncanny in mind, two inseparable questions come into full view. On the one hand, if one thinks of the 'uncanny' in relation to a play of concealment and revelation, one must ask what does the constant appeal to anthropological knowledge by Salvador's tourist industry reveal and conceal that seems to prompt an outright unwillingness to examine it on the part of anthropologists. On the other, one must also address the question of what that unwillingness may point to insofar as anthropologists' knowledge practices in themselves go.

At this point, it is worthwhile calling upon a somewhat different etymology for the word uncanny, that is, the uncanny as antonym to 'canny' or 'knowing [how to].' The latter sense brings us to more comfortably familiar terrain and, one might add, to a more comfortably familiar kind of problem. For whereas it is clear that the tourist industry recognises the benefits it can accrue from invoking the expertise of anthropologists in their attributions of value to Salvador's heritage, anthropologists have failed to 'measure' the distance between what is invoked and the effects that invoking it produce—or, in other words, the scales that inform the two forms of knowledge (Strathern 1999). What knowledge is, in fact, called upon by the tourist industry, and to what effect? A closer look provides us with an interesting puzzle: Bahian tourist authorities invoke anthropological knowledge, but anthropological knowledge of the kind that describes neatly what their agents deem to have always been there, that is, to what makes Salvador *Salvador*. And, in fact, with noticeable exceptions (for example Verger, whose work is widely known by Bahians of all kinds), it does so with little concern for specific scholars, texts or arguments. It neither enumerates nor qualifies. Rather, it lists the general themes of interest to a general 'community' of anthropologists. This brings forth another puzzle: if becoming a World Heritage Site or attracting tourists—something Salvador has clearly shown to be capable of doing on its own merits—is no longer a concern, what might explain the constant reiteration of anthropological interest? This question invokes yet another sense of scale, that is 'scales as orders of knowledge, where particular orders of relations (of economy, or religion, kinship, etc.) are measured up against other scales' (Corsín-Jiménez 2005: 3). One might ask, for example, what image of 'culture' is conjured when the president of Salvador's council says 'culture is also economy' (Cit. in Muricy 2001: 182)? Taking my cue from Strathern (1999) and, to a lesser extent, Battaglia (1995), I would argue that the appeal to anthropological interest works, at once, as a public test of authenticity, and as a display of capacity, the capacity to effectively attract scholarly attention. Attracting scholarly attention thus becomes a means and an end in itself.

With this in mind, we come to the second question. At the risk of repeating myself and drawing on the problems posed by the concept of a 'double', I would suggest that the unexamined sense of an 'echo' or a 'double' must not become so 'uncanny' as to blind us to the distance between the knowledge practices at play—yet another kind of scale. That some of the people anthropologists study should understand themselves through terms and categories that seem to replicate those of the social scientist can be seen as an instantiation of what Lash (1994) aptly calls 'modernisation's doubles'—a common enough phenomenon after all. My point here has been to show that this 'echoing' or 'doubling' may not be quite what it seems. Anthropological literature, in Salvador, has been appropriated as expert knowledge on culture. And herein lies the problem: that anthropologists are not, strictly speaking, expert *knowers* of culture. That other people treat ethnographic practice otherwise, if and when our work fits the purpose, is a matter for serious analytical inquiry, not a cause for retreat into the more familiar terrain of a methodological device that cannot yield serious critique. And one of the unwarranted and most problematic consequences of anthropologists' resistance to address the deployment of their work as expert knowledge, in the case of Salvador, has been to keep them from engaging with the multiple, complex and uneven effects tourism has had on the city. In the meantime, Salvador and its people can only become more exotic, as ethnographic practice itself unwittingly echoes the verdicts of a tourist industry that can often seem too canny for comfort.

References

Agier, M. 2000 *Anthropologie du Carnaval: la Ville, la Fête et lÁfrique à Bahia.* Marseille and Paris: Parenthéses, IRD.

Battaglia, D. 1995 On Practical Nostalgia: Self-Prospecting among Urban Trobrianders. In *Rhetorics of Self-Making*, (ed.) D. Battaglia. Berkeley: University of California Press.

Bourdieu, P. 1984 *Distinction.* Cambridge, MA: Harvard University Press.

Collins, J. 2008 'But What if I Should Need to Defecate in Your Neighbourhood, Madame?': Empire, Redemption and the 'Tradition of the Oppressed' in a Brazilian World Heritage Site. *Cultural Anthropology* 23(2): 279-328.

Corsín-Jiménez, A. 2005 Changing Scales and the Scales of Change: Ethnography and Political Economy in Antofagasta, Chile. *Critique of Anthropology* 25(2).

Ferry, E.E. 2006 Memory as Wealth, History as Commerce: A Changing Economic Landscape in Mexico. *Ethos* 34(2): 297-324.

Freud, S. 2003 [1899] *The Uncanny.* London: Penguin.

Guerreiro, G. 2005 A Cidade Imaginada: Salvador Sob o Olhar do Turismo. *Revista Gestão e Planejamento* 6(11): 6-22.

Guimarães, Luís E.B. 2009 *A Baianidade Como Elemento Diferencial na Atração Turística da Bahia: uma Análise da Estratégia da Bahiatursa.* MA Dissertation, Salvador, UFBA.

Harvey, D. 1990 *The Condition of Postmodernity.* Cambridge MA \ Oxford UK: Blackwell Publishers.

Herskovits, M. 1937 African Gods and Catholic Saints in New World Negro Belief. *American Anthropologist* 39(1): 635-43.
Herskovits, M.J. 1942 The Negro in Bahia: A Problem in Method. *American Sociological Review* VII: 294-404.
Hill, M.J. 2007 Reimagining Old Havana; World Heritage and the Production of Scale in Late Socialist Cuba. In *Deciphering the Global: Its Scales, Spaces, and Subjects*, (ed.) Saskia Sassen. London and New York: Routledge.
Johnson, P. 2002 *Secrets, Gossips and Gods: The Transformation of Brazilian Candomblé*. Oxford: Oxford University Press.
Landes, R. 1947 *The City of Women*. New York: Macmillan.
Lash, S. 1994 Reflexivity and its Doubles: Structure, Aesthetics, Community. In *Reflexive Modernity: Politics, Tradition, and Aesthetics in the Modern Social Order*, (ed.) U. Beck, A. Giddens and S. Lash. Stanford: Stanford University Press.
Lowell Lewis, J. 1992 *Ring of Liberation: Deceptive Discourse in Brazilian Capoeira*. Chicago: University of Chicago Press.
—— 1995 Sex and Violence in Brazil: Carnaval, Capoeira and the Problem of Everyday Life, *American Ethnologist* 26(3): 539-77.
Matory, J.L. 2005 *Black Atlantic Religion: Tradition, Transnationalism and Matriarchy in the Afro-Brazilian Candomblé*. New Jersey: Princeton University Press.
Muricy, I.T. 2001 O Éden Terrestre: O Consumo da Cidade como Mito. *Bahia Análise and Dados* 11(2): 180-93.
Pierson, D. 1942 *Negros in Brazil: A Study of Race Contact in Bahia*. Chicago: The University of Chicago Press.
Pinho, Osmundo de Araújo. 1998 A Bahia no Fundamental: Notas para uma Interpretação do Discurso Ideológico da Baianidade. *Revista Brasileira de Ciências Sociais* 13(36): 1-16.
Queiróz, Lúcia Aquino de. 2002 *Turismo na Bahia: Estratégias para o Desenvolvimento*. Salvador BA: Secretaria da Cultura e Turismo.
Ribard, F. 1999 *Le Carnaval Noir de Bahia: Ethnicité, Identité et Fête Afro à Salvador*. Paris: Harmattan.
Riles, A. 2000 *The Network Inside Out*. Ann Arbor: University of Michigan Press.
Sá, Natália Coimbra de. 2006 A Baianidade Como Produto Turístico: Uma Análise da Ação dos Orgãos Oficiais de Turismo na Bahia. Presented to the 24th Brazilian Congress of Communication Studies. Available Online: http://www.bahia.com.br/site/institucional/teses.jsp?filterteses.page=1.
Strathern, M. 1988 *The Gender of the Gift: Problems with Women and Problems with Society*. Berkeley, CA: University of California Press.
—— 1999 *Property, Substance and Effect: Anthropological Essays on Persons and Things*. London and New Brunswick, NJ: Athlone Press.

Futher information available from:

http://whc.unesco.org/en/list/309
http://www.bahia.com.br/site/institucional/teses.jsp?filterteses.page=1
http://www.braziltour.com/site/gb/home/index.php
http://www.emtursa.ba.gov.br
http://www.ibge.gov.br/english/
http://www.ipea.gov.br/default.jsp

Interface 2

Re-discovering the Variety of Discovering Pasts

Aspects of Alain Schnapp and Stephanie Koerner's Conversation, May 2009

Stephanie Koerner (SK): Thank you very much for organising our getting together today, and especially for our writing to one another over the last months. Otherwise it would have been impossible to gather materials that might be useful to refer to in writing up aspects of our conversation. Do you think it would be useful to begin by trying to recall something of where and how we met?

Alain Schnapp (AS): Yes very. And do let us note some of topics we touch upon, as well as how these might be further pursued in the near future.

SK: I think that the first time that we spoke about some shared interests was at the 2004 European Association of Archaeologists (EAA) meeting in Lyon. You were developing connections between AREA (Archives for Research on European Archaeology) and discussions of defining and conserving 'intangible' as well as 'tangible heritage in relation to the UNESCO World Heritage Congress (for instance, Schnapp 1982). Laurent Olivier and I had organised a session on implications of conceptions of time for approaches to such theoretical issues as those of human agency, the diversity of humanity's history and conditions of archaeological knowledge. The panel marked a turning point in Olivier and my work together by providing a context for exploring the bearing that deeper historical backgrounds of these themes might have upon concerns with changes taking place in the dynamics of so-called 'academic' and 'public archaeology', and in processes of 'globalisation' and 'risk society'.

AS: The work of Walter Benjamin (1940) played important roles?

SK: Yes definitely. But much of the session concerned the ways in conceptions of human agency and history grounded in systems of supposedly synonymous dichotomies (nature-culture, materiality-ideas, reality-historical contingency) impede appreciating the relevance of the diversity of the past (including, with regards to past societies' experiences and representations of the past) for fresh approaches to 'lived heritage'. For example, how might research on the diversity of 'the past in the past' (Schnapp 1983; Olivier 2004) relate to concerns expressed in *History and Ethnicity* (Tonkin, MacDonald, and Chapman (eds) 1989: 1) about the ways in which such dichotomies are perpetuated by still very influentially opposed approaches to: 'How did the past lead to the present? And how does the present create the past?'

With regards to Benjamin, we have long been interested in such themes in Benjamin's work as:

- the extent to which 'state of emergency' marked contemporary social life, not in an exceptional sense, but as one of its 'ruling principles',
- the ways in which some of the most influential 'meta-narratives' of modern times render in visible the 'barbarity' of what powerful political ideologies refer to as 'civilising' processes,
- the relevance of issues by (a) and (b) for efforts to 'go against the grain' of forces that threaten conditions of possibility for variability of human life-worlds.

For Olivier, (2004: 210) elements of Benjamin's work can bring light not only to the diversity of the past in the past, but also the embeddedness of the present in the past—'the present is not what is uniquely happening at this very moment, but on the contrary what has always been happening.'

AS: Do you still have a copy of the panel's programme?

SK: Yes, here it is. A number of participants are contributors to the *Unquiet Pasts: Risk Society, Lived Cultural Heritage, Re-designing Reflexivity* volume—which reminds me that one topic of our 2004 conversations was that of implications of deeper historical backgrounds of archaeological methods and theory for controversies centring on questions of for and by whom heritage is managed (cf. Skeates 2000; Carman 2002; Cleere 1984, 2006; Darvill 2007).

AS: Then—as more recently—at the 2008 meeting of the Theoretical Archaeology Group (TAG) in Southampton we compared materials Amos Funkenstein used to investigate 'the revolution in historical reasoning' for his book, *Theology and the Scientific Imagination From the Middle Ages to the Seventeenth Century* (1986) and those explored in *The Discovery of the Past. The Origins of Archaeology* (Schnapp 1993). How did you come across these similarities? Would it be useful to include reference to Funkenstein's work in our discussion's write up?

SK: Those two questions are related. With regards to your first question, I think it was when I was trying to write a section on the history of Scandinavian archaeology alone on the basis of approaches such as those of David Clarke (1968, 1973) and Bruce Trigger (1984, 1989). Their approaches have numerous advantages for research on the importance to the 19th and 20th century histories of archaeology as a specialised academic profession of (1) techniques developed by C.J. Thomsen (1837) for classifying and chronologically ordering artifact assemblages, and (2) the implications of adopting geological historians' methods for adding previous unimagined time depth to human history. But especially the history of Scandinavian archaeology—as your work explores in detail—reflects much deeper and heterogeneous historical roots, including roots explored by Funkenstein. More recently, I found further similarities by considering implications of materials discussed in Funkenstein's writings on the contextual circumstance of the 'revolution in historical reasoning' and in your *Discovery of the Past* (1993) for appreciating the extent to which controversies over 'heritage' pose fundamental 'life quality issues' (Darvill 2007). Examples include your work's concerns with questions of:

- Why did men of the classical age find it so difficult to escape from the biblical chronology, and why did the weight of Scripture continue to confine them to a short chronology—so short that they were forced into the most complicated contortions in order to conceptualise the ancient history of man and therefore the earth? (Schnapp 1993: 221),
- the bearing upon (a) of inquiries into the deeper histories of dualist conceptions of nature and culture, and (b) societies with and without written histories,
- contributions 'landscape anatomists' made to what Funkenstein calls the 'revolution in historical reasoning' by investigating not only the fragmentary nature of texts, but also the embeddedness of artifacts in the extraordinary depth and diversity of the histories of physical, organic and human cultural realms.

AS: Emmanuel Le Roy Ladurie may have been focusing on similar elements of the work when he stressed, in the French edition's Preface, its concerns to underscore that:

> Ever since antiquity observers, thinkers and philosophers in china and in Greece, just as in the East, have had an intuition of the very long history of the world and of humanity. For over a millennium in Europe (from Saint Augustine until Darwin) specialists, learned societies and the ruling powers in particular, refused to allow that human history ran to hundreds of thousands of years, and that it was the heterogeneous prolongation of a still older adventure: the history of nature (Le Roy Ladurie, in Schnapp 1993: 9).

The passage relates in some interesting ways to those you suggest as prefixes for our discussion's summary.

SK: It does indeed. But that is one of the topics we need to pursue in the project with Tim Ingold and Gísli Pálssen around the idea of 'anthropology and archaeology after simplicity'. And themes of Funkenstein's work that compare with the elements of your approaches mentioned earlier will be incorporated into that project. For Funkenstein—as for Hans Blumenberg in *The Legitimation of the Modern Age* (1983)—a useful point of departure for exploring why authoritative mediaeval cosmology could not dispend with representations of human history based on Scriptural texts is to consider the importance of that latter to claims (rarely explicitly stated) that the world was created with human (salvation) purposes in Mind. Perhaps especially relevant for our present topics, such orientations also were involved in played central roles in arguments that people, places and things of historical significance could be expected to be documented by texts.

AS: A recurrent argument against expressions of concerns that Scriptures' accounts of human history clash with evidence of 'antiquities and buildings' that may be 'millions of years old' was that the latter lack 'a book concerning which a multitude of people held the same opinion' but are associated with 'idols, talismans, and witchcraft' (Judah Halavi, Kazari, 12th century Spain, cited in Popkin 1987: 35, and Schnapp 1983: 224).

SK: For Funkenstein, such arguments were often motivated by the notion that history was '*simplex narratio gestarum*' the simple story of things that happened as they really happened, and recorded by the 'eyewitness.' Interpretation, in this view, was the task of theology—of the *spiritualis intelligence.* Isidore of Seville's *Etymologies or Origins*—the most widely used encyclopaedic reference between the 8th and 12th centuries—contributed importantly to standardisation of such views. For Isidore:

> History is the narration of events by which we learn about what happened in the past. The word is derived from the Greek terms meaning to see or recognize: among the ancients, namely, no one wrote history unless he was present and saw the events to be written down (translated by Funkenstein 1986: 206).

In such views, since every generation can be trusted to commit to writing the events which are 'worthy of memory', history was to be seen as a continuous and unbroken chain: *erat enim continua historia mundi* (Funkenstein 1986: 206-207). Such views relate to some of the very diverse roots of archaeology explored in *Discovery of Antiquity* (1993), which variously relate to factors responsible for the 'revolution in historical reasoning' and its implications for arguments that:

> historical facts are meaningful only in their context; that this context has to be reconstructed painstakingly; often by alienating words or institutions from their present connotation and function, lest we fall into anachronism; that the eyewitness is not at all the best historian, because even if subjectively sincere, he is captive of his vantage point; that, indeed every period inevitably reinterprets history from its own vantage point and a unique cannon of questions born out of its own experiences (Funkenstein 1986:208).

AS: Important contributions to such developments were made by late mediaeval and Renaissance humanists, who objected to hitherto authoritative cosmology's disregard of the historical significance of contemporary times—the *modo*—'now'. But much of these scholars' efforts centred on comparing texts or evaluated texts in light of material culture documentation. By contrast, regions not well and/or indirectly documented by texts, saw the development not only of interest in the direct historical importance of ancient objects, structures and landscapes, but also of methods for what might be called the earliest systematic problem oriented excavations: 'Taking a look around at the world around him, the antiquary discovered in the present the material remains of the past and in so doing this he freed himself, partially but decisively from tradition' (Schnapp 1983: 163).

SK: You stress the importance of activities and orientations of 'anatomists of landscape' to early manifestations of the idea of stratigraphy?

AS: The expression 'anatomists of landscape' helps describe the nature of the contributions to this idea made by antiquarians working in areas of northern Renaissance that lacked text based continuity with the distant past. John Aubrey's (1626-1697) 'comparative archaeology' centred on 'combining the observation of the past and present, ethnology with written records analysis of landscape and the anatomy of monuments…His ambition was not to enrich an enthusiast's collection or to construct a universe after the fashion of the *Museum Wormianum*, but to restore antiquity in a palatable form by marrying the rigor of a naturalist with the passion of the historian' (Schnapp 1983: 192). Two chapters of the *Discovery* (1983) book concern the complex antiquity of the ideas about, methodological responses to, and drawings of the embeddedness of artifacts and monuments in geological stratigraphy. Already by the 17th century, in many areas of northern Europe,

> archaeology had progressed as much in the methods of field survey as in the appraisal of sources and the application of botanical and geological knowledge. But the principle progress came from excavations…This progress did not rely solely on recourse to excavation to support [historical] reasoning, but was linked to the attention given to detail, to the composition of layers, the analysis of the of

> the context of traces in the soil—in short, the underlying idea was that the earth was composed of remains of different kinds which allowed the reconstruction of its history… if one may say it, the idea of stratigraphy (Schnapp 1983: 200).

SK: A key theme in Funkenstein's work is the range of contributions to the 'revolution in historical reasoning' (as well as to mathematical and mechanical natural philosophy and experimental science) made by new interpretations of the very ancient expression, 'The Scriptures speak a human language'—interpretations that stressed their having been written in a highly context specific language and style. He illustrates this with, amongst others, examples from the 12th century philosopher, Ibn Ezra's, contextualist exegesis of Genesis I.

Time and time again, Ibn Ezra emphasises that Genesis is not a scientific account of the creation of the world *ex nihilo*, but rather an account of sublunar realms and natural processes…[For Ibn Ezra] even celestial bodies appear in the narrative of creation only from an historically contingent vantage point, not with reference to timeless essences…The narrative of creation is, according to Ibn Ezra, the narrative of the creation of objects immediately perceived in proportion to the ways in which they are perceived. If not to give an adequate cosmology, what is its purpose?

For Ibn Ezra, as for Spinoza at the beginning of modern biblical criticism, one of Scripture's purposes was ultimately to indicate that it does not 'prove' that the whole universe was created with human purposes in Mind. Another purpose may relate to traditions, which you describe as having 'held that the history of man went back several thousands of years…and that there were worlds much more ancient than ours' (Schnapp 224-25). This purpose was to indicate that (contrary to traditions that equate historical significance with measures of written documentation) Scripture addressed questions about the antiquity and circumstances of human origins metaphorically.

AS: Maimonides echoed these traditions in the 12th century. Perhaps informed by documents, which attributed to the Sabean inhabitants of ancient Arabia the belief that the history of man went back several thousand years, he wrote:

> The Sabeans allowed the eternal nature of the world, because according to them the sky was God. They held that Adam was a person born of a man and a woman, like other human beings, but they glorified him saying that he was a prophet and an apostle of the moon, that he encouraged the cult of the moon and that he wrote books on agriculture (Maimonides, quoted in Schnapp 1983: 225).

SK: It probably bears noting, Funkenstein stressed that, for Ibn Ezra and Maimonides, not only scholarly matters were at issue. One of the most controversial questions—long associations with arguments that 'Scriptures speak a human language'—of their times was that of whether 'laws and customs' are determined by some form of necessity, or historically contingent. Could we conclude with a passage relating to implications for archaeology under such circumstances of research on the field's deeper historical backgrounds?

AS: Gladly. How about:

> Contemporary archaeology has never ceased to debate the contradiction between human and natural sciences. In so doing it has detached itself from antiquarianism, but it has a long way to go in order to become a social science complete in itself. Modern prehistory, as part of a necessary critical movement tends to deny the physiological and ecological determinisms in vogue since the beginning of the twentieth century, and discovers how close these often were to ideas known since antiquity (Schnapp 1993: 324).

References

Benjamin, W. 1994 [1940] Theses on the philosophy of history, in H. Arendt (ed.), *Illuminations: works of Benjamin*, 245-55. London: Fontana Press.

Carman, J. 2002 *Archaeology and Heritage: An Introduction*, Leicester University Press: London.

Clarke, D. 1968 *Analytic Archaeology*. London: Methuen.

—— 1973 Archaeology: the loss of innocence. *Antiquity* 47: 6-18.

Cleere, H. (ed.) 1984 *Approaches to the Archaeological Heritage*. Cambridge: Cambridge University Press.

—— 2006 The World Heritage Convention. Management for and by whom?, in R. Layton, S. Shennan and P. Stone (eds) *A Future for Archaeology*, 65-74. London: Routledge.

Funkenstein, A.1986 *Theology and the Scientific Imagination: from the Middle Ages to the Seventeenth Century*. Princeton: Princeton University Press.

Le Roy Ladurie, E. 1996 Preface to the French Edition of *The Discovery of the Past*, translated by I. Kinnes and G. Varndell, 8-10. London: British Museum Press.

Olivier, L. 2004 The Past of the Present. Archaeological memory and time. *Archaeological Dialogues* 10(2): 204-213.

Nicholis, G. and Prigogine, I. 1989 *Exploring Complexity: an introduction*. New York: W.H. Freeman and Company.

Prigogine, I. 1997 *The End of Certainty. Time, chaos and the new laws of nature*. London: Free Press.

Schnapp, A. 1982 France, in H. Cleere (ed.) *Approaches to the Archaeological Heritage*, 48-53. Cambridge: Cambridge University Press.

—— 1993 *The Discovery of the Past*, translated by I. Kinnes and G. Varndell. London: British Museum Press.

Skeates, R. 2000 *Debating the Archaeological Heritage*. London: Duckworth.

Tonkin, E., MacDonald, M. and Chapman, M. 1989. *History and Ethnicity*. London: Routledge.

Trigger, B. 1984 Alternative Archaeologies: nationalist, colonialist, imperialist. *Man* (NS), 19, 355-70.

—— 1989 *A History of Archaeological Thought*. Cambridge: Cambridge University Press.

PART 3
Re-designing Reflexivity, or Can there be a Cautious Prometheus?

Introduction to Part 3

Interdisciplinarity and Ethics

Stephanie Koerner

The last decade has seen much interest in themes relating to deliberative democracy in political philosophy (Benhabib (ed.) 1996; Brandom 1996; Bohman 2003; Habermas 2003; Rancier 1999; Newman 2008) and many areas of anthropological and archaeological specialisation emerging on interstices of so-called 'pure' and 'applied research'. Deliberative democracy defends an ideal of equality as political efficacy (Bohman 2003: 85). This ideal hinges precisely upon equal valuation of highly discrepant perspectives on what counts as crucial matters of public concern, on what human beings can aspire to, and in what sort of world. However, in views of many powerful global elites, 'difference' (ethnic, linguistic, religious, cultural) is antithetical (even the primary obstacle) to a defect—an obstacle to universalisation of democracy through instrumental use of science as means to put the future at the service of the present (Bernstein 1996). For some, solutions lie in new cosmopolitan treatments of disagreements in terms of notions of 'incommensurate conceptual schemes,' 'world views' and even ideas about 'alternative realities.' One problem with this idea is that it hinges upon equations of consensus with mutual intelligibility. These impede appreciating the value of democratic means to address social conflict and existential problems, which do not hinge upon unreasonable expectations of consensus (Bohman 2003). Wars are not fought over so-called 'alternative realities' but over different experiences of what matters in the world that we occupy together. One of the most difficult challenges may be that of encouraging appreciation of the democratising value of 'neither fearing the absences of consensus…nor harboring the fantasy that conflictive situations may ever achieve a final equilibrium' (Lazzari 2008: 647; Laclau 1990; cf. Deleuze1990; Meskell 2002).

At the outset we noted that, until quite recently, few anthropologists and archaeologists are likely to have been receptive to the idea that fundamental change in theoretical orientations would come from areas of specialisation emerging on interstices of so-called 'pure' and 'applied research'. Likewise few are likely to have expected these fields to initiate arguments for the importance to reflexive approaches—indeed for re-designing reflexivity—of revisiting 'elementary timbers of the modern cosmopolis' (Toulmin 1990).

The contributions collected into this section variously illustrate fresh orientations towards reflexive research, teaching and diverse forms of applied anthropological and archaeological practice. The broad topic of this final section introduction is the interdisciplinarity of developments facilitating alternatives to programmes based on presuppositions about the world's reducibility to simplicity, and such dichotomies as those of nature-culture, pure-applied research, global-local, risk-ethics, and the real versus the historically contingent. Emphasis falls upon developments relating to themes running through Toulmin's *Cosmopolis* (1990):

- the relevance for appreciating the long history of alternative to dualist pedagogical and political ideals of shifting foci away from 'crises over representation' towards highly contradictory settlements,
- the centrality amongst timbers of modern dualist forms of reasoning and practice of deterministic conceptions of nature,
- ways ahead against the grain of such supposed settlements.

Themes stressed by Philippe Descola and Gísli Pálssen (*Nature and Society. Anthropological Perspectives* (1996)) will help to structure our considerations of three broad developments.

Not Just Another Analytic Category

Deconstructing the dualist paradigm may appear as just one more example of the healthy self-criticism which now permeates anthropological theory. After all, burning conceptual fetishes has long been a favourite pastime of anthropologists and very few domains have escaped this iconoclastic trend. If such analytic categories as economics, totemism, kinship, politics, individualism, or even society, have been characterised as ethnocentric constructs, why should it be any different with the disjuncture between nature and society? The answer is that this dichotomy is not just another analytic category belonging to the intellectual tool kit of the social sciences. It is the key foundation of the modernist epistemology (Descola and Pálssen 1996:12).

Whether interpreted as a triumph or tragedy, 'science and modernity' have long figured paradoxically amongst both the most and the least historicised of all topics in the humanities and social sciences. The expression 'the most and the least historicised' comes from an article entitled 'The History of Science as Self Portraiture' that Lorrain Daston wrote on the occasion of the award of the 2005 Erasmus Prize for work in the history and philosophy of science to Simon Shapin and Steven Schaffer, especially for their jointly authored study, *Leviathan and the Vacuum Pump Hobbes, Boyle and the Experimental Life* (1985). For Daston, the work marked an epoch of change in the kinds of questions historians posed and the ways in which they went about answering them:

Perhaps the simplest ways of describing the change is that what had previously been regarded as self-evident now demanded historical explanation: 'What is an experiment? We want our answers to be historical in character [Shapin and Schaffer 1985: 3]. Moreover, the historical explanation in question linked fundamental innovations in science such as the emergence of the experiment as a method of inquiry to coeval political and social events' (Daston 2006: 523).

Despite the remarkable variety amongst approaches, we can discern several shifts in orientations that have made especially important contributions to appreciating the remarkable 'historiality' of such dichotomies as those of nature-culture, the global versus the local, pure—applied research, the real versus the historically contingent, and so on. It is useful to treat these as three broad shifts in foci.

Shifts in foci from notions of science, rationality, objectivity, etc. as 'givens' towards the variety of forms such entities have taken

Few developments have contributed more significantly to radical change in orientations towards highly problematical dualist categories than inquiries that have moved the centre of research foci from the levels of abstraction at which much intellectual history operated to concrete circumstances, practices and entities (for instance, Galison 1987, 1996, 2002, 2007; Galison and Hevly 1992). Until this development got under way, Daston says, the categories that divided the most influentially opposed objectivist and subjectivist, or essentialist and relativist paradigms were taken largely as 'givens'. By contrast, approaches that she refers to as 'applied meta-physics' explore:

> how whole domains of phenomena—dreams, atoms, monsters, culture, morality, centers of gravity, value, cytoplasmic particles, the self, tuberculosis—come into being and pass away as objects of scientific inquiry. The echo to the title of Aristotle's treatise On Generation and Corruption is deliberate: this is a meta-physics of change, of the 'perpetuity of coming-to-be.' If pure meta-physics treats the ethereal world of what is always and everywhere from a God's-eye viewpoint, then applied meta-physics studies the dynamic world of what emerges and disappears from the horizon of working scientists (Daston 2000: 1).

Shifts in foci from intellectual history (the history of ideas) towards practices, instruments, cycles of credibility

'Applied meta-physics' has contributed to major change in methodological orientations. Instead of having to chose between approaches to intellectual history (the history of ideas) and approaches to social contexts (or between macro- and micro-analytic scales), applied metaphysics concerns the contextual circumstances (practices, instruments, cycles of credibility, social relations, and so on) under which 'epistemic things or entities' ('experimental objects and systems') emerge, are transformed and/or endure in states far from equilibrium (Rheinberger 1997). For Hans-Joerg Rheinberger (1997), one of the most interesting things about such entities (contrary to notions that 'the experiment' is universally concerned with eliminating contingency) is precisely their lack of determinate completeness. They are objects of inquiry, as well as open-ended sites where further novel objects, instruments and practices emerge. Instead of being closed, experimental objects are open to question, controversy and projection. Instead of owing their significance to reduction (as the Occam Razer story about the simplicity of the world goes), the significance of 'experimental things' (or 'matters of concern') hinges upon adding to their complexity—and to their implications for widening the diversity of possible futures (see also, Callon 1986; Latour 1986, 1993, 1999).

The terms representing and intervening were introduced by Ian Hacking (1983) in order to shift focus away from traditions centring on abstract ideas to inquiries into experimental instruments and practices (cf. Kuhn 1962). Notions of representation, for Hacking, perpetuate the realist/antirealist debate. For Hacking, much of the tenacity of debates over foundationalism and relativism has been a consequence of ignoring the jointly epistemic and political dimensions of representations. By exploring practices of representing and of intervening in physical, organic and social realms, we can appreciate the concrete historiality of the instruments, activities and relationships whereby we intervene in the world, which shapes how our surrounding impact us. In this view, the primary aim of experimental practice is not to create representations but to change the world and especially our social relationships within it. As Latour and Steve Woolgar (1979, 1982) might have put it, we don't formulate our representations of objects of knowledge production first, and then start to think about audience persuasion afterwards. Concerns with persuasion are always there from the start, over evidence, adequacy, salience, consistency, and especially over audience's capacities (or lack thereof) to believe what we supposedly know is true.

Whifts in orientations replacing assumptions that consensus is determined by necessity by concerns with how consensus is achieved, transformed and endures

By treating such epistemic things as nature, culture, reality, and tradition, as simultaneously real, historical, and contingent upon political and social events, applied metaphysics contributed to radical change in orientations towards so-called 'standard accounts' of science and modernity. Especially important has been replacing assumptions that consensus is somehow determined by necessity (and that continuity requires no explanation) by questions about how consensus (a 'common world') is achieved and endures in states far from equilibrium. This change has been encouraged by shifting foci from preoccupation with 'crises over representation' towards highly contradictory supposed settlements (Toulmin 1990; Hale 1993; J.L. Koerner 1998; Schaffer 2002; Latour and Weibel 2002, 2005). The relevance of this change for revisiting the contextual circumstances of dualist 'timbers of the modern cosmopolis' (Toulmin 1990; cf. Flyvbjerg 2001) is difficult to overstate. For example, 'standard accounts' of science and modernity conventionally stress 17th century economic prosperity, the withering of religious restrictions on social mobility and intellectual life, expansion of secular culture, the political centrality of the nation-state, and the overturning of pre-modern worldviews by the new mechanical experimental science and natural philosophy (Toulmin 1990; Koerner 2004). But the deeper we delve into materials eclipsed by these accounts, the more light is thrown on the circumstances under which highly problematical caricatures of 'others' and 'publics' played key roles in contradictory conflict settlements. Writing on an example associated with the worsening conditions if the Thirty Years War (1618: 16-48), Toulmin notes that:

> The longer the bloodshed continued, the more paradoxical the state of Europe became…For many of those involved, it ceased to be crucial what their theological beliefs were, or where they were rooted in experience, as 16th-century theologians would have demand. All that mattered, by this stage, was for supporters of Religious Truth to believe, devoutly in belief itself. For them, as for Tertullian long ago, the difficulty of squaring a doctrine with experience was just one more reason for accepting this doctrine that much more strongly (Toulmin 1990: 54).

What bears stressing, is that it was not until those who claimed to be peace negotiators had developed devout belief in the beliefs of 'others' that they were able (without sensing any contradiction) to (a) reduce issues of trust to matters of expert competence, (b) believe that it is possible to start altogether from 'a clean slate', (c) eclipse the importance of acknowledging disagreement to anything like a 'common good'.

New Alliances between Human Life Ways and Natures' Exploratory Adventures

It is realistic to assume that the environment matters and that to understand both humanity and the rest of the world anthropology, ecology and biology need new kinds of models, perspectives and metaphors. Such a realisation may necessitate a fundamentally revised division of academic labour; in particular, the removal of disciplinary boundaries between the natural and the social sciences. We may have to abandon the current separation of physical and biological anthropology, giving new life to the old philosophical, anthropological project which focused on the unity of the human being (Descola and Pálssen 1996:14).

To the best of my knowledge, few have made more important contributions to appreciating the relevance for political and ethical reflexivity of interdisciplinarity in anthropology and archaeology than John Barrett (1999 [1988], 1994, 2000) and Tim Ingold (1993, 1996, 2000). In the 1980s John Barrett (1999 [1988], 1994, 2000) examined impacts of such divisions on predominant conceptions of the archaeological record and human agency. He stressed the bearing upon problem of Linda Patrik's exploration of the question, 'Is There an Archaeological Record?' (1999 [1985]). For Patrik (1999: 119), archaeology's influentially opposed paradigms could be characterised in terms of two models ('physical and textual') of what they 'do to or with the 'archaeological record''. One model envisages the record as fossilised (static) evidence of past (dynamic) natural and social processes. Exponents of this 'physical model' archaeologists 'apply scales to, give meaning to, extract information from, carry out experiments to test hypotheses against the record' (Patrik 1999; Binford 1982; Watson et al. 1971; Schiffer 1976). One problem with this model is its assumption about direct mechanical relationships between 'imprints in the archaeological materials [and] dynamic past processes' (Barrett 1999: 5). In the 'textual model…the record consists of physical objects and features that are material signs or symbols of past concepts' (Patrik 1999: 127), and what archaeologists 'do to or with the record' is said to depend on cultural conventions for interpreting the past (Hodder 1982) and on aims to 'discover' meanings, 'translate' ideas and 'read' symbols (Patrik 1999: 140-141). For Barrett (1999 [1988], 1994) and quite a number of other archaeologists and anthropologists (Rosaldo 1989; Dobres and Robb (ed.) 2000; Gardner (ed.) 2004), both options imply problematic conceptions of human agency. Amongst other difficulties, agency is envisaged paradoxically both as a crucial cause of social change and as mere node through which 'systems' supposedly pursue states of equilibrium (Koerner 2004). Writing on the relevance of appreciating the indeterminacy of physical, organic and social processes for fresh approaches to such difficulties, Gregiore Nicholis and Ilya Prigogene note that:

> Our everyday experience teaches that adaptability and plasticity of behavior, two basic features of non-linear dynamic systems capable of performing transitions in far from equilibrium conditions, rank among the most conspicuous characteristics of human societies…A basic question that can be raised is whether, under those conditions, the overall evolution is capable of leading to some kind of global optimum, or, on the contrary, whether human [behaviors, communities and histories involve] complex stochastic processes whose rules can in no way be designed in advance. In other words, is past experience sufficient for predicting the future, or is a high degree of unpredictability of the future the essence of human adventure, be it at the level of individual learning or at the collective levels of history making? (Nicholis and Prigogine 1989: 238).

Barrett long argued for the relevance of the indeterminacy of human life ways and histories for fresh approaches to problems with influentially opposed 'physical and textual models of the archaeological record' (Patrick 1999 [1985]). For Barrett such approaches might explore 'the range of contextual mechanism by which different forms of agency have gained their various historical realities' (Barrett 2000: 61). Human agency, Barrett says:

> …is not something which lies behind the material residues of an 'archaeological record', to be recovered (literally 'dug out') by the archaeologist who explains that 'record' as being a 'record' of something' ('ideas', 'actions', etc.). Instead, agency lies in front of [its historical and material conditions as]…the means of understanding and reworking them in an interpretive cycle, and it is this

> interpretive cycle which can be glimpsed through archaeological analysis. [And an] archaeological engagement with the past now becomes an attempt to understand how, under given historical and material conditions, it may have been possible to speak and act in certain ways and not in others, and by so doing to have carried certain programmes of knowledge and expectation forward in time (Barrett 1994: 4-5).

Tim Ingold has never ceased to examine complex historical connections between problematic disciplinary divisions and caricatures of 'others'. He has thrown important light on connections between schemes, which divide the biological evolution of 'anatomically modern humans' from cultural history, and problems with hitherto predominant conceptions of 'cross-cultural translation' of supposedly 'incommensurable world views'. The former assume that humanity supposedly stepped out of 'nature' through 'reason.' The latter assume that science and modernity 'represents the culmination of the potentials common to humanity' (Ingold 1993: 215-117; see also, Asad 1986). Both depend on dichotomies of nature versus culture, 'universal capacities to reason versus tradition' and 'the real and total versus the local and historical' (Ingold 2000). These dichotomies perpetuate conceptions of cross-cultural translation that oppose traditional 'views in the world' versus a supposed universal 'view of the world'—a view supposedly alone able to disclose 'others' 'as world views, as alternative ('emic') modeling of supposedly immutable ('etic') reality' (Ingold 1993: 224). Much of Ingold's recent work centres on critical and constructive approaches to interlinked problems with:

- deterministic approaches to molecular biology that eclipse the importance of emergent novelty to processes of growth and maturation, and the diversity of life forms and their relational capacities (see also, Johannsen 1911; Gould and Lewontin 1979; Kauffman 1993, 1995; Fox-Keller 1995; Jablonka and Lamb 1995; Kay 1996; Jablonka 2000),
- approaches to cognitive psychology, which depart from dualist characterisations of the mind versus the world (see also, Habermas 1979, 2003; Barrett 1988, 1994, 2000; Pribram 1997),
- versions of cultural theory that attribute human behaviour to designs that are passed from one generation to the next as the content of traditions (Ingold 1996, 2000; see also, DeMarrais et al. 2004; Gamble and Gittens 2004; Gamble 2008).

This work has led to several constructive hypotheses, including that:

- Much of what we call cultural variation may consist of 'variations of skills—not just techniques of the body, but capabilities of action and perception of the whole organic being.'

- 'Becoming skilled in the practice of a particular form of life is not a matter of furnishing a set of generalised capacities, given from the start as compartments of universal human nature, with specific cultural content.' Rather than being transmitted from generation to generation by tradition, they are 're-grown, incorporated into the modus operandi' of the developing human organism person through training and experience it the performance of particular tasks.
- One of the mistakes of anthropology has been to insist upon a separation between the domains of technical and social activities, a separation that has blinded us to the fact that one of the outstanding features of human technical practices lies in their embeddedness in the current of sociality. It is to the entire ensemble of tasks, in their mutual interlocking, that I refer by the concept of taskscape (Ingold 2000: 6, 298).

Complexity and Reflexivity

Going beyond dualism opens up an entirely different landscape, one in which states and substances are replaced by processes and relations. The main question is not any more how to objectify closed systems, but how to account for the diversity of the processes of objectification (Descola and Pálssen 1996: 12).

Complexity and reflexivity are key components of Ingold's (2000: 5) approach to 'the human being as a locus of creative growth within unfolding fields of relationships.' For Ingold:

- Intentionality and functionality are immanent in the practice itself rather than being prior properties, respectively of an agent and an instrument.
- Skill is not an attribute of the individual body in isolation but of the whole system of relationships constituted by the presence of the artisan in his or her environment.
- Rather than representing the mere application of a mechanical force, skill involves care, judgement and dexterity.
- It is not the transmission of formulae that skills are passed from generation to generation, but through practical, 'hands on' experience.
- Skilled workmanship serves not to execute a pre-exiting design, but actually to generate the forms of artifacts (Ingold 2000: 291; see also, Ingold 1996).

For Ingold and Elisabeth Hallam (2007) and quite a number of others (for instance, Van der Leeuw and Torrence 1989; Toren 1999; Edgerton 2007) much light can be thrown on the complex indeterminacy of human life forms by shifting attention from approaches to change centring on 'innovation' towards processes of 'improvisation.'

> The difference between improvisation and innovation…is not that the one works within established convention while the other breaks with it, but that the former characterises creativity by way of processes, the latter by way of its products, To read creativity as innovation is, if you will, to read it backwards, in terms of its results, instead of forwards, in terms of the movements that gave rise to them (Ingold and Hallam 2007: 2-3).

Like skills, improvisation is not about the already made, finished and given, but rather:

- generative—'rather than the realisation of an *a priori* design,'
- temporal—inherent in the 'onward propulsion of life rather than being broken off, as a new present, from a past that is over,'
- relational—'the creativity of our imaginative reflections is inseparable from our performative engagement' with the dynamics of our surroundings (Ingold and Hallam 2007: 3).

Importantly envisaging life ways and activities along lines suggested by Barrett, Ingold and Hallem and several contributors to the present volume, call for strongly relational conceptions of time—one similar to that developed by Prigogine, which envisages time as an operator—as a property of dynamic relationships rather than an external parameter (see also Kubler 1962). Amongst other things, Hans Joerg Rheinberger (1997) says, especially useful about such operator (or internal) time is that it can illuminate dynamic relationships analogically.

Let us assume that with respect to the movement of material systems, systems of things, or systems of actions, time can be viewed as an operator and not simply as a chronological axis of extension in a system of coordinated. In this sense, time is structural, local and intrinsic characteristic of any system maintaining itself in at least some respects enduring state—though far from equilibrium and reaching from time to time, as a result of turbulences, points of bifurcation…Thus every system of material entities and therefore any system of actions of such entities can be said to possess its own intrinsic time a process (Rheinberger 1997: 180).

Ingold says that such relational approaches are intrinsically reflexive. They situate 'the practitioner from the start in the context of active engagement with his or her surroundings' (Ingold 2000: 5). Further, reflexive orientations need not be restricted to human or even only to animate realms:

> The rhythms of human activities resonate not only with those of other living things but also with a host of other rhythmic phenomena—the cycles of the day and night and of the seasons, the winds and the tides, and so on…[L]ife is not a principle that is separately installed inside individual organisms, and which sets them in motion upon a stage of the inanimate. To the contrary…life is a name for what is going on in the generative field within which organic forms are located and 'held in place' [Ingold 1990: 215]…This means that in dwelling

> in the world, we do not act upon it, or do things to it; we move along with it. Our actions do not transform the world; they are part and parcel of the world's transforming itself. And this is just another way of saying that they belong to time (Ingold 2000: 200).

Similarly, writing on the indeterminacy of the 'metamorphosis of nature,' Stengers recalls Michel Serres evoking the respect that peasants and fishermen have for the world in which they live:

> They know that no one has control over time and that one cannot rush the growth of the living, the autonomous transformations that the Greeks called *physis*. In this sense, science may be at last on its way to becoming a physical science since it has to finally accept the autonomy of things, and not just living things…As with the development of plants, the development of this new nature, peopled by machines and technology, the development of social and cultural practices, the growth of cities are continuous and autonomous processes in which no one can certainly intervene to modify or organize them, but whose intrinsic time must be taken into account, under threat of failure (Stengers 1997: 56-7).

In these lights, she argues that the 'time has come for new alliances, which have always existed but for a long time have been ignored between the history of humankind, its societies, its knowledges, and the exploratory adventure of nature' (Stengers 1997: 59).

Re-designing Reflexivity

Writing on the growing interest in themes variously relating to notions of 'design' and, especially 're-design' (rhythms, enskillment, improvisation, and so on), Bruno Latour (2008) notes that conditions summarised by expressions like 'globalisation' and 'risk society' may be useful to envisage in terms of a:

> disconnect between two great narratives—one of emancipation, detachment modernisation, progress and mastery, and the other, completely different, of attachment, precaution, entanglement dependence and care—then the little word 'design' could offer an important touch stone for detecting where we are heading and how well modernism (and also postmodernism) has been faring. To put it more provocatively, I would argue that design is one of the terms that has replaced the word 'revolution'! To say that everything has to be designed and redesigned (including nature), we imply something of the sort: 'it will neither be revolutionized, nor will it be modernized'… (Latour 2008: 2).

What bears stressing is that until quite recently it was unthinkable to envisage 'design' as means to address large scale problems. Amongst others, connotations of 'design' include:

- 'a humility that seems absent from…the Promethean sense of what it means to act' (Latour 2008: 2) and from what Toulmin (1990) calls the 'myth of the clean slate',
- an attentiveness to details, and to the indeterminate multi-dimensionality of enskillment and creativity (Ingold 2000; see also Bourdieu 1990) that is eclipsed by the 'Promethean, dream of action—Go forward, break radically with the past and the consequences will take care of themselves!' (Latour 2008: 3),
- emphasis on historically contingent meanings ('matters of concern') being as real as brick, rather than in terms of dichotomies of nature-culture, reason-tradition, and the global, total and real versus the local and illusory' (Ingold 2000),
- awareness that no process begins from scratch—to design is always to re-design, to experiment is always to revisit pasts that differ to the present and indeterminate futures (re-designing and cautious experimentality may be an 'antidote to colonialising, establishing, or breaking with the past'—'to the search for absolute certainty, absolute beginnings, and radical departures'),
- emphasis on ethical issues, as suggested by the expression 'good or bad design' and the question of for and by whom experiments are pursued.

For Latour,

> The expanding concept of design indicates a deep shift in our emotional make up: at the very moment when the scale of what has to be remade has become infinitely larger (no political revolutionary committed to challenging capitalist modes of production has ever considered redesigning the earth's climate), what it means to make something is also being deeply modified. The modification is so deep that things are no longer 'made' or 'fabricated', but rather carefully 'designed', and if I may use the term, precautionary designed. It is as though we had to combine the engineering tradition with the precautionary principle; it is as though we had to imagine Prometheus stealing fire from heaven in a cautious way! (Latour 2008: 4).

Although there is great diversity amongst contributions to this volume, they may variously touch upon the bearing that fresh rethinking and re-designing archaeological and anthropological reflexivity can and has had upon the 'needs of a world in which simplicity is a memory of a bygone age' (Funtowicz and Ravetz 1997). Further, as Toulmin (1990) notes, such orientations are not at all new but have predecessors. One of our aims with this volume has been to explore the possibility that insights which bear upon these needs might arise out of realms that go against the grain of the long history of dualist characterisations of nature:culture, reality:historical contingency, universal:particular, and pure:applied research.

The phrase, 'public grounds of truth', is a rough translation of the Italian expression, *publici motive del ver*. It reflects the antiquity of associations of things with publics (cf. *res—publici* = things public; Latour and Weibel (eds) 2005) and human sociability with intentionality (cf. Wittgenstein 1958; Brandom 1994). Perhaps most influentially developed by Giambattista Vico in *The New Science of the Common Nature of the Nations* (1948 [1744]), the expression has roots in the long history of writings that argue for poet orators rather than Platonist and Aristotelian philosopher kings as pedagogical and political ideas. Much of what we know about the earliest roots of such 'hidden' or eclipsed agendas (cf. Toulmin 1990) comes from such founders of the 'arts of memory' as Horace (65-8 BC) (1928) and Cicero (106-43 BC) (1942). For these and other contributors to alternatives to pedagogical and political ideals grounded in images of science hinging upon presuppositions about the 'given' reducibility to simplicity of a 'common world', appreciation of plurality of public grounds of truth is intrinsically reflexive and hermeneutical. It involves attention to the importance to the emergence and endurance (in states far from equilibrium) of interpretations of things unknown by analogy with familiar things, or (to use Geertz's 1979 terms) interpretations of 'the experience far' by analogy with 'the experience near'. Indeed, the most ancient public grounds of truth were aspects of the earliest human beings' (*primi oumini*) surroundings that they singled out as exemplary (the sky, a mountain range, animals, and other concrete images) and employed to express social aspirations, accountabilities, disagreements, and diverse senses of 'we.'

Such a 'concrete logic' is not at all 'irrational'. We are mutually susceptible, accountable and intentional creatures whose relationships and practices form the 'family resemblances' between our practices, anchor these to matters of shared concern, and inform us that we do not know and agree on everything (cf. Barnes 2000). What Vico (1948 [1744]) might have called 'poetic wisdom' or what many now refer to as 'material culture' figures amongst the plurality of embodied and socially embedded means we require, as such beings, to use our capacities for *poesis* (creativity) for rationally making our implicit experiences explicit (for transform 'knowing-how' into 'knowing that'), and for logically showing one another how and why to do so. For Vico—as for Cicero and Horace—such insights were important not only amongst conditions of possibility for historical understandings, but also for democratic political leadership—leadership that began with the most rather than the least tractable difficulties by appreciating that there is no such thing as a context independent public problem. In such views, distinctions between such things as nature:culture, reality:contingency, science:values, risk:ethics are normative judgements, and making their public roles explicit is one of the tasks of democratic pedagogical and political authority.

Cicero's (*De Oratore* 1942, II, lxxxvi: 351-354; Yates 1966; Carruthers 1990) paradigm exemplar of such was the legend of how the poet orator, Simonides of

Ceos, invented the 'art of memory'. The legend recounts his being commissioned to present a lyric poem at a celebration given by Scopas, a noble of Thessaly. Simonides' oration was not restricted to praising his patron, but spoke of the virtues of the twin gods, Cator and Pollux. This infuriated Scopas, who says he will pay only half. But before he could respond, Simonides is summoned to meet two travellers. He searches for them, but he finds no one. On returning he finds the palace's roof has collapsed and crushed hosts and guests beyond identification. Now as the only survivor, Simonides is summoned to chant a memorial of the event and the guests. 'How can I do this?' he asks, since he knew none of them personally. The answer offered by the legend is that Simonides walked through the architectural remains and, using these as mnemonic, performed the oration he did at the ill-fated banquet again—but now with diversions to the names and honours of the people lost in the event by their living families and neighbours.

For Cicero, the legend showed the importance of the 'art of memory' not only for understanding the conditions of possibility for historical knowledge, but also for democratically achieved political leadership. The legend showed how concrete things and places (*loci*) act as images in the recollection of words, people, places and events, and in the very constitution of history and a common world (*sensus communis*). For Cicero, the legend illustrated the formal dimensions of the art of memory, namely: (a) a sequential framework and structural conventions, and (b) elements of dialogue and explanation. The former are structured around poetic (*phronesis* or ethical) practices (or tropes, *verba translata* = words with transferred meanings), the most elementary of these being transfers by:

metaphor	from one thing to something similar,
metonymy	from cause to effect or vice versa,
synecdoche	from whole to parts,
irony	from one thing to its opposite.

The later are structured around dialogical practices required to critically engage and constructively change the *sensus communis* of a particular situation, the practices most important for pluralist dialogue being:

invention	finding images to make different views on matters of concern explicit,
dispositi	arranging them,
elocution	considering the language of communication most appropriate,
memoria	memorising,
pronuncia	presenting one's own views in what issues count.

These practices are not restricted to words, but involve sensible objects as concrete metaphorical vehicles of communication and as sites for further metaphorical image creation. Images and objects do not 'stand for' (represent) but express meaning relationships; their significance hinges upon the contextual contingencies of their use. Hence, Latour's (2000) arguments that images count but not because they somehow possess powers of their own with sources somewhere 'above and beyond' the contingencies of everyday human affairs:

> They are [also] not mere tokens, and not because they are prototypes of something away, above, beneath; they count because they allow one to move to *another* image—to new means to make agreements and disagreements over matters of concern explicit—no less frail and modest as those before but *different* (Latour 2002: 36).

One of the central themes in the writings of contributors to alternatives to pedagogical and political ideals, which variously hinge upon claims that 'state of emergency' is the norm for contemporary times (if not the human condition in general, and what Toulmin 1990 calls the 'myth of the clean slate'), is that of irony. Irony:

- resupposes consciousness of distinctions between true and false,
- makes deception and self deception possible, including about equations of intelligibility with consensus, and mutual unintelligibility with disagreement,
- can facilitate reflective awareness of the scope and limits of ideological hegemony, as well as means not only for resistance but for changing the circumstances that allowed hegemony to arise.

Writing under conditions where it was difficult to 'assume that anything can be assumed' (cf. Sloderdijk 2005), Vico argued that irony is crucial not only for illuminating contradictory ideologies but also for 'going against the grain' (cf. Benjamin 1940) to challenging their hegemony. For Vico, metaphor, metonymy, synecdoche, and irony form means whereby everyday people can and have successfully addressed glaring discrepancies between how things are and ought to be. Vico stressed examples of situations where plurality of public grounds of truth enabled communities to recognise that a critical barrier to democratic participation in public affairs are ideologies, which ruling elites both truly believe in and use to their own advantage. Initial responses to this recognition, Vico said, centre on creating counter-ideologies, which help resist hegemonic value judgements and exploitation. But where ruling ideologies are self-contradictory and unjust, counter-ideologies are of restricted use for engaging the forces out of which ruling powers arise. Irony, Vico (1948 [1744]: 916-918, 925-927), said, facilitates such forms of public reflection as those evidenced by demands in antiquity on the part of the 'plebs' to have their public significance recognised as human beings with human rights. Then, perhaps as now, plurality of interpretations of 'lived heritage':

- emerge around particular events that cannot be accounted for or determined by pre-existing conditions (Newman 2008),
- can and often do variously involve 'dis-identification' with pre-exiting identities and interests (Vico 1948 [1744]; De Beauvoir 1948 [1994]; Rancier 1999) on the very scales on which historically salient meanings and values are generated (Husserl 1936),
- may be about creating some 'common sense' of 'universality' under historically contingent circumstances (Laclau 1993; Latour 2004).

Importantly though, since a 'common world' is unlikely ever to be 'given', efforts to re-design reflexivity must be continuously made anew.

References

Asad, T. 1986 The Concept of Cultural Translation in British Anthropology, in G.E. Marcus and J. Clifford (eds) *Writing Culture*, 141-164. Berkely: University of California Press.

Barnes, B. 2000 *Understanding Agency: social theory and responsible action.* London: Sage Publications.

Barrett, J. [1988] 1999 Fields of Discourse: reconstituting a social archaeology, in J. Thomas (ed.) *Interpretive Archaeology: a reader*, 23-32. Leicester: Leicester University Press.

—— 1994 *Fragments from Antiquity: an archaeology of social life*. Oxford: Basil Blackwell.

—— 2000 A Thesis on Agency, in M.A. Dobres and J. Robb (eds) *Agency in Archaeology*, 61-68. London: Routledge.

Benhabib, S. 1996 *Democracy and Difference: contesting the bounderies of the political*. Princeton: Princeton University Press.

Bernstein, P. 1996 *Against the Gods: the remarkable story of risk.* New York: Wiley.

Binford, L. 1982 Objectivity—Explanation—Archaeology, in C. Renfrew, M. Rowlands and B.A. Segraves (eds) *Theory and Explanation in Archaeology: the Southampton Conference*, 125-138. New York: Academic Press.

Bohman, J. 1991 *New Philosophy of Social Science: problems of indeterminacy*. Oxford: Polity Press.

Bourdieu, P. 1990 *The Logic of Practice*, translated by R. Nice. London: Polity Press.

Brandom, R. 1994 *Making It Explicit: reasoning, representing and discursive commitment.* Cambridge, MA: Harvard University Press.

Callon, M. [1986] 1999 Some Elements of a Sociology of Translation: Domestication of the Scallops and the Fisherman of St. Briec Bay, in Biagioli (ed.) *The Science Studies Reader*. London: Routledge.

Carruthers, M. 1990 *The Book of Memory. A study of memory in medieval culture*. Cambridge: Cambridge University Press.

Cicero, M.T. (106-43 BC) 1942 *De Oratore*, Books 1 and 2, translated by E.W. Sutton, book 3, translated by H. Rackham. London: Loeb Classical Library.

Daston, L. (ed.) 2000 *Biographies of Scientific Objects*. Chicago: University Chicago Press.

—— 2006 The History of Science as European Self Portraiture. *European Review* 14: 523-536.

De Beauvoir, S. 1994 [1948] *The Ethics of Ambiguity*. New York: Citadel Press.

Deleuze, G. 1990 *Negotiations: 1972-1990*. New York: Columbia University Press.

DeMarrais, E., Gosden, C. and Renfrew, C. (eds) 2004 *Rethinking Materiality: the engagement of the mind with the material world*. Cambridge: McDonald Institute of Archaeological Research.

Descola, P. and Pálssen, G. (eds) 1996 *Nature and Society. Anthropological perspectives*. London: Routledge.

Dobres, M.A. and Robb, J. (eds) 2000 *Agency in Archaeology*. London: Routledge.

Edgerton, D. 2007 *The Shock of the Old: technology and global history since 1900*. Oxford: Oxford University Press.

Flyvbjerg, B. 2001 *Making Social Science Matter: why social theory fails and how it can succeed again*, translated by S. Sampson. Cambridge: Cambridge University Press.

Fox-Keller, E. 1995 *Refiguring Life. Metaphors of twentieth century biology*. New York: Columbia University Press.

Funtowicz, S.O. and Ravetz, J.R. 1997 The Poetry of Thermodynamics. Energy, entropy/exergy and quality. *Futures* 29(9): 791-810.

Galison, P. 1987 *How Experiments End*. Chicago: University of Chicago Press.

—— 1996 Computer simulations and the trading zone, in P. Galison and D.J. Stump (eds), *The Disunity of Science: boundaries, contexts, and power*, 119-157. Stanford: Stanford University Press.

—— 2002 Images scatter into data. Data gather into images, in B. Latour and P. Weibel (eds) *Iconoclash. Beyond the image wars in science, religion and art*, 300-323. London: MIT Press.

—— 2007 *Einstein's Clocks and Poincaré's Maps: empires of time*. London: Sceptre.

Galison, P. and Hevly, B. (eds) 1992 *Big Science: the growth of large-scale research*. Stanford, Calif.: Stanford University Press.

Gamble, C. 2008 Hidden Landscapes of the Body, in B. David and J. Thomas (eds) *A Handbook of Landscape Archaeology*, 256-262. Walnut Creek: New Left Coast Press.

Gamble, C. and Gittens, E. 2004 Social Archaeology and Origins Research: a Paleolithic Perspective, in L. Meskell and R.W. Preucel (eds) *A Companion to Social Archaeology*, 96-119.

Gardner, A. (ed.) 2004. *Agency Uncovered: archaeological perspectives on social agency, power and being human*. London: UCL Press.

Geertz, C. 1973. *The Interpretation of Cultures. Selected essays*. New York: Basic Books.

Gould, S.J. and Lewontin, R.C. 1979 The Spandrels of San Marco and thhe Panglossian Paradigm: a critique of the adaptionalist programme. *Proceedings of the Royal Society of London* 205: 581-598.

Habermas, J. 1979 *Communication and the Evolution of Society*, translated by T. McCarthy. London: Heinemann Educational.

—— 2003 *Truth and Justification*, translated and edited by B. Fulner. Cambridge: Polity Press.

Hacking, I. 1983 *Representing and Intervening. Introductory topics in the philosophy of science*. Cambridge: Cambridge University Press.

Hale, J.R. 1993 *The Civilization of Europe in the Renaissance*. London: Simon and Schuster.

Hallam, E. and Ingold, I. (eds) 2007 *Creativity and Cultural Innovation*. Oxford: Berg.

Hodder, I. (ed.) 1982 *Symbolic and Structural Archaeology*. Cambridge: Cambridge University Press.

Horace (65-8 BC) 1928 *The Art of Poetry*, translated by H.R. Fairclough. Cambridge, MA.: Harvard University Press.

Husserl, E. 1970 [1936]. *The Crisis of European Science and Transcendental Phenomenology*. Evanston, IL: Northwestern University Press.

Ingold, T. 1993 The Art of Translation in a Continuous World, in G. Pálssen (ed.) *Beyond boundaries. Understanding, translation and anthropological discourse*, 210-229. Oxford: Berg Publishing.

—— 1996 Situating Action V: The History and Evolution of Bodily Skills. *Ecological History* 8(2) 171-182.

—— 2000 *The Perception of the Environment. Essays in livelihood, dwelling and skill*. London: Routledge.

Jablonka, E. 2000 Lamarkian Inheritance Systems in Biology: a source of metaphors and models in technological evolution, in J. Ziman (ed.) *Technological Evolution as an Evolutionary Process*, 27-40. Cambridge: Cambridge University Press.

Jablonka, E. and Lamb, M. 1995. *Epigenetic Inheritance and Evolution*. Oxford: Oxford University Press.

Johannsen, W. 1911 The Genotype Conception of Heredity. *American Naturalist* 45: 129-159.

Kauffman, S. 1993 *The Origins of Order: self-organisation and selection in evolution*. Oxford: Oxford University Press.

—— 1995 *At Home in the Universe. The search for laws of self-organisation and complexity*. Oxford: Oxford University Press.

Kay, L. 1996 Life as Technology: Representing, Intervening and Molecularizing, in S. Sarkar (ed.) *The Philosophy and History of Biology*. Dortrecht: Kluwer, 87-100.

Koerner, J.L. 1998 Heironymus Bosch's World Picture, in C. Jones and P. Galison. P. (eds), *Picturing Science and Producing Art*, 297-325. London: Routledge University Press.

Koerner, S. 2004 Agency Against the Grain of Privatised Ethics and Globalise Indifference, in A. Gardner (ed.), *Agency Uncovered: archaeological perspectives on social agency, power and being human*, 211-240. London: UCL Press.

Kubler, G. 1962 *The Shape of Time: remarks on the history of things*. New Haven: Yale University Press.

Kuhn, T. 1970 [1962] *The Structure of Scientific Revolutions*, second edition. Chicago: Chicago University Press.

Laclau, E. 1990 *New Reflections on the Revolution of Our Time*. London: Verso.

Latour, B. 1986 Visualization and Cognition: drawing things together. *Knowledge and Society* 6: 1-40.

—— 1993 *We Have Never Been Modern*. Cambridge, MA: Harvard University Press.

—— 1999 *Pandora's Hope: essays on the reality of science studies*. Cambridge, MA.: Harvard University Press.

—— 2004 Whose Cosmos? Which Cosmopolitics? *Common Knowledge* 10(3): 450-462.

—— 2008 A Cautious Prometheus: A few steps toward a philosophy of design (with special attention to Peter Sloterdijk). Keynote lecture for the 3 September 2008 meeting of the Design History Society ('networks of Design'), Cornwall, UK.

Latour, B. and Weibel (eds) 2002 *Iconoclash. Beyond the image wars in science, religion and art.* London: MIT Press.

—— (eds) 2005 *Making Things Public. Atmospheres of democracy*. Cambridge: MIT Press.

Latour, B. and Woolgar, S. 1979 *Laboratory Life: the social construction of scientific facts*. London: Sage.

Lazzari, M. 2008 Topographies of Value: ethical issues in landscape archaeology, in B. David and J. Thomas (eds) *A Handbook of Landscape Archaeology*, 644-658. Walnut Creek: New Left Coast Press.

Meskell, L. 2002 Negative Heritage and Past Mastering in Archaeology. *Anthropological Quarterly* 75(3): 557-574.

Newman, S. 2008 Control, Post-politics and the Invisibility of the People. Paper presented in workshop on *Futures: Engaging Technologies of Protection and Control*. Lancaster University, June 2008.

Nicholis, G. and Prigogine, I. 1989 *Exploring Complexity: an introduction*. New York: W.H. Freeman and Company.

Patrik, L.E. [1985] 1999 Is There an Archaeological Record?, in J. Thomas (ed.), *Interpretive Archaeology: a reader*, 118-144. Leicester: Leicester University Press.

Pribram, K.H. 1997 Interfacing Complexity at the Boundary between the Natural and Social Sciences, in E.L.K. Khalil and Boulding, K.E. (eds), *Evolution, Order and Complexity*, 40-60. Cambridge: Cambridge University Press.

Rancier, J. 1999 *Disagreement—Politics and Philosophy*. Minneapolis, MN.: University of Minnesota Press.

Rheinberger, H.-J. 1997 *Toward a History of Epistemic Things: Synthesizing Proteins in the Test Tube*. Stanford: Stanford University Press.

Rosaldo, R. 1989 *Culture and Truth. The remaking of social analysis*. Routledge: London.

Schaffer, S. 2002 The devises of iconoclasm, in B. Latour and P. Weibel (eds) *Iconoclash. Beyond the image wars in science, religion and art*, 498-516. London: MIT Press.

Schiffer, M. 1976. *Behavioral Archaeology*. New York: Academic Press.

Shapin, S. and Schaffer, S. 1985 *Leviathan and the Vacuum Pump. Hobbes, Boyle and the experimental life*. Princeton: Princeton University Press.

Sloderdijk, P. 2005 Forward to the Theory of Spheres, in M. Ohanian and J.C. Royoux (eds), *Cosmographs*, 223-241. New York: Lukas and Sternberg.

Stengers, I. 1997 *Power and Invention*. Minneapolis: University of Minnesota Press.

Toren, C. 1999 *Mind, Matter and History: explorations in Fijan ethnography*. London: Routledge.

Toulmin, S. 1990 *Cosmopolis. The hidden agenda of modernity*. Chicago: University of Chicago.

Van der Leeuw, S.E. 1989 Risk, Perception, and Innovation, in S.E van der Leeuw and R. Torrence (eds) *What's New? A closer look at the process of innovation*, 300-329. London: Unwin Hyman.

Vico, G. 1948 [1744] *The New Science of the Common Nature of the Nations of Giambattista Vico*, unabridged third edition translated by T.G. Bergin and M.H. Fisch. London: Cornell University Press.

Watson, P.J., Leblanc, S.A. and Redman, C.L. (eds) 1971 *Explanation in Archaeology: an explicitly scientific approach*. New York: Columbia University Press.

Wittgenstein, L. 1958 [1955] *Philosophical Investigations*, translated and edited by G.E.M. Anscombe. Oxford Blackwell.

Yates, F. 1966 *The Art of Memory*. London: Routledge and Kegan Paul.

Chapter 15
Memory Practices and the Archaeological Imagination in Risk Society: Design and Long Term Community

Michael Shanks and Christopher Witmore

Disciplinary Crisis

A broad disciplinary crisis was recognized and embraced across the humanities and social sciences from the late 1960s and early 1970s. Gouldner notably signaled the crisis in sociology in his book of 1970; it was associated with a revaluation of the classics of Durkheim, Weber and Marx, and an assimilation of hermeneutic and phenomenological critiques of social science-based methodologies and theories. Debate has circulated around their confident positivist premise that essentially deterministic processes can account for the complexity of social and cultural life and history.

In anthropology the questions have been about the coherence and scope of a four-field anthropology instituted in the US as cultural, biological and linguistic anthropology, plus archaeology, with challenges from the 1960s to structural functionalist anthropology and then cultural evolution, tied to programmes of structuralist, marxist and interpretive anthropologies.

More broadly, we note the rise of critical theory, derived from a strain of western Marxian thinking from Lukács and through the Frankfurt School in the 1920s and 1930s, across the humanities and social sciences. After Mannheim's sociology of knowledge (1936, 1952) and Kuhn's concept of paradigm (1970), the rise of science studies has endeavored to situate science in its constitutive practices, with the absolute authority of science itself under question as a set of historically located cultural practices in institutions such as the academy, and with their ramifications through funding agencies, state and corporate policy, down to laboratories, instruments and everyday practices (Biagioli 1999).

In the early 1970s in archaeology, David Clarke (1973) famously highlighted a loss of disciplinary innocence and the emergence a critical self consciousness, an awareness of archaeology as a mature field in the sciences and humanities. This sense of crisis has not dissipated, though there is now a kind of neoliberalist orthodoxy in the academy that recognizes a market of competing ideas, even while it champions the behaviourist norms of a certain conception of humanity, one taking its essential nature from the natural selection of a market economy (Canaan and Shumar 2008).

At the heart of this disciplinary crisis there regularly appear the following characteristics: challenges to the authority of science; questions raised of agency and the motor of history (just what forces are driving change?); debates about the applicability of universal and global systems of knowledge in relation to local interests and specific contexts; different approaches to a human habitat now seen as inherently 'complex'; deep and radical reframing of theory and methodology, when we can almost no longer assume that anything can be assumed (Sloterdijk 2009b).

In the debates around research in academic archaeology we still read much of the contrasts between social science and interpretive approaches (however see Rathje, Shanks and Witmore (2010) on the dynamics of archaeology over the last 30 years). Nevertheless, the sense of crisis in the profession of archaeology is now centred upon threats to the material past, the challenges of managing the loss, of mitigating impacts, of policing the looting and the illicit trade in antiquities, of regulating, of recognizing local claims on historical roots, on past legacies, when history may become conflated with heritage. We now look upon an ethics and politics of property, responsibility, threat and stakeholder interests (Brodie and Tubb 2002, Renfrew 2000).

This discourse of threat and loss of course goes back to the 19th and 18th centuries and a modernist sensibility attuned to the decline of tradition, to the broad rise of historical sensibilities, of appreciation of the remains of the past and of other cultures, housed in the new architectures of the museum and increasingly protected by institutions of the nation state (Olsen and Svestad 1994). This modernist historicity, or sense of the historical past, is not simply about the flow of events and personalities, social change, and individual agency, or indeed human mortality in the face of the tide of history. The modern disciplines of history, archaeology and anthropology offered means of holding on to what was changing, of curating the past and of connecting with the future, through visions and narratives of progress, of improvement, or of simply understanding where we had come from (see Thomas 2004).

Though there are these deeper roots, two centuries old, Cultural Resource Management has seen exponential growth since the 1970s. At first growing perception of threats to the remains of the past prompted programmes of 'rescue' and 'salvage' archaeology, mostly in the US and UK, with visions of the remains of the past rescued from beneath the developers' bulldozers. Since then, and aided by various legal and legislative instruments, there has been a shift of emphasis to the management or stewardship of valued sites and artifacts under a sense of ethical responsibility to future generations. This is now a global phenomenon, with prominent recognition coming from the likes of UNESCO and indeed the World Bank's acknowledgement of the crucial role of culture in economic development. Of note also is the convergence of concern about cultural as well as natural resources, indeed a convergence that pertinently questions the very distinction.

Archaeology in Risk Society

How are we to understand this trend towards a sense of 'crisis'? Clearly, the quickening pace of urban and industrial development over the last 40 years has had a tremendous impact on archaeological remains, prompting the legislative and state interventions. The past itself seems under threat; at the minimum it needs protection. Archaeology, as a mode of appropriation and engagement with the past, is a component of a now global and hegemonic academic culture industry, with comparable curricula and discourse found in every university the world over. The archaeological past of sites, monuments and works housed in museums is at the heart of the tourist industry and features prominently in popular mass media (Clack and Brittain 2007). Archaeology and archaeological awareness is more than ever obviously wrapped up in contemporary (post)modernity.

Bruce Trigger's lifelong project of writing a history of ideas of archaeology successfully connected the discipline with broad modernist trends in nationalism, imperialism and colonialism (1984, 1989). We can go much further by considering not just ideas but also practices, locales and things (Olsen et al. 2011). We suggest that it is useful to treat this sense of crisis as an archaeological manifestation of what is being called 'risk society'.

The term 'risk society', associated with the pioneering thinking of Ulrich Beck (1992) and Anthony Giddens (1991), is shorthand to describe escalating shifts in modernity centred upon concern with manufactured risks and threats. Giddens emphasizes changes that involve an end of tradition, in the sense of the past no longer being guarantor of contemporary security, in the sense that individuals are increasingly held responsible for their own security in a world experienced as more and more subject to risks to self, family and community. Sloterdijk (2009a) and Serres (1995) flag up growing senses of threat to global humanity's very habitat and survival—atmosphere, food and sustenance, water and housing. We are no longer simply subject to fate and nature, but the cumulative effect of certain behaviours, policies and values is having deleterious effect on the stability of our human cultural ecology. Considerable attention is given to the implication of individuals, institutions and corporations in changes that seem to threaten the very core of human being: the engineering of genetic change, environmental change, the instabilities of a global monetary economy, international security in the face of terrorism and nuclear proliferation. And the loss of the past, associated with changes in the way history itself is conceived and experienced.

It is most useful to locate archaeology in such analysis because it helps us understand the changes in archaeology over the last few decades and to evaluate our options, as professionals, academics, stakeholders and coinhabitants of earth. Connecting archaeology with an analysis of risk society locates archaeology primarily outside the academy; we suggest that changes in archaeological thought are a fundamental component of contemporary memory practices that have a particular manifestation in the academy and its discourses. Consider the growth of garbology and the archaeology of the contemporary past, stretching archaeological interests in directions inconceivable, or at best marginal, a few decades ago. Consider how there is a new shape to the scope of archaeological inquiry, with the crucial relevance of long term archaeological process to the understanding of climate change, human interventions in the environment, to the development of state and empire, to communication, travel and mobility (what Giddens (1984, chapter 4 and 5) subsumes under his term time-space distanciation). We are beginning to see some bold broad-brush thinking in archaeology and cognate fields again (see Redman 2001, Tainter 1990, Diamond 2005 on the aforementioned topics) where our counter measures need to be equal in scope to the perceived risks threatening humanity. Archaeologists have never had to work harder to understand the past in the present.

We will outline the characteristics of this extended field of archaeological practice and sensibility as a component of risk society.

Thirty years ago, when one of us was starting out in an archaeological career in the academy, there simply was no scope or agenda for questioning the value-freedom of (social) science, for debating the politics of the archaeological past, for treating pasts forged in the present in association with cultural struggles for genuine local identity in an increasingly globalist and neoliberal world. Quite appropriately, archaeology and CRM were embedded in long-standing agendas to establish a coherent time-space systematics for both the management of endangered sites and finds as well as for academic research, and to harness the power of quantitative social science for modelling social change. They were, and still largely are, part of a broad modernist programme instituted in the early 19th century and typically involving abstract expert systems that permit disembedded comparison and calculation across indefinite time and space.

The development from the 18th century of such large scale abstract systems of knowledge acquisition and management related to the monitoring and direction of everyday life, has been amply explored by the likes of Foucault. Focused on everything from medicine to criminology, economy and environment, state directed and coordinated through bureaucracies, markets and all kinds of research agencies, including the academy, colossal resources have been given over to surveillance, measurement and analysis with the aim of regulation and control. The building of these knowledge-based systems has involved the development of instruments and techniques of observation and measurement such as cartography and photography, standards and infrastructures that facilitate comparison and analysis, statistics operating upon databases, as well as institutions and management structures that allow the translation of observation into data into information into policy into execution—witness a control society (Deleuze 2002).

If all of human life and experience is in principle calculable and subject to knowledge, that we might understand better the likely outcomes of particular actions, attention is thrown onto the future because some sort of assessment of likely risks can be made for virtually all habits and activities. These abstract and comparative knowledge systems throw suspicion on traditional answers and precepts in favour of research and analysis oriented on the future and implying assessment of threats and opportunities, hence the notion of risk is central in a society which is taking leave of the past, of traditional ways of doing things, and which is opening itself up to a problematic future.

This is part of the wholesale reorientation of temporality so central to modernity and which, of course, encompasses the likes of cultural resource management and archaeology. Great voids in the antiquity of humankind came with the challenges to senses of history based upon religious teaching, biblical chronologies and Graeco-Roman historiography. Archaeology has worked so successfully over two centuries to populate the past with sites and artifacts in a global time-space systematics of timelines and distribution maps rooted in universally applicable systems of classification and categorization. While this inventory of archaeological remains has become the foundation and instrument of the management of the past in ministries of culture and planning departments the world over, it has nevertheless, indeed necessarily come with a growing awareness of threats both to the remains of the past and to the possibility of creating any kind of meaningful knowledge of what happened in history, if access to sources is overly restricted, if contextual information is lost or never acquired.

Here we experience a new kind of threat or risk to the past itself as well as to the potentiality and richness of pasts in the future, based upon new modern dynamics of presence (of the remains of the past) and absence (of past lives themselves as well as future memories and histories). The past is conspicuously not a datum, but subject to contemporary interests and concerns, infused now with the interests of knowledge and also with erosive threatening interests. Just as the natural environment is now seen as a thoroughly socialized and institutionalized habitat, a hybrid that includes threats, culpability, and responsibility on the part of humanity to care and curate, so too the past is a matter of concern, a matter of foresight, another risk environment affecting whole populations' needs and desires for history, heritage, memory.

The paradox or contradiction is that the control that knowledge affords, for example, in managing the impact of development or of the trade in illicit antiquities on the possibility of a past in the future, comes at the cost of a sense of security. It is not just that the past is threatened; senses of personal and community identity are threatened. The growth of these systems of calculation and control is intimately connected with growing political, social, cultural and indeed ontological insecurity.

The security threat which individuals face is, at base, a threat to their very identity because of the ways in which these abstract systems of knowledge work. When who you are, including your history, is no longer given by traditional institutions and cultures, but is constantly at risk, if who and what you are is subject to changing expert research, the challenge to individuals is to constantly construct and reconstruct their own identity. The growing absence of traditional sources of authority in answering who we are accompanies a growing emphasis upon the individual to take responsibility for self and decisions, to monitor their self, to self-reflect and to assert their own agency, exercise discipline in being who they are. This responsibility is, of course, full of risk, and the possibility of asserting individual agency is seriously circumscribed by horizontal and vertical divisions in society.

Neoliberal thinking upholds the principle of the individual taking responsibility for their choices in a rational market based upon contractual relationships. Its orthodoxy since the 1980s in political science and governance questions the role of the state, of institutions and corporate bodies in promoting such a principle in the context of a society as divided and rooted as ever in class inequalities. This again exacerbates senses of risk, threat and insecurity.

Abstract systems of knowledge are meant to operate everywhere and to allow comparative assessment and judgement to be made. They are global in reach, lifting patterns out of local contexts and simultaneously reaching into our most intimate depths. Distant happenings, wars, environmental damage, the decisions of faceless investment bankers thousands of miles away, ozone depletion over the Antarctic can have profound influence on events close-by, in our everyday lives, and on the intimacies of the self. This is a folding of time and space, shifting time-space distanciation, as the times and places of social and cultural life, the zones or locales that give sense and identity to experience, constantly shift.

And worse. Any authority that science may claim is based upon its skeptical and questioning attitude; this has now been turned against science itself. Risk assessment demands an assessment of knowledge claims: are we to believe, for example, those who deny the evidence for a human origin to global warming? Is the threat to global heritage as serious as some would lead us to believe? Scientists have not only been wrong in the past, but culpably wrong. Science is conspicuously now an ethical field where we find no easy answers, in spite of the expansion of systems of information management, analysis and interpretation.

This new moral economy of assessment of culpability and blame involves a distinctively forensic attitude. At a scene of crime anything might be evidence in the search for coherent account, in the attribution of blame, in the pursuit of the guilty. How are we to know where to look? And anywhere might be a scene of crime: threat is ubiquitous. Forensics is best based upon reason and careful work upon evidence, drawing upon general principles, but we should also recall that the field is a legal one of inquiry and advocacy, case-based and so specific or situated, dealing in human motivation and desire, as well as matters of right and wrong, good and evil. The burden of proofs demands we open up every step of the process to close public scrutiny.

A positive side to this indeterminacy has been the development of sciences of complexity. It is no longer feasible to treat the human habitat as one of discrete systems and determinable parts such as material culture and natural environment, animal and plant species, lithosphere and troposphere. Instead we need to deal with distributed and underdetermined processes which meld human and machine, nature and culture, challenging all those Cartesian dualisms at the heart of much modernist thinking.

One particular implication of archaeology in risk society is that, as part of the pervasive construction of risk objects, it is a system of practice and knowledge, a discourse of dealings with otherness, alterity, the abject (Shanks 1992, 2010). We refer to the potential anxiety elicited in dealing with other cultures and times that present questions of difference, challenges to establish understanding, to translate and establish common ground. The centrality of entropic processes in archaeology, decay and loss, the erosion of order and form, makes this dealing with otherness particularly sharp and challenging. It is not just that an archaeologist may raise questions of historical and cultural continuity in their research, asking 'is this the way we were?'; but the rot and ruin, the debris of humanity in the decaying garbage heap that is history, may mean that we may never know, that no sense may ever be established. The ruin and loss may even tend to nausea, an aspect of the abject: the loss of the past may be sickening; mortal flesh rots; and, without a past, we may never know who we were. This struggle in the face of perpetual perishing is a distinctively archaeological dimension to contemporary threats to ontological security.

Organizations like the World Archaeological Congress have raised awareness of the ethical implications of archaeological research and of the impact of globalization, particularly in relation to non-western pasts, histories and cultures. The proliferation of disembodied knowledge has involved a counter tendency that emphasizes local knowledge, situated in the way described above in connection with the forensic 'case', but also meaning knowledge associated with a particular standpoint, and/or held by stakeholders, those with a specific and motivated interest in a case. Issues of 'indigeneity' are prominent in these agendas that focus on the relation between the global and the local and embrace the need to re-embed knowledge. We note a new premium on what interpretive anthropologists and sociologists call 'local knowledge'.

What Role forArchaeologists in Risk Society?

It is because of the threats facing humanity, because of the potentially catastrophic risks, that it becomes urgently necessary to clarify or even rethink our relationships to the past and inevitably, therefore, what archaeology has to offer.

These changes in our historicity and senses of time and temporality, this orientation on the future, yet growing attachment to the past mediated through our actions now to recover, curate and conserve, the patent elision of materiality and immateriality with the development of digital archives makes of archaeology a critical component of new and contemporary memory practices.

Archaeology, we suggest, has an extended historiographical scope, a much broader scope than history (consider Foucault's use of the terms archaeology (1972, 1973) and genealogy (1980, 1986)), encompassing the mundane and the material, with archaeology as the tangible mediation of past and present, of people and their cultural fabric, of the tacit, indeed the ineffable. Archaeology offers rich resources for building alternatives in a risk society, reframing matters of common and pressing human concern. Indeed because, in the era of future orientation and short term thinking, nothing is guaranteed memory, archaeology's work on the material remains of the past provides a route to the vital insertion of pasts into the present.

We have three propositions. (1) Archaeology, as a genealogical mediation of past in present, offers a distinctively mundane and tacit, yet rich and nuanced, non-teleological basis to the human archive, to collective memory practices. (2) We call this work done on the past to connect its remains with current concerns and a respect for future concerns, a mode of unforgetting, rooted in rich reservoirs of experience and information that are not primarily narrative in character. (3) This work of archaeology can and should be modelled as craft and design.

Archaeology as genealogy

We suggest that archaeology be thought as a genealogical field. This is to distinguish it from the varieties of historiography that implicate the writing of historical narrative or construction of analytical models. Genealogy recognizes that we would not be here but for the past, but also that there is no necessary coherent narrative that leads from past to present, or a neat unitary model to which the past may be fitted. In genealogy there are affiliations and radical discontinuities, a primacy of contingencies in an underdramatized past, characterized by complex, mundane and inglorious origins, contingent and indeterminate flows and turns, no great or grand stories to the history of humankind. Fundamentally this is to question the modernist celebration of progress, that there is progression to history, even in the notion of a greater well-being to humankind now as contrasted with the past. Instead of narrative and progress or development, archaeology as genealogy highlights distribution and complexity—topologies and foldings, flows and eddies. Liquid history; weather-bound.

Take the example of optics. More specifically, consider any lens in your proximity—a pair of eyeglasses, a digital camera, or even the small video lens in your computer. We readily recognize how each lens is made of glass or some other transparent material. We also understand how each has either one or two curved surfaces which deflect rays of light in a consistent manner. Beyond this basic description of a lens we often fail to appreciate the trials and tribulations that went into the making of these things.

As things, each of these lenses gather achievements distant in both space and time so that they are nonetheless simultaneously aggregated in the lens itself (Webmoor and Witmore 2008). In other words, folded into each lens are designers, engineers, optical tables, mathematical calculations, opticians, experiments in refraction, glassmakers, silica, kilns, pyrotechnological skill, and so forth. In fact, the mathematical calculations translated into the curvature of the lens can be traced to the labours of Kepler and al-Haytham, to the optical tables of Hero of Alexandria and to the geometry of Thales (Authier 1995). A sociotechnical genealogy of the glass in each lens is marked by events, incremental shifts, where transactions between humans, silica, fire, ovens, and other things made of silica (techniques) in Egypt, the Near East or Venice are translated into things which either establish a foothold or are lost to Lethe. Mention should even be made of potential connections with Chinese glass and optics as exemplified in the labours of astronomer Shen Kua (Zielinski 2006: 84).

There are several subsidiary theses associated with this repudiation of modernist progress. For archaeology we question the oft-held premise that the discipline of archaeology grew out of a pre-disciplinary tradition of antiquarian thought. Instead we note continuities of concern with the manifestation and documentation of the past, of sites, locales, monuments and portable artifacts in the present.

In an assertion again that we have never been modern, the past is always contemporary with us, presenting opportunities of investigation, of reworking and reuse. So long as we hold to the ill-founded notion of the unswerving march of progress, the idea that the past holds alternatives to the present will be regarded as reactionary and backward, as regression. When such seemingly 'reverse' movements are no longer regarded as the antithesis to a modernist image of progression, the past can once again be regarded as a locus of insight and innovation. Engaging with the past and pooling up a reservoir of alternatives is the key to securing our global future. Archaeology has never been more relevant than in a risk society. In other words, the seemingly out-of-date, the obsolete, the discarded, can once again become futuristic. Examples of such 'reversals' are not new; indeed, they are pervasive. Consider two vignettes from the roots of the modern, industrial era.

Exposed to the onslaught of the English Channel south of Plymouth, the Eddystone Rocks had hosted two different lighthouses before John Smeaton (who would later describe himself as a 'civil engineer', the first to do so) was commissioned by the President of the Royal Society to design a third in 1756. Smeaton's lighthouse would continue to do work for 122 years, outlasting the previous two combined by 75 years (they had lasted 5 and 47 years respectively). Unlike his predecessors, Smeaton experimented with various mortars to produce a 'hydraulic lime', a form of mortar that set even under the onslaught of the unruly sea in the English Channel. In the course of this development, no *Homo faber* myth will do—no form was imposed on shapeless matter (Ingold 2009). For without the presence of Roman cement, without the texts of Vitruvius, Smeaton would have been limited to the limes of Britain and not privy to the properties of pozzolana (Rankin 1916: 749; Smeaton 1793). The fact that such associations intervened in Smeaton's case is captured in the label subsequently given to hydraulic lime, 'Roman cement'.

Such polychronic exchanges abound. Ralph Wood's design of the Tanfield bridge in the industrial north east of England (also known as the Causey arch, the oldest standing railroad bridge in the world) was most probably suggested by the remains of Roman arches so prevalent in this region (Greene 1990: 42). This instance of co-present exchange between an ancient and an industrial arch may seem incidental and trivial; they are quickly overlooked when we conceive of the process of design, design as the imposition of form upon matter—the *Homo faber* myth again. No entity can ever entirely contain another in this way (Harman 2009: 30); it is the negotiation between them that may lead to new translations, foldings, aggregates and assemblages in a percolating temporality.

In *The Shock of the Old* David Edgerton (2007) has documented how some of the most touted technological innovations of the 20th century, for example in the aero-space industries, have actually been costly failures, little adopted. In contrast, major changes have been associated with mundane unheralded technologies, like corrugated iron, often well-known and old, given new use and purpose in modernity.

We now want to sketch a new context for memory practices like archaeology: to signal the contemporary and digital basis of the changes, we call this new condition 'Archive 3.0' and it involves the reanimation of storehouses of experience and cultural works. As Sloterdijk (2009b) would have it, from now on, curation must accompany technical science.

Most generally, archives are the storerooms of humanity, what has come down to the present. An archive is a place where records are kept, a record so preserved. Archives are an architecture of access to the remains of the past. The verb is to place or store (in an archive); archival practices are the principles of how archives are organized and how they work.

The prefix 'arche' (found in archive and architecture and archaeology) is Greek for beginning, origin, foundation, source, first principle, first place of power, authority, sovereignty. It represents a starting point or founding act in both an ontological sense ('this is whence it began') and a nomological sense ('this is whence it derives its authority'). Archaeology's 'archaia' are old things, where 'things', to bear witness and carry meaning, are inseparable from their context, from their find-spot, their connections with other things. Archives are material artifacts and systems all about narratives of origin, identity and belonging, and the politics of ownership, organization, access and use. To recognize how we begin in the midst of things is to acknowledge the primacy of the archaeological.

We see ourselves moving into a new phase in the history of archives. We don't have the space to give a full outline of our proposition; a caricature holds the changes as follows.

Archive 1.0 refers to bureaucracy in the early state, the temple and palace archives and storerooms, with inscription as an instrument of management and the redistribution of goods. With Archive 2.0 comes the mechanization and digitization of archival databases under an aim of fast, easy and open access based upon efficient dendritic hierarchical classification and retrieval. From the early 19th century Archive 2.0 is associated also with statistical analysis performed upon data.

Archive 3.0, coming in the last two decades, brings a reemphasis on personal affective engagement with cultural memory. Rather than static depositories, we see archives increasingly as active engagements with the past, animated archives, where the holdings of museums, libraries, galleries and public collections are opened up to personalized use, where curation and information management increasingly aim to tailor services to different needs and desires. This shift is visible in the efforts made by museums to capture visitor interest by assuming less of them and providing more in the way of supporting information and narrative, as well as by recruiting attractive media affect, literally replacing the static vitrine with a media-rich experience. There is a new focus here on interface, but not solely as a human factor of ergonomics or of behavioural efficiency and conceptual transparency: interface has become an issue now of richness of engagement.

But there is much more. At the heart of the shift to Archive 3.0 is a complete turn in the conception and design of information based upon digitization. Databases have, until recently, required clear conceptualization and hierarchical structure to facilitate efficient access and retrieval. Powerful electronic processing and cheap digital data storage have all but eliminated the need for data directories or look-up tables based upon clear metadata standards; quick and efficient searching of free-form data is now commonplace. While the traditional library required careful and expert recording of book metadata (author, date, title, keywords) according to a standardized system such as Dewey Decimal, Google Books offers customized searching through the complete texts of its holdings of several million works, with the results freely available for individual reuse.

At the heart of such vast digital mediation is fungibility. Digitization allows the gathering of moving image, still image, music, text, 3D design, database, geological survey, graphic detail, architectural plan, virtual walk-through etc., into a single environment. These may be infinitely manipulated and re-mobilized without loss in that space. The eventual output as video, photograph, CD ROM, DVD, paper based printed text, web page, broadcast, archival database, live event, exhibition, site specific installation, 3D model, building etc., is only weakly constrained by limiting factors inherent to the 'originals' being reworked.

Numerous attributes of digital practice—cutting, pasting, undoing, reformatting, layering, mixing, and so on—belong to an arena in which design decisions have become ubiquitous and even the simplest of tasks can take on a speculative, investigative, critical, and/or creative character. And this character, in turn, is inflected by the new associative and collaborative opportunities, the novel ways of moving ideas, communications, and culture around, provided by digital networks. Potentially this raises issues about differences of power and influence between centre and periphery, between the urban and the rural, traditionally privileged and newly empowered classes. There is enhanced potential for small-scale and locally-based artisanal and pre-, post- or non-industrial modes of operation. Digitally mediated culture may imply a re-negotiation of the relationship between the global and the local, the physical and the virtual. The 'virtual', as an ever-expanding experiential, cognitive, and socio-cultural domain, moves alongside or into competition with the physical environment and, as is already the case in certain youth subcultures, mixed reality experiences become not the exception, but the rule.

To this volatile and still somewhat inchoate mix must be added what is perhaps best described as the digitally enabled de-territorialization of data. Vast amounts of cultural, social, and other information, valuable or not, organized or random, information that was once mined exclusively either by restricted circles of specialists or by eccentric data dumpster divers, has become widely available thanks to efforts extending from Brewster Kahle's Internet Archive to Project Gutenberg to the digital repositories of the world's great libraries, gathered in the likes of Google Books, to an archipelago of private undertakings. The ongoing efforts of a variety of interests to treat such information as private property and to restrict its circulation find themselves regularly thwarted by the sheer ubiquity of means to promote their uncontrolled circulation and proliferation. Wikipedia, the collaborative encyclopaedia-for-all, and news blogging similarly challenge proprietorial attitudes towards information. The battle intensifies the closer one gets to contemporary cultural production, but it encompasses the entire cultural field, from prehistoric relics in the possession of the world's most venerable museums to yesterday's detergent advertisement. Whatever its outcome, there can be little doubt that this process of de-territorialization will continue.

We suggest that these distinctive features of contemporary digital culture create an expanded and intensified 'poetic' space; for archaeology, this space is that of the 'design' of the past and memory. Anyone with a personal computer may author, appropriate, share, rework, and publish works in this new political economy of media. The facility offered by digital technologies to exchange, locally rework and remix is the basis for the conflicts over intellectual and cultural property, over matters of creativity, authenticity and ownership of both the means of cultural production as well as its goods. The social and collaborative media of Web 2.0 add to the collaborative possibilities of such remix or re-association.

At the heart of Archive 3.0 are new mixed realities (digital and analog) or prosthetic architectures for the production and sharing of archival resources. The past circulates in ways that it never has. So the implications of Archive 3.0 encompass some crucial components of contemporary memory practices:

- how data and information might circulate and be managed, as described above; an elision of research, pedagogy and publication, as non-experts gain access to what was controlled by experts;
- historiography—how we make history using archives, without the necessary mediation of academic expertise;
- the future of the book and of publishing in the new library—the cybrary that extends way beyond books and shelves;
- the figure of the cybrarian—mediating engagement with the archive rather than organizing its components;
- the creative reuse (re-mix, re-generation, re-collection) of archival re-sources.

We pick up a key point, in this nexus that runs through many of the concerns raised by archaeology in a risk society; it concerns authority and agency. In Archive 3.0 we see not just recirculation of the past, but the collaborative co-production of the past. We will elaborate this point below. Here we simply mention some political and ethical issues. What some see as a democratization of information, or at least the opening up of digital databases on the world wide web, intersects with senses of threat to the past as more people realize the ability to collect and contribute to their own and to community pasts, in their own online photo albums, through online genealogy searches, while also being alerted to events such as the looting of the Baghdad Museum, the destruction of the Bamiyan Buddhas, or the bombing of historical monuments in the former Yugoslavia. Peer to peer sharing of music and movies raises the same questions:

- who owns what is left of the past and what is its value?—The politics of cultural property;
- who should be able to use what archival resources?—The politics of access and stakeholder interests;
- who organizes what for whom?—The politics of authority, authenticity, and trust.

In this cultural politics of the constant recirculation of the past, we suggest that archaeology has a distinctive role to play in offering awareness of precisely the possibility of a return of the past in a new guise, creatively remixed and offering innovative view, a concern reframed through connection with the remains of the past.

Consider the pressing concern of sustainable food production. The way we grow our food is connected to several crises in the US and beyond: health, energy and ecology. Health, because most chronic diseases can be linked to diet. Energy, because agriculture and food processing accounts for a significant percentage of greenhouse gas emissions in the United States (a recent UN study estimated rearing livestock produced more greenhouse gas emissions than driving cars (Steinfeld et al. 2006)). Ecology, because high levels of nitrate from industrial agriculture, whether with chemical fertilizers liberally deployed on Iowa farms or with bird cities along the Chesapeake, these industrial sites are creating large dead zones in the oceans.

Drawing attention to these crises, Michael Pollan describes living off the modern industrial food system as akin to 'eating oil and spewing green house gases' (2008). Add up all the pesticides and fertilizers, all the petrol required for farm machinery and long-distance transportation, not to mention all the energy that goes into food processing and contemporary industrial farming. Pollan estimates that it requires ten calories of fossil fuels to generate one calorie of food energy. The message is that soft solutions in the form of cheap calories come with hard consequences.

While this scheme is unsustainably costly, it is, for better or worse, given weight and support by a vast infrastructural, sociotechnical network. Travelling west along Interstate 40 from the east coast of the United States one may note the regionalization of US agriculture. Polyculture farms, formerly a common backdrop in North Carolina and Tennessee have been replaced by tree farms, suburbs and industrial parks. As one moves into Arkansas and Oklahoma huge livestock and monoculture farms become more prevalent. The town grain elevator, set up to move soybeans or corn only, becomes a recurrent feature.

Pollan's answers are centred on a shift back to polyculture farms, which practice careful crop rotation on a solar energy base—of course, with the help of government investment. These are excellent and necessary measures. Still, without addressing the broader question of the shape of sustainable societies over the very long term, these solutions might be reckoned to stopping a full-speed freight train with but a few meters of track (Serres 1995).

It is not the case that archaeological history will be able to find easy answers in the way of a gardenworld economy such as Pollan's. But archaeology does have much to say about early farming and its changes with state management and through into feudalism and beyond that lends considerable nuance and sophistication to any discussion of alternatives to monocultural agribusiness. This is that reservoir of alternatives to which we have alluded.

Archaeology as unforgetting

Perpetual perishing is a universal principle. Through entropy, life, information, and relations drain away. Bequeathed to future generations are radically transformed, heavily fragmented, and partial remains of what was. But we should not think of entropy primarily as a uniform process. Entropy is relational; that is, entropy is replete with utterly specific occasions and events. Entropy does not flow evenly everywhere. Given the right conditions, the right set of relations, ephemera, that would have otherwise wasted sometime ago, may pool in places. At a cave in Armenia, Areni-1, extreme aridity, consistent temperature and a hardened carbonate crust have conspired to perpetuate dried prunes, grape husks, cloth, rope, reeds, wood, and even a child's brain, all ephemeral details of the Chalcolithic some 6,000 years ago (Areshian 2009). In entropy's flow toward a uniform distribution of energy there are spiral eddies, countercurrents and reverse fluxes that maintain more or less coherent forms of what was, even though they may be subject to radical sea-change. Entropy's flow is turbulent and percolating, bringing together different times and separating what once was connected. This, in spatial terms, is a topological folding, just as a landscape, through inhabitation and entropy, comprises a mélange of pasts, presents and future aspirations: an ancient hillfort by a neglected coppice by a new barn storing genetically modified seed.

Archaeology involves lived relations with these nodes and flows through folded temporality, immediately material and intangible (in the implication of all that went into the life of the past, the past's absent presence). Archaeologists work with what remains of what once was above all to explicate, that is to unfold, in the sense just implied. But this work, it should be emphasized, is occasioned by a very particular orientation.

Archaeology's work on what remains of what was may appear sometimes as a discovery or unveiling of a past become hidden. But the percolation of entropy defines archaeology as a field of practices encapsulated by the notion of unforgetting. Unforgetting is not merely the unveiling of a hidden aspect of what was; unforgetting is a risky, continuous, and laborious struggle against entropy. Unforgetting is not only to toil against forgetting; it is to circulate, articulate, crucially to reconnect neglected aspects of what was in order to enrich the present and potentially avert risk in the future. We suggest that there need be more to unforgetting than translating what remains of the past into forms that circulate. In risk society archaeology begins with issues of how the things of the past are caught up within relations today; it then tentatively and cautiously designs stories of what the past was like; archaeology can feed such episodes of unforgetting into reservoirs of alternatives that enable us to think outside the box on matters of common and pressing human concern, as we just tried to illustrate. We have also already mentioned how long-term archaeological thinking is being applied to big questions such as societal stability, environmental change and sustainable living.

But if unforgetting may fill reservoirs that give what has passed renewed presence, how are we to maximise their potential to help creative reframing, and to ensure longevity? How are we to ensure those pasts are open, flexible, malleable, should future generations deem it necessary to deploy them to other ends? This is a classic question of the archive, and one that has been partly addressed in definitions of good archaeological archival practice and publication of, for example, fieldwork. We add to these professional recommendations two characteristics of unforgetting, of archival practices that promise the maintenance of dynamic forging of pasts in the present and for the future, of memory as lived transactions.

The first characteristic is substantial empirical detail that is free from narrative overdetermination. Archaeologists have always delivered catalogues of collections and assemblages that exceed analytical and interpretive efforts to reduce them to coherence, particularly narrative coherence. Consider here Henry Gee's arguments (2002) for evolutionary cladistics in deep time that do not admit narrative treatment. We encourage a celebration of the fascination of collage and assemblage that goes beyond and subverts narrative. This is something very evident in modernist and contemporary art (see, for example, Belknap (2004) on the list in modernist literature, Merewether's edited collection (2006) on the archive in contemporary art). To maintain this empirical richness we should not shy away from indeterminacy and ambiguity (consider Shanks 2004), mess and complexity. Connected with an appreciation of the complexity of the human habitat, we should embrace the multiple ontology of things (Shanks 1999, Olsen et al. 2011).

We suggest an expansion of this classic component of the archaeological project by connecting it with our other characteristic of rich memory practices in the effective archive.

This second characteristic concerns the tacit and the ineffable. Much of our experience and knowledge is tacit, that is, unvoiced, and most is ineffable, that is, irreducible to words alone. We are all aware of the power of sensory and involuntary memory, the way a smell or sound can unexpectedly evoke so powerfully a complete past experience. We suggest that it is not only possible, but crucial to the future of the archive to document our archaeological work and finds in ways that encompass the tacit and ineffable. Antiquaries and archaeologists have always been early adopters of new media technologies in order to capture the qualities sites and artifacts, in the illustrated book, the map, the photograph, the spreadsheet, the electronic database, the 3D CAD reconstruction. We suggest an extension of this to explicitly document the qualities of archaeological experience, the qualities of the human habitat through history. The challenge is not a trivial one of making more videos of an excavation, to be filed away in order that we might experience the site again some day. The circulation that is at the core of unforgetting, the need to constantly reconnect, suggests that we fully acknowledge that any definition of even something as apparently self-evident as a potsherd requires connecting that thing to the milieu from which it has emerged (the terms context and matrix apply here too, though they already carry heavy archaeological connotation). No signal makes sense unless it is decoded as part of a semiotic system, and, we emphasize, except as it is distinguished from background noise. We notice things only as they stand out against the mundane, the everyday, the unexceptional, what is normally overlooked and left tacit, even unrecorded and discarded in archaeology. This noise of history is what makes it live, even as it belongs with an underdramatized history, one that is instead mundane and ambient, rather than constituted by a few familiar historical characters and plots or narratological paradigms (consider Hayden White (1973) on history and Shanks and Tilley (1987) on archaeology; examples of experiment can be found in Stanford University's Metamedia Lab—http://metamedia.stanford.edu). Ambient noise is the source of ambiguity, of an alternate view, of a challenge to the conventional decoding or account (see Serres (1985) on noise and multiplicity).

As an example of a long-term matter in the qualities of the human habitat, consider light pollution. The circadian rhythms of light and dark, day and night, the tones of which were fine tuned over the course of countless millennia. Since the 19th century they have been altered by the proliferation of artificial light. High exposure to light at night has been linked to accelerated tumor growth in women with breast cancer (Chepesiuk 2009). Sea turtles, disoriented by nearby lights, fail to find their way to the sea, and every year in the US alone millions of migratory birds die from collisions with skyscrapers, thrown off course by their bewildering luminosity (Rich and Longcore 2005). In the face of new risks brought on by light pollution, we are now faced with a situation where legion lighting systems must be re-designed to reclaim what was once plainly given (Sloterdijk 2009b). Every new technology modifies subsequent relations with the world and, as Marshall McLuhan put it, 'the message of the electric light was total change' (1994, 52). Light pollution has lead directly to the explication of the formerly taken-for-granted, very long term relations which now need to be conserved and protected. Matters of archaeological concern and endeavor, formerly understood differently or not in the least, enter the fold.

An attention to the qualities of things involves signal-noise relationships, distinguishing figure from ground, an attention to quiddity and haecceity, the whatness and this-ness of things, to appropriate those technical philosophical terms for our purposes here. As we have indicated, this is not new to archaeology; we note with encouragement the growing acceptance of for example, anthropologies of the senses (Stoller 1990, Jones and MacGregor 2002), of notions like ancient soundscapes, of the performative and embodied character of human practice past and present that demands rich and nuanced appreciation and representation (Pearson and Shanks 2001, Hamilakis, Tarlow and Pluciennik 2001). The media and tools are ready to hand: curators and museologists, for example, have long worked with the theatricality of the museum in designing rich modes of embodied engagement involving visualization, diorama, reenactment, annotation (labelling and interpretive text), multiple interfaces, new archival architectures.

New digital media, with their ubiquity and the fungibility noted above, offer multimedia authoring environments that can complement well this quest for rich documentation and explication as well as the recirculation of work that we suggest should lie at the core of an animated archive. The deterritorialization of information and ease of editing and reuse of digital work, also noted above, has prompted debates about a 'creative commons', movements to challenge restrictive copyright and establish a richer public domain of cultural creativity rooted in the reworking or remixing of past works (Lessig 2002 and 2005). Certainly a major feature of recent developments in digital networking, including Web 2.0, involves participatory, peer-to-peer and social software, large scale content management and collaborative authoring environments such as blogs and wikis, or modular systems such as Drupal. This could be treated as a third characteristic of unforgetting: collaboration.

Archaeology as design and craft

What kind of work is archaeology, this work on what remains, this unforgetting, this explication of the deep genealogies of humanity?

Elsewhere we have suggested that archaeology may profitably be seen as craft (Shanks 1992, McGuire and Shanks 1996). The likes of Sennet (2008), Brown (2009) and Latour (2003), have recently written about craft and design in the context of late modernity and we find support a good deal of what they propose. We suggest that a view of archaeology as craft is even more apposite when associated with notions of design practice and thinking. To conceive archaeology as design and craft encompasses much of the argument we have made in this chapter.

Craft implies embodied knowledges applied with an intense awareness, acquired through iterative engagement with materials and making, of the qualities of materials. Design we define broadly as a field of integration, of pulling together whatever is necessary to attend to a problem needing solution, of application of diverse fields of skill and expertise, (typically engineering, psychology, materials science, anthropology) with the interests, needs or desires of an individual or group, of management of this process of making. It is also a field of rhetoric, where arguments are made for a particular solution, where what is designed is frequently an implicit argument for the good life.

Though 'Design' has been conspicuously associated with charismatic designers and sometimes quite grand design philosophies, the actual processes of design thinking and practice are much more mundane. The terms raise all the right questions about modern(ist) attitudes towards materials, labour and artifacts, including those concerning agency and alienation, our involvement with the world of goods today, the role that archaeologists share in designing the past. As craft and design archaeological agency is founded upon engagements with the past, and with others who share such an interest. Both craft and design imply local attention, working in a humble way in a process that will deliver an artifact, but an open process of constant and iterative improvement (we have elsewhere elaborated this point with reference to 'agile management': Shanks 2007).

There is a crucial epistemological point. To conceive archaeology as craft and design frees the archaeologist from striving primarily to achieve an epistemological correspondence between their representations and accounts and the past. Craft and design involve a very different kind of accuracy that nevertheless is founded upon detailed empirics, expertise and knowledge. It is an accuracy that usually involves constantly putting out and sharing trials, seeking feedback and listening to others, rather than striving for a final and definitive account.

Crisis and Actuality

When he wrote his theses *On the Concept of History* in a threatened Paris of 1940, Walter Benjamin was certainly experiencing a profound sense of crisis and danger, one that eventually led to his attempt to flee to the United States and then suicide when it appeared he would not be able to cross the Spanish border. The theses convey in Benjamin's prescient and hybrid imagery of Jewish mysticism and Marxism a good deal of what we have attempted to elaborate in this chapter. At the heart is a historiographical temporality of *actuality*—the conjunction, in a moment of crisis, of past and present, with a view to changing the future. This is *Jetztzeit*, now-time, a conjunctural moment when the continuum of history is blown apart, when we take a stand against empty homogenous time in constructing a unique relationship now with the past. So Thesis VI opens: 'Articulating the past historically does not mean recognizing it the way it really was. It means appropriating a memory as it flashes up in a moment of danger' (2003, page 391), when 'the true image of the past flits by' (Thesis V 2003, page 390)—when historical truth depends upon the work of connection we are calling unforgetting. Historical articulation of this kind requires constant work, because the line of least resistance is for the past to be assimilated into familiar and comforting stories of progress: 'every age must strive anew to wrest tradition away from the conformism that is working to overpower it' (2003, page 391), that is working to forget the power of the past to prompt reflection and action to redeem erstwhile hopes that threaten to be lost in a tide of so-called progress. It is in this context that Geoffrey Bowker, in his book on memory practices in the sciences, cites Yosef Yerushalmi, who wonders whether 'the antonym of 'forgetting' is not 'remembering,' but justice' (2005, page 25).

References

Areshian, G.E. 2009 Breaking Away from Dominant Paradigms: An Unknown World of Near Eastern Chalcolithic Organic Remains from the Cave Areni-1, Armenia paper given at the annual meetings of the AIA in Philadelphia.

Authier, M. 1995 Refraction and Cartesian 'forgetfulness.' In Serres, M. (ed.) *A History of Scientific Thought*. Blackwell Publishers, Oxford, pp. 315-43.

Beck, U. 1992 *Risk Society: Towards a new modernity*. London: Sage.

Belknap, R.E. 2004 *The List: The uses and pleasures of cataloguing*. New Haven: Yale University Press.

Biagioli, M. (ed.) 1999 *The Science Studies Reader*. London: Routledge.

Bowker, G.C. 2005 *Memory Practices in the Sciences*. Cambridge MA: MIT Press.

Brodie, N. and K.W. Tubb 2002 *Illicit Antiquities: the theft of culture and the extinction of archaeology. One World Archaeology* 42. London: Routledge.

Brown, T. 2009 *Change by Design: How design thinking transforms organizations and inspires innovation*. New York: Harper Business.

Canaan, J.E. and W. Shumar (eds) 2008 *Structure and Agency in the Neoliberal University*. London: Routledge.

Clack, T. and M. Brittain (eds) 2007 *Archaeology and the Media*. Walnut Creek, CA: Left Coast Press.

Clarke, D. 1973 Archaeology: the loss of innocence. *Antiquity* 47:6-18.

Deleuze, G. 2002 Postscript to control societies, in *CTRL SPACE: rhetorics of surveillance from Bentham to Big Brother*. Edited by T.Y. Levin, U. Frohne and P. Weibel. Cambridge MA: MIT Press.

Diamond, J. 2005 *Collapse: How societies choose to fail or succeed*. Penguin.

Edgerton, D. 2007 *The Shock of the Old: Technology and global history since 1900*. Oxford; New York: Oxford University Press.

Eiland, H. and M.W. Jennings (eds) 2003 *Walter Benjamin: Selected writings*. Volume 4: 1938-1940. Cambridge MA: Harvard University Press.

Foucault, M. 1972 *The Archaeology of Knowledge*. London: Tavistock.

—— 1973 *The Order of Things: An archaeology of the human sciences*. New York: Vintage Books.

—— 1980 *Power/Knowledge*. Brighton: Harvester.

—— 1986 Nietzsche, genealogy, history, in *The Foucault Reader*. Edited by P. Rabinow. Harmondsworth: Penguin.

Giddens, A. 1984 *The Constitution of Society: Outline of the theory of structuration*. Cambridge: Blackwell Polity.

—— 1991 *Modernity and Self-identity: Self and society in the late modern age*. Cambridge: Polity Press.

Gouldner, A.W. 1970 *The Coming Crisis of Western Sociology*. New York: Basic Books.

Greene, K. 1990 *The Archaeology of the Roman Economy*. Berkeley: University of California Press.

Harman, G. 2009 *Prince of Networks: Bruno Latour and Metaphysics*. Melbourne: re.press.

Ingold, T. 2009 The textility of making. *Cambridge Journal of Economics*. Online version.

Jones, A. and G. MacGregor 2002 *Colouring the Past: The significance of colour in archaeological research*. Oxford, UK; New York: Berg.

Kuhn, T. 1970 *The Structure of Scientific Revolutions*. Chicago: University of Chicago Press.

Latour, B. 2008 A cautious Prometheus? A few steps toward a philosophy of design (with special attention to Peter Sloterdijk) in *Networks of Design. Design History Society*. Falmouth: Cornwall UK.

McLuhan, M. 1994 *Understanding Media: The Extensions of Man*. Cambridge, MA: The MIT Press.

Mannheim, K. 1952 *Essays on the Sociology of Knowledge*. London: Routledge and Kegan Paul.

Mannheim, K., L. Wirth, and E. Shils 1936 *Ideology and Utopia; An introduction to the sociology of knowledge*. London, New York: K. Paul, Trench, Trubner; Harcourt, Brace.

Merewether, C. 2006 *The Archive*. London; Cambridge MA: Whitechapel; MIT Press.

Olsen, B., M. Shanks, T. Webmoor and C.L. Witmore 2001 *Archaeology: The discipline of things*. Berkeley, CA: University of California Press.

Olsen, B. and A. Svestad, 1994 Creating prehistory. Archaeology museums and the discourse of modernism, *Nordisk Museologi* 1994(1), 3-20.

Pollan, M. 2009 Farmer in Chief. *The New York Times Magazine*. Available at: <http://www.nytimes.com/2008/10/12/magazine/12policy-t.html>.

Rankin, G.A. 1916 Portland cement. *Journal of the Franklin Institute* 181, 747-84.

Rathje, W., M. Shanks, and C. Witmore 2010 *Conversations through Archaeology*. Berkeley, CA: Chicago University Press.

Redman, C. 2001 *Human Impact on Ancient Environments*. University of Arizona Press.

Renfrew, C. 2000 *Loot, Legitimacy and Ownership: The Ethical Crisis in Archaeology*. Duckworth.

Rich, C. and T. Longcore (eds) 2005 *Ecological Consequences of Artificial Lighting*. Island Press.

Sennett, R. 2008 *The Craftsman*. New Haven: Yale University Press.

Serres, M. 1995 *The Natural Contract* (translated by E. MacArthur and W. Paulson). Ann Arbor: The University of Michigan Press.

Shanks, M. 1992 *Experiencing the Past: On the character of archaeology*. London: Routledge.

—— 1999 *Art and the Early Greek State: An interpretive archaeology*. Cambridge: Cambridge University Press.

—— 2004 Three rooms: archaeology and performance. *Journal of Social Archaeology* 4:147-80.

—— 2007 Digital media, agile design and the politics of archaeological authorship, in *Archaeology and the Media*. Edited by T. Clack and M. Brittain. Walnut Creek, CA: Left Coast Press.

—— 2010 *The Archaeological Imagination*. Walnut Creek, CA: Left Coast Press.

Shanks, M. and C. Tilley 1987 *Social Theory and Archaeology*. Cambridge: Blackwell Polity.

Sloterdijk, P. 2009a *Terror from the Air* (tanslated by A. Patton and S. Corcoran) Semiotext(e).

—— 2009b Excerpts from *Spheres III: Foams*. *Harvard design magazine* 29, 38-52.

Smeaton, J. 1793 *A narrative of the building and a description of the construction of the Edystone Lighthouse with stone: to which is subjoined, an appendix, giving some account of the lighthouse on the Spurn Point, built upon a sand*. London: H. Hughs.

Steinfeld, H., P. Gerber, T. Wassenaar, V. Castel, M. Rosales and C. de Haan 2006 *Livestock's Long Shadow: Environmental issues and options*. Rome: Food and Agriculture Organization of the United Nations.

Thomas, J. 2004 *Archaeology and Modernity*. London: Routledge.

Trigger, B. 1984 Alternative archaeologies: nationalist, colonialist, imperialist. *Man* 19:355-70.

—— 1989 *A History of Archaeological Thought*. Cambridge: Cambridge University Press.

Webmoor, T. and C.L. Witmore 2008 Things Are Us! A Commentary on Human/ Things Relations under the Banner of a 'Social' Archaeology. *Norwegian Archaeological Review* 41:1, 53-70.

White, H. 1973 *Metahistory*. Baltimore: Johns Hopkins University Press.

Zielinski, S. 2006 *Deep Time of the Media: Toward an Archaeology of Hearing and Seeing by Technical Means*. Cambridge, MA: The MIT Press.

Chapter 16
Archaeologies and Geographies of Value

Gail Higginbottom and Philip Tonner

> The value of rhetoric as a meta-critical perspective *resides* in its care for *cultural plurality*, its modest, pragmatic willingness to recognise irreconcilable cultural personae, critical genres, and affective stimuli as elements of a *cultured understanding*…(R)hetoric enables us to ask questions about the state of political discussion and public-sphere conversation, to critique the discursive *quality of historical understanding*…
>
> Curthoys 2001

Philosophical Base

This chapter illustrates, along with Curthoys, Deleuze and Guattari, 'that every mode of thinking, every school of thought, needs to account for the 'plane of immanence' upon which it operates', for it is through this account that a clearer and more workable philosophy emanates (Curthoys 2001: § 'Deleuze', Deleuze and Guattari 1994). The 'plane of immanence' is the unstructured plane out of which an individual's or group's thought and concepts are created. The plane of immanence does not produce concepts that are determined in advance; yet developing a knowledge of the various elements that came into prior-existence to allow for the Formation of any thought or School of Ideas in the first place is possible only by reference to such a plane. 'Planes of immanence' contain 'many voices'; they are 'composed of speeds and slownesses, movements and rest' (Deleuze and Guattari 2004: 298). In archaeology today we use the term 'Reflexive Archaeology' (Hodder 1999 and Salzman 2002). Reference to 'places of immanence' imply that a competent form of Reflexive Archaeology will recognise that 'conceptual thinking needs to retain a multifarious 'sense' of what it is doing, the kinds of problems it addresses and the cultural context it seeks to influence and is influenced by' (Curthoys 2001: § 'Deleuze'). In this way, we recognise a reflexivity that, as Hodder describes it, involves 'recognizing the value of multiple positions, and multivocality (2003: 58)'. *Reflexivity is social.*

We feel that it is essential to seek an understanding of those 'planes of immanence' connected to notions of heritage in order to comprehend the many forms heritage takes and to gain an appreciation of how we may face the challenges of cultural heritage management in the future.

We begin our search, in order to understand this, by way of the discussion below.

Where Are We?

The issue

In the 'risk society' everything is in question, not least the complex series of relationships intersecting the debates surrounding cultural and natural heritage, archaeology itself and the manifold issues bound up with the notion of a 'common world'. We argue that a comprehensive understanding of heritage pivots upon the fundamental belonging together of cultural and natural heritages and that human/communal relations to such heritage 'objects' will be complex and plural. For this reason the geographer of values must begin from the recognition of a plurality of valuations of the manifest phenomena of the world, and in doing so, will question the very foundation of a possible common world as a future possibility for humanity. Further, such plurality of values is temporally underscored and thus for the 'risk society' the complexity of human appropriations regarding the past, the present and the future can seem to be an interminable barrier to political unity and consensus. Given these facts, the project of charting the multiple *landscapes of values* becomes a philosophical and political necessity, not to create the common world, if that is taken to be something homogenous, but to aid in the understanding and establishing of some common goals.

A divided world

In this chapter, we look at division. Specifically, we look at the divide between cultural and natural heritage, and suggest that perhaps it does not actually exist. Along with the analytic and dialectic traditions of critical reflection we recognise that there is an incorrect application of dualistic notions leading to erroneous decisions and behaviours, whether as researchers, managers or operating in the everyday.

In the history of philosophy, broadly speaking, discussions surrounding the issues of nature and culture have emerged as commonplace. Some of the earlier European influences include Plato's *Gorgias* and *Protagorus*, Aristotle's ideas on the teleology of Nature (*passim*), St. Augustine's *City of God*, Aquinas' *Of Being and Essence* and *The Principles of Nature*, and Descartes' *Discourse on the Method of Rightly Conducting the Reason* and *Principles of Philosophy*. In the German tradition this can be seen from the Enlightenment onwards with Herder's focus on culture (out of which anthropology is said to develop), to the Romantics, Idealists and the master of suspicion Nietzsche and his *Birth of Tragedy* of 1872 (Denby 2005 and Kauffman 2000). These ideas continue into the 20th century with Heidegger's *The Origin of the Work of Art*, reworking Nietzsche's forces of duality (Dionysus and Apollo) with his World and Earth (Kauffman 2000, Heidegger 1993 and Polt 1999).

Today, work specifically questioning the unnecessary divisions of nature and culture have continued to be *moralised* and can be quite heavily politicised, as emphasised by publications like the journals *Environmental Ethics* and *Ethics and the Environment*. In the late 20th century works were written that emphasised both the place and importance of Nature and/or critiqued those traditions that have devalued it or felt that its 'form' has been misunderstood (Loy 1999 and Vogel 1996). More importantly, for our work, are those papers that emphasise the relevance of humans as part of that nature, such as in Ferré's 'Persons in Nature' (1996) and Kato's 'Community, Connection and Conservation' (2006). Like Nagel's work, these works emphasise that there is no *View from Nowhere* (1986).

Demonstrating the Fundamental Belonging Together of Cultural and Natural Heritages

What is cultural heritage?

In order to recognise the plural and perspectivistic human cultural universe which is composed of a series of more or less complete cultural wholes, we take our point of departure from the *contemporary* anthropological notion that the term 'culture' should really be replaced by 'cultures' (see Barnard and Spencer 1996, Herder *passim* and Rapport and Overing 2000: 92-102).

We also argue that any 'culture' is, in opposition to quasi-structuralist synchronic analysis, fundamentally *historical* (Barnard and Spencer 1996, 138). That is, any culture is at once composed of past, present and futural temporal modalities. Further, from a Heideggerian perspective, cultural communities remain vibrant by engaging in creative activities that produce artworks that establish self-referential circuits of interpretation (Polt 1999, Tonner 2006). From the point of view of a phenomenological-hermeneutic understanding of human being, human beings are understood as *historical* agents who produce diverse ideological and material artefacts. By agents' appropriative acts these artefacts become constitutive of the heritage of particular groups. Thus cultural 'heritage' is produced by the general structure of cultural production: agents create culture by their productive activity and by appropriative acts they constitute their shared heritage. Such cultural creation, from a functional point of view, attempts to establish a point of stability from which individual and communal interpretation can project into a shared future. It is within such a temporal network that the meaning of the term 'heritage' must be decided. To this extent, *cultural production constitutes cultural heritage*.

There can be no separation, then, of the notion of 'heritage' from the notion of the culturally produced and it is a fundamental assumption of our approach that, from a phenomenological-hermeneutic point of view, cultural production and the very essence of human existence is temporally underscored (Heidegger 1962). Thus, there can be no division of these things when we realise that all that we are and do is bound by the temporality of human existence and the temporality of action; where past, present, and future interact in the production of the meaningfulness of things.

What is natural?

It is our view that human existence has to be understood in the context of being-in-the-world in the Heideggerian sense. Heidegger's basic starting point for his hermeneutic phenomenology was an understanding of the agent's basic state as being-in-the-world (Heidegger 1962). In this chapter, we take it that being-in-the-world is the basic state of human existence and that, as such, there can be no ontological separation of 'being', 'in' and 'world'. As *existence*, human beings are being-in-the-world. Given this, what is taken as 'natural' and 'cultural' has to be approached from this point of view.

Interestingly, a complementary theoretical move has recently been made by Tim Ingold in his 'The Temporality of the Landscape' when he states that 'social or cultural anthropology, biological anthropology and archaeology form a necessary unity…they are all part of the same intellectual enterprise' (Ingold 2000: 510). The reason this move is complementary is that it insists on a holistic approach in what we can call generic human science. Similarly, our approach calls for a holistic approach that takes its point of departure from the agent's basic state of being-in-the-world. Being-in-the-world is the basic state of what Heidegger calls Dasein (being-there). The term Dasein, as we are employing it, can be applied to any human agent and its constitutive state of being-in-the-world insists that in order to gain an understanding of 'the agent' you must also gain an understanding of its generic 'context'. That is, to understand any agent it is also necessary to engage with the socio-historic and natural environment within in which that agent acts.

Ingold's move, from our point of view, constitutes an insistence that the generic context of Dasein is pursued through a mutually supporting and complementary enterprise by placing the agent together with its cultural, social, historical and biological milieu in question. And to this extent, there can be no *a priori* separation of what is 'culture' or 'cultural' and what is 'nature' or 'natural'. Any approach to 'the human' will from the outset problematise these divisions.

Nietzsche provides an early clue to our problematic. He suggests in his *Homer's Contest 1872:*

> When one speaks of humanity, the idea is fundamental that this is something which separates and distinguishes man from nature. In reality, however, there is no such separation: 'natural' qualities and those called truly 'human' are inseparably grown together. Man, in his highest and noblest capacities, is wholly nature and embodies its uncanny dual character. Those of his abilities which are terrifying and considered inhuman may even be the fertile soil out of which alone all humanity can grow in impulse, deed, and work (in Pearson and Large 2006: 95).

Thus, humans, and all they are and create, the idea of the 'cultural' itself, can not be divided from the idea of 'Nature', they and theirs are a part of Nature, and being Natural, they co-exist on the same material plane as all other Natural 'Things' and 'Events'. Therefore, you *can't* usefully explain human beings in abstraction from their environment and this includes what people often refer to as the 'natural environment'. Recognition of this is recognition of our point of departure, the primordial unity of being-in-the-world.

It follows from this that natural and cultural heritage have to be regarded as fused. Cultural and natural heritage are fundamentally joined and not only that, both are bound up as a '*site*' where a historically configured/negotiated people dwell. Thus, when approaching the question of heritage there are immediately three variables in play: nature, culture and 'people' and all three are historically negotiated, and all of this can be approached as part of the Natural.

Application of Cultural and Natural Heritages

We emphasise these important ideas because they are fundamental to the way we choose to think and deal with what we now call *Cultural* Heritage, and note that whilst Nietzsche recognised this interplay in 1872, it wasn't until 1992 that UNESCO could acknowledge that 'significant interactions between people and the natural environment have been recognized as cultural landscapes' (http://whc.unesco.org/en/criteria/; accessed 5/08/06).

Though such statements show an improvement in understanding something of humanity's heritage or what could be considered as heritage, UNESCO are still failing to see what heritage *is* in relation to people and their environment. By doing so they further limit their capacity to consider or *explain* what is possibly important or has possible relevance or is possibly of value to people. Such a statement has ignored, or put aside, the issue that humanity and their environment and their interactions *are all in fact parts of the cultural landscape*; and this includes people. This is very perplexing given the great amount of time, effort and application UNESCO has put into the research and effect of the recognition of such things as 'intangible' and 'tangible' heritage (http://portal.unesco.org/culture/en/; last accessed 24/03/07). The effect of such lack is evident when the management of heritage comes into play.

Being, experience, and the meaning of place

Kato's 2006 paper is pivotal in the recognition of the ontological identification of being and experience in heritage and highlights the connectivity of people with place on the existential and phenomenological levels.

Shirakami-Sanchi the mirror: Where nature is, we are

Kato voices her concern of the protected so-called natural areas, like World Heritage Sites, which essentially exclude human presence; according to Kato, 'in this rhetoric, 'Natural Heritage' tends to be situated in the context of dichotomous nature-culture or nature-contemporary human' (2006: 459). This paper tells us, that due to the strong connection the local community of Shirakami-Sanchi had with their surroundings, they fought to obtain World Heritage status for the area in order to prevent further destructive incursions. The irony is that through the local communities' successful acquisition of a World Heritage area, they were eventually denied *their* own *place* within it. These very same people, whose sense of connection can be described as one of mutuality and reciprocity with Shirakami, could no longer carry out traditional, minor subsistence activities in the protected area, such as the harvesting of naturally occurring produce (2006: 464). We say minor, as this is the technical descriptor for activities not essential for basic biological survival. However, such terms in no way connect to or cover the issues of cultural importance and, possibly, not even well-rounded nutrition (and again we see those troublesome divisive adjectives that separate the well-being of humans).

Specifically, for example, the social grouping called the *matagi*, who spend four years learning all of the practices, related rituals, languages and taboos associated with their position, could no longer practice their self-regulatory systems of living with the mountains; these include restoration and regeneration activities—planting, seeding, weeding, as well as the hunting of six or seven bears a year in total (2006: 464). Oral reports tell us that *matagi* and community members felt that the mountain needed them (2006: 465). The loss of interaction is clearly evident:

> We fought to save the area from destruction, but now the area we saved excludes us. For us not being in the mountain is like being told not to go outside at all. Shirakami has never been untouched. People always lived with it. It was always our ordinary mountains (2006: 466).

It is obvious that people have been left out of the equation resulting in an erroneous calculation and management of a place. For, naturally, incorrect equations result not only in the mismanagement of the variables that were left out (people) but also the remaining dependent variables and the entire outcome. A case in point is the inappropriate care and consideration of the other animals in the area and the interactions between them and peoples. Once the protection status was applied 450 bears were killed in 2003 'under a pest-control scheme in Akita when they came down to the residential areas.' Previously bears were not known to come into the villages and the number taken was very low indeed (six or seven as mentioned above). Now the animals that were once considered as part of the ecology of the place (both by the World Heritage status and the population) are now 'pests' to be dealt with and not an animal community that has an interactive part to play with the environment and the people. As reported by Kato, a *matagi* is quoted as saying this form of management is not only 'outrageous but seriously worrying (2006: 466)', for according to this man the population had been quite stable over the last 20 years.

How does mismanagement happen?

In a practical sense, it is as Kato states—Shirakami is now a place regulated by a number of external managing bodies with their regulations which restrict traditional practices but allow in tourists; we would add that they are managing without the consideration of the tripartite whole (culture-nature-people) that makes up heritage. Further, it is not clear why World Heritage committee did not offer or insist on giving the Shirakami area (whether or not it was asked for in the original application) the added criteria of cultural inclusion (as defined by them) as part of its status, namely:

> to be directly or tangibly associated with events or living traditions, with ideas, or with beliefs, with artistic and literary works of outstanding universal significance (Criterion vi, http://whc.unesco.org/en/criteria/, last accessed 24/03/07).

Indeed this criterion (number six) could have been used with criterion seven, the criterion the area did gain; for the Committee considers that this criterion (six) should preferably be used in conjunction with other criteria. Criterion seven is:

> to be outstanding examples representing significant on-going ecological and biological processes in the evolution and development of terrestrial, fresh water, coastal and marine ecosystems and communities of plants and animals (criterion vii, http://whc.unesco.org/en/criteria/, last access 24/03/07).

Obviously people were not considered as part of this landscape, despite the intangible cultural heritage that had played a vital role in the nature conservation of Shirakami-Sanchi. They were denied the very access that gave them the sense of connection with their surrounding nature—to quote one woman: 'we feel so grateful because it has always been there' (Kato 2006: 465).

And whilst the area now meets the UNESCO criterion number seven, inappropriate management has made the designated area largely culturally defunct for the area no longer so clearly adheres to criterion six. The world for these people has been turned upside down—and a once *functioning* 'place' is lost. What can be considered lost for these people includes:

- an environment tangibly associated with events or living traditions,
- knowledge,
- appropriate local management of land and animals,
- balance of human and animal behaviours.

We have observed, then, the fundamental problematic of attempting to divide the tripartite entity of culture or heritage by management 'reaffirming (the) externality of nature with, and within, which societies are inextricably intermeshed' (Kato 2006: 467).

Recognition of unity and plurality

Stated again, though differently, this is the result when people do not treat the world and its inhabitants as an organic, interactive whole (and by this we mean specifically the culture-nature-people paradigm). We are not saying that the world is a closed system. What we are saying is that the world is composed of completely intermingled, beneficial and non-beneficial reciprocating elements.

Thinking of the world in such reciprocal terms allows us to understand that each and every thing forms part of a unity. However, we must be cautious for such an understanding might fool us into thinking that because we have this unity, we may be able to act and think with a common voice and create a united actuality, a recognised 'one voice' of humanity. But whilst the world is a unity, we, its component actors, are divided. The very foundation of a possible 'common world' as a future possibility for humanity may not be possible.

The Geographer of Values

What are geographies of values?

Nevertheless, this does not mean we should fall into some nihilistic universe of self- or universal-loss. Rather this gives us the understanding of *what* and *where* we are, to look outwards, which in turn helps us to understand *who* we are and what we can do, to look inwards. Such knowledge gives us the ability to comprehend our situation and ourselves; that is our past, present and future, and the possibility of communicating this to others. As well, it allows us to comprehend, somewhat, aspects of the same predicament in others, in our own area or another. In this way, we can work towards comprehending the 'plane of immanence' upon which we operate as 'encultured' beings. Such an awareness and appreciation are fundamental if we are to create, chart and use the landscapes of values. Such a map will allow us to explore what heritage is in general and relate this to the specific case or cases in question.

Creating the *geography* of values here involves negotiating the cultural-historical and the geo-political, and our schematic of nature—culture—people has to be *multiplied* to accommodate each interest group for whom heritage matters. In agreement with Shanks, heritage can include virtually 'anything' thus this threefold determination of the question surrounding heritage, must be fundamentally malleable (2005: 221). That is, what counts as 'natural' heritage and 'cultural' heritage has to remain open to question, to which there are possibly an infinite number of answers. This is so, too, for the identity of the 'people' in this tripartite-looking structure for whom these questions can also be posed.

As *values* are the fundamental issue in public archaeology (as Shanks also suggests), in order to *engage in the debate* over heritage, a comprehensive *geography* of values must be 'mapped-out' using these descriptions. By recognising and acknowledging that there is a *multiplicity* of 'peoples', 'places' and 'objects' constituting this cartography we are then faced with the recognition that it is the plural '*geographies* of values' that we are mapping.

By allowing us to look inward and outward and to give some understanding of others, the provision of such a map is the prerequisite, or the condition of possibility, of raising the question of the possibility of the emergence of an agreed notion of heritage, never mind of a common world. Thus, the 'geographies of values' project is fundamentally practical and political. The geographer of values begins from the *recognition* of a plurality of valuations of the manifest phenomena of the world and raises the question of the very foundation of a possible 'common world' as a future possibility for humanity.

And whilst we problematise this possibility for the reasons stated in the section above, we feel it is important to offer a technique for thinking the numerous possibilities. Through this way of thinking and through the acceptance of plurality and public debate, another answer might be born that defies our qualified one in this chapter.

Geographies of values, ad infinitum

Part of what is required, then, is a methodology for charting these geographies of values, and for this methodology to be ampliative, as required by such a multiplicity of outcomes, the descriptive method is best; specifically, the *description* of each set within our schematic of nature—culture—people. These contain the infinite answers to those questions raised by this threefold determination of heritage: those questions of who and what we are that metamorphose into things like 'what does this place mean to me, or to them, or for whom does this place have meaning?'

From these things, a broadly phenomenological approach seems to be a candidate early on. With the multiple views that geographers must collect, and those that they have themselves, there is no 'view from nowhere' (Nagel 1986). So we must ask ourselves from whose perspective should this phenomenology of the landscape of values be written?

Turning somewhat to our own state of affairs in the 21st century, Garden (2006) has argued that the most visible, accessible and tangible manifestations of heritage are 'heritage sites', and holds that there is some kind of meaningful consensus as to what is a 'Heritage site' today. However, things have to be meaningful to a community and to the geographer as intangible heritage before there is any meaningful consensus about what is a heritage site/object. Thus, the 'transcendental' question that we are raising is fundamentally important since it provides for the possibility of this meaningful discourse.

If it is impossible to understand what is going on, or what is 'up for discussion', then it is impossible to know what past choices and understandings have had a hand in creating the culture that constitutes the 'heritage site'. Further to this, just because a site has little relevance now does not mean it wont have any for future generations, who could well be astonished as to why we allowed the destruction of places that presently 'do not appear to have any value to anyone'. To these extents, any projective evaluations are rendered *a priori* impossible and once more we feel the force of the temporal-historical dimension to the geographies of values and that the consideration of heritage contains all tenses: past, present and future.

In this regard the work of Keith H. Basso proves instructive. In his 1996 study he reminds us that, with reference to the Western Apaches:

> Losing the land is something the Western Apaches can ill afford to do, for geographical features have served the people for centuries as indispensable mnemonic pegs on which to hang the moral teachings of their history (Basso 1996: 62).

Here, geography, people, history, time and values are intertwined. In this connection also, Basso invokes Bakhtin's notion of the chronotope. As Bakhtin puts it, chronotopes can be understood as:

> points in the geography of a community where time and space intersect and fuse. Time takes on flesh and becomes visible for human contemplation; likewise, space becomes charged and responsive to the movements of time and history and the enduring character of a people…Chronotopes thus stand as monuments to the community itself, as symbols of it, as forces operating to shape its members images of themselves (Bakhtin 1981 in Basso 1996: 62).

At once the geographer of values task is set and made infinitely complex. That is, to map the diverse *chronotopes* representative of the manifest 'cultures' of the contemporary world. However, just such considerations of the possible future evaluations of sites/objects continue to render the coherence of a common world problematic. This is so since questions over the possible future value of sites/objects can, in principle, go on forever.

Having thought about the multiple views that geographers must collect, we can see that the geography of value also requires a hermeneutic (interpretive) rendering—making explicit the prejudices (pre-judgements) of the geographer as an integral part of the project. These are all part of our *cultured understanding* as mentioned by Curthoys (2001). To do this we need to come to grips with our own prejudices, rendering them transparent or explicit, thus allowing us the possible opportunity to move towards a consensus of understandings of *others*, of what constitutes a heritage site and hopefully the appropriate management of the tripartite 'site'.

Cultured understanding: A philosophical and political necessity

Without the ability to look inward and outward, the lack of recognition of our own and others' cultural heritage and the loss of communicating our own cultural heritages to others is assured. The possibility of creating, charting and using the geography of values would likely be denied if this were the case, and certainly any international decisions of assigning World Heritage status and the management of areas could inadvertently under-value such heritages' worth, as we think has happened at Shirakami-Sanchi and no doubt in many other places. Such activities, we believe, are allowed because of the lack, or imbalance, of looking outwards and inwards: of critically evaluating who we are and what we do and the resultant cultural forms these create. It is with these creations, and each-other, that we must engage as individuals and communities, and recognise that *plurality is an essence of Unity*. Be these variations in religious or political attitudes, they must be identified and mapped in order that we and others can have the opportunity to be aware of and know them as a first step towards them being more fully understood. The actual existence and use of this resource by others may help them to reconsider their own understanding of themselves and others in 'this world' and finally help us reconsider the ways of administering a 'site'.

In following Curthoys' critique of rhetoric, we see that the geography of values resides in its care for *cultural plurality*; *that it contains* cultured understanding. Further, the creation and use of the geography of values will not only assist us in the articulation and communication of heritage but will allow us 'to ask (informed) questions about the state of political discussion and public-sphere conversation, to critique the discursive *quality of historical understanding*' of what is heritage (Curthoys 2001 § 'Benjamin'). In this way, we can constantly reassess the state of play of heritage as is inherent in the model of geography of values where 'questions over the possible future value of sites/objects can, in principle, go on forever'. All of this sits comfortably with Deleuze and Guattari's concept of knowing of the 'plane of immanence' which, they argue, 'is the complex *ongoing conversation*, the dilemma, the received history of fraught questions that one intuitively recognises as a formative background for one's own critical enunciations' (Deleuze and Guattari 1994: Chapter 2 in Curthoys 2001).

Plurality of values: The actuality and the good

In this vein of recognising and accepting plurality we, similar to Nietzsche, *desire the affirmation of difference* for public good. And like Nietzsche, we hold that a vibrant community can only exist if it engages in rhetorical debate (Nietzsche 1895 in Curthoys 2001). According to Curthoys, Nietzsche, 'in his celebration of the sociably sympathetic condition of genuine communication and intellectual comportment' theorises a cultured, discursive 'completion' of the concept as a public-intellectual task (Nietzsche 1887, 1888 in Curthoys 2001). It is in the spirit of this, recognising and embracing difference and encouraging public debate, that we propose the geographies of values.

Ultimately, it is in cultural production that human beings first glimpse themselves and it is in cultural production that 'we' co-perform and verify 'who we are'. Such realities are projected into the future and can be caught by all of us as heritage. And it is heritage that can begin to be understood by mapping the tripartite Unity of culture-nature-people. At the end of the day, it is the identity of the people that is co-performed in the production of the heritage site.

References

Aquinas, T. 1944 *Basic Writings of Saint Thomas Aquinas*, Vol. 1, (ed.) A.C. Pegis, London: Random House.
—— 1961 *Commentary on Aristotles Metaphysics*, trans. J.P. Rowan, Notre Dame, Indiana: Dumb Ox Books.
—— 1965 *Selected Writings—[The Principles of Nature, On Being and Essence, On the Virtures in General,* and *On Free Choice*]. trans. R.P. Goodwin, Indiana: Library of Liberal Arts, Bobbs-Merrill.
—— 1998 *Selected Writings*, (ed.) and trans. R. Mcinerny, London: Penguin.
Aristotle 1961 *Physics*, trans. R. Hope, Nebraska: University of Nebraska Press.
—— 1963 *Categories and De Interpretatione*, trans. J.L. Ackrill, Oxford: Clarendon Press.
—— 1998 *The Metaphysics*, trans. H. Lawson-Tancred, London: Penguin.
Augustine 2003 *City of God*, trans. H. Bettenson, London: Penguin.
Bakhtin, M.M. 1981 *The Dialogic Imagination: Four Essays*, cited in, Basso, K.H. 1996 *Wisdom Sits in Places*, Albuquerque: University of New Mexico Press.
Barnard, A. and Spencer, J. 1996 *Encyclopaedia of Social and Cultural Anthropology*, London: Routledge.
Basso, K.H. 1996 *Wisdom Sits in Places*, Albuquerque: University of New Mexico Press.
Curthoys, N. 2001 Future Directions for Rhetoric—Invention and Ethos in Public Critique, in *Australian Humanities Review*, Available at: <http://www.australianhumanitiesreview.org/archive/Issue-April-2001/curthoys.html>/. Accessed: 2 September 2009.
Deleuze, G. and Guattari, F. 1994 *What is Philosophy?*, trans. G. Birchill and H. Tomlinson, New York: Verso Books.
Denby, D. 2005 Herder: culture, anthropology and the Enlightenment in, *History of the Human Sciences*, No. 1, 55-76.
Descartes, R. 1968 *Discourse on Method And The Meditations*, trans. F.E. Sutcliffe, London: Penguin.
Ferré, F. 1993 Persons in Nature: Toward an Applicable and Unified Environmental Ethics, in *Zygon*: *Journal of Religion and Science*, Vol. 28, No. 4, December 1993, pp. 441-53.
—— 1996 Persons in Nature: Toward an Applicable and Unified Environmental Ethics revised from *Zygon* 28:4, in *Ethics and the Environment,* Vol. 1, No. 1, Spring 1996, pp. 15-25.

Garden, M.C.E. 2006 The Heritagescape: Looking at Landscapes of the Past in, *International Journal of Heritage Studies*, Volume 12, Issue 5, 2006, pp. 394-411.

Heidegger, M. 1962 *Being and Time*, trans. J. Macquarrie and E. Robinson, Oxford: Blackwell.

—— 1971 The Origin of the Work of Art, in *Poetry, Language, Thought,* trans. A. Hofstadter, Harper and Row.

—— 1978 *Basic Writings*, (ed.) D.F. Krell, Routledge.

Hodder, I. 1999 *The Archaeological Process: An Introduction*, Blackwell, Oxford.

Ingold, T. 2000 The Temporality of the landscape, in *Interpretive Archaeology, a Reader*, (ed.) J. Thomas, Leicester: Leicester University Press.

Kato, K. 2006 Community, Connection and Conservation: Intangible Cultural Values in Natural Heritage—the Case of Shirakami-sanchi World Heritage Area in, *International Journal of Heritage Studies*, Vol. 12, 5, 2006, pp. 458-473.

Loy, D. 1999 *Nonduality: A Study in Comparative Philosophy*, Amherst, New York: Humanity Books.

McIntosh, E. 1964 *Concise Oxford English Dictionary*, Oxford University Press, Oxford.

Nagel, T. 1986 *The View from Nowhere*, Oxford University Press, USA.

Nietzsche, F.W. 1994 *Basic Writings of Nietzsche* (Modern Library), trans. W. Kaufmann, Random House USA Inc; Modern Library Ed edition.

—— 1999 *The Birth of Tragedy and Other Writings*, (ed.) R. Geuss and R. Speirs, Cambridge: Cambridge University Press.

—— 2006 Homers Contest (1872), in, *The Nietzsche Reader*, (ed.) K. Ansell. Pearson and D. Large, Oxford: Blackwell.

Plato 1997 *Complete Works*, (ed.) J.M. Cooper and D.S. Hutchinson, Hackett.

Polt, R. 1999 *Heidegger, An Introduction*, UCL Press.

Rapport, N. and Overing, J. 2000 Culture, in *Encyclopedia of Social and Cultural Anthropology. The Key Concepts*, Routledge.

Salzman, P.C. 2002 On Reflexivity in, *American Anthropologist*, 104(3):805-813.

Shanks, M. 2005 Public archaeology/Museology/Conservation/Heritage in C. Renfrew and P. Bahn (ed.) *Archaeology: The Key Concepts*, Routledge.

Tonner, P. 2006 The Univocity of Being, with special reference to the doctrines of John Duns Scotus and Martin Heidegger, PhD. The University of Glasgow.

UNESCO 2009 The Criteria for Selection Available at: <http://whc.unesco.org/en/criteria/>. Accessed: 2 September 2009.

—— 2009 Intangible Heritage, Available at: <http://portal.unesco.org/culture/en/ev.php-URL_ID=34325&URL_DO=DO_TOPIC&URL_SECTION=201.html>. Accessed: 2 September 2009.

Vogel, S. 1996 *Against Nature: The Concept of Nature in Critical Theory*. Albany: State University of New York Press.

Chapter 17
The Stuff of Dreams: Archaeology, Audience and Becoming Material

Angela Piccini

> [username1]: Tell us more about Warburton, [username2]. What did you find? [username2]: 'Roll Eyes' Well you know better than to ask me that one!!! I am obviously going to say wait and see…It was a great three days and the programme will be all the better without any clues before broadcast. 'Valentine'
>
> Having a *Time Team* programme about my site was a dream come true…

Introduction

Questions of 'public engagement' in the context of archaeology and the historic environment have been transformed over the past decade by what has been termed the 'New Managerialism' (Deem 2004). Numerous initiatives commissioned by the UK's Department for Culture, Media and Sport have sought to audit the 'public value' of the past by compiling and analysing visitor information for museums, historic sites and other cultural attractions (eg., DCMS 2006; English Heritage 2006; Travers 2006). *Heritage Counts 2006*, English Heritage's annual look at England's historic environment, concludes that major barriers to public engagement persist in the historic environment (English Heritage 2006). Despite the museums sector's Herculean efforts to attract more diverse audiences, Black and other visible ethnic minorities, young people outside formal education and people of all ages and ethnicities from lower socio-economic groups continue to feel excluded by those glass cases (Bourdieu 1984; Macdonald and Fyfe 1996; Merriman 1991). However, recent research indicates that the situation is slowly changing: 'between 2002-2003 and 2003-2004 participation by socio-economic groups C2, D&Es and by black and minority ethnic groups who have traditionally not been active users of or visitors to museums increased by 15.2 percent and 60 percent respectively' (quoted in Travers 2006: 35). At the same time, archaeologists themselves continue to call for a more active and creative engagement in the production of archaeological narratives and images (e.g., Carman 2002; Holtorf 2006; 2007; Piccini 1996; 1999; Russell 2006a; 2006b; Shanks and Tilley 1992 [1987]).

Within archaeology and heritage studies there continues to be a tension between, on the one hand, our belief in archaeology's affective power within the population as a whole (Holtorf 2007) and, on the other, the anxieties expressed by the official guardians of archaeology and the historic environment about barriers to that engagement. This chapter perhaps stems in part from Stephanie Koerner's observations that:

1. The efficacy of symbolic (discursive) practice is relational and actualised only through its exercise.
2. The forces (or in Foucault's terms, relations of power and knowledge) essential to the efficacy of symbolic forms are productive and enabling, not merely prohibitive.
3. Symbolic communication is unintelligible from perspectives that envisage individuals as atomistic parts. At the very heart of language and human communication, as such, are mutually susceptible and mutually accountable intentional creatures whose engagement with one another and the world in which they live hinges upon recognising each other as such (2006: 214).

That is, beyond the evidence for or against the importance of archaeology in people's lives and their relative access to and participation in archaeology, we might attend to the mechanisms of that engagement to consider how the intertwining of heritage media and people materialises archaeology *itself*. That materialising involves key questions about memory, perception and mobility. As Matt Matsuda has detailed (1996), the history of 'the modern' is characterised by the proliferation of tools designed to transmit memories, what Alison Landsberg terms 'prosthetic memories' (2004). Each transmission of these memories, or what Lyotard calls intensities (1989: 164), produces its own medium. Archaeology is not, therefore, something that has an effect *on* people and needs to be protected *from* people. Rather, what we recognise as archaeology is continually transformed through co-constitutive relationships between people and media.

In this chapter, I consider the plays between broadcast factual television and other screen-based media. Specifically, I attempt to explore questions around matter, memory and dreaming things, around those moments during which encounters among people and things generate the archaeological. To do this I discuss quantitative and qualitative data produced around archaeological television viewing. Rather than argue from a position in which I use these data to say something stable and concrete about the nature of archaeological audiences and the quality of their learning experiences, however, I am attempting to think around processes of perception, creation and obliteration via what may seem to be largely immaterial media. This moves some of the discussion around media and archaeology away from an ocularcentric view that, following the work of Bruno Latour (1986), sees media's relationship to all sciences as that of transforming the material world into flat inscriptions (Witmore 2006: 268).

Without wishing to get ahead of myself, the questions I ask rely on a number of key points. Firstly, I have an affinity with those who write about materiality as the relationality of people and objects, as a process of bringing 'things' into existence (Badiou 2007). That sense of 'bringing forth' has clear roots in Heidegger's elaboration of *techne* (very loosely, technology) as a form of *poeisis* (just as loosely, art) (1977). Rather than simple instrumentality, *techne* as *poeisis* requires that human and non-human agents gather together in an act of bringing forth. Mitch Rose (2005) draws on both Heidegger and Derrida to write eloquently about gathering dreams of presence in the landscape in an attempt to avoid both extremes of humanistic and environmental determinism. Humans do not simply create landscape, nor do the shapes, textures, breezes and folds of the landscape determine human responses. Rather, the interplay between the two sees people and things presencing place: this shoreline is where I fished with my father; I remember camping out with my friends on that hill in 1990 when the rain didn't stop. Gathering is another of Heidegger's key, contested notions. Irigaray critiques what she argues is its masculinist drive to presence, the filling of the void with things, the fear of what is visually and materially 'absent' (1999). However, Rose attempts to argue that this gathering is not an ordered process governed by human whim. The presence brought forth is co-produced via acts of memory reliant on more symmetrical relationships between things and people (see Webmoor and Witmore 2005).

A similar resistance to the extremes of materialism and phenomenology appears in the writings of Henri Bergson (1912). For him, the relationship between matter and memory emerges in acts of perception. It is not simply either that matter has the power to produce representations in humans, nor is it that matter can be reduced solely to the representations that humans have of it. Akin to Rose's dreams, Bergson's perceptions rely on the shivering vibrations of encounters between human and non-humans (1912: 276). Where Bergson's cone of perception appears to over-determine the role of the individual perceiving subject, the hierarchy of the vertical axis over the horizontal and the difficulty that we have in accepting qualitative as well as quantitative multiplicities. That is, we can readily understand the concept of a single perceiving subject able to imagine geographies that contain all simultaneous objects in space, even where that reality exceeds our perception. However, it is more far more challenging to imagine the multiplicity of perceiving subjects through time. Despite Bergson's call to us to do exactly this (183-85) in the tip of the cone that encounters matter in the durational present we perhaps too readily locate an all-knowing subjectivity.

Some of the ideas expressed under the umbrella term Actor Network Theory might help out here. As Bruno Latour has written, objects are 'more interesting, variegated, uncertain, complicated, far reaching, heterogeneous, risky, historical, local, material and networky than the pathetic version offered for too long by philosophers (2005: 11). Latour writes about things in terms of 'immutable mobiles' (1986: 7-14), which emphasises a node-network model. Perhaps more useful in this chapter are John Law and Vicky Singleton's 'mutable mobiles': 'we can't understand objects unless we also think of them as sets of present dynamics generated in, and generative of, realities that are necessarily absent' (Law and Singleton 2003: 8). Intensities are transformed into transportable media, yet these media themselves are mutable, shifting, 'radioactive' (ibid.). The sense in which things persist and yet change through time and space returns me to Bergson to consider the potential implication of his graphic model of perception if we consider the symmetries between human and non-human is that things occupy the multiplicity of the vertical time axis while memory's relationship to processes of materialisation involve a topology. As Bergson himself argues, 'matter resolves itself into numberless vibrations, all continuous and travelling in every direction' (1912: 276).

By considering the dynamism of things I am of course also addressing a particularly resilient thematic within the discipline: that 'archaeology as modern science asserts to society that there is an ascertainable and tangible reality of the past' (Russell 2006b: 361) and that 'awe at science and the results of the photographic and archaeological process inspired belief in the two processes as quests for visible and tangible evidence of human agency' (Russell 2006a: 6). As Koerner has so convincingly argued, 'archaeological materials' are central to such foundational elements of the 'modern episteme' as: the artefactual route to via Doubt of Rene Descartes (1596-1626) Doubt, the artefactual Nature of Isaac Newton's (1642-1724) laboratory (1934 [1687]), the artefactual Society of Francis Bacon's (1561-1626) 'Atlantis' (1909) and Thomas Hobbes' (1588-1679) *Leviathan* (1962 [1651]), and the artefactual Value of Isaac Newton's Mint and Market (e.g., Schaffer 2002; Koerner S. 2006) (see Koerner 2006: 200).

That tension between duration and fragmentation, which we know from these Enlightenment scholars, is also what drives modernity's technologisation. Matsuda's discussion of the importance of memory and the forensic (1996) is focused through archaeology in Pearson and Shanks (2001) and it is this fragmentation of duration and the subsequent reconfigurations into a 'whole' that marks a key link between Bergson, archaeology and cinematographic endeavour. Although Bergson argues that 'the moment that recollection is linked with the perception, a multitude of events continguous to the memory are thereby fastened to the perception—an indefinite multitude' (1912: 218), the process of reflection works to obliterate that multiplicity via attempts to cut reality into fragments.

Documentary film and video similarly 'extract what is useful' (ibid.) from the multiple. Despite continuing debates within documentary studies about documentary's relationship to 'reality', from the earliest years of cinema filmmakers and writers have also understood John Grierson's definition of documentary as the 'creative treatment of actuality' (1926). An extensive literature in anthropology and archaeology (see Piccini 1996; Webmoor 2005) and in documentary studies (see Cowie 1999) argues that films and photographs literally frame and are framed by lived experience and do not unproblematically provide 'evidence' for what really happened. The transformation of the intensity of event into material representation obliterates multiplicity. At the same time, these media are perceived, perception necessitates recollection and the 'thingness' of the media as material opens out into multiple relationships. As much as film fragments duration to isolate and materialise specific images and then stitch those fragments together into the semblance of a coherent whole, it operates within the multiplicity of existence to fold back into the dynamic relationships between matter and memory discussed above.

In short, in this chapter I discuss the ways in which humans, archaeology and media are entangled in a productive relationship from which material intensities are brought forth. To make this argument, I discuss two examples of research into audience practices of engaging with archaeological media in order to think through the materialising practices of perception, the production and obliteration of memory and the co-constitution of 'archaeology' itself as thing. To return to the exchange quoted at the beginning of this chapter, that constitution is not located primarily in the field of media's visuality. Rather, in dreaming archaeology into presence we hear, see, smell and touch it. This literally takes place via the tensions between the materialities of television and computer screen—the whine of electricity, the crackle of static across the surface of the screen, the relative comfort of our chairs, the interruptions of children, partners, advertising, the play of electrons across the fluorescent screen of the cathode ray tube television, or the electrically excited phosphors of the plasma screen or the LCD screen's dance of liquid crystals within millions of backlit pixels—and the surprisingly resistant reality of the luminous and sonorous illusions that these technologies and settings produce. 'Television' is both co-terminous with its material specificities and stretches far beyond. The box in the room links to other boxes in other rooms and they all reaffirm both physical presence and virtual absence—the 'now-here, now-gone' of the photographic rather than painterly image (Cowie 1999: 22).

Audience Research

In 2006, I was commissioned to undertake a piece of research for English Heritage and the Council for British Archaeology on archaeological television audiences (Piccini 2007b). The aim was to produce baseline television audience research data for archaeology programming that would be comparable with the data being produced through the *Heritage Counts* and *Taking Part* initiatives. That is, English Heritage wished to compare patterns of television viewing with heritage visiting behaviours. Moreover, they wanted to look at how archaeological television viewing compared with more general television viewing patterns, with a focus on gender, age, ethnicity and social class. The research was not designed to drill down into individual programmes or individual transmissions of programmes across the schedules.

I had some reservation about my involvement in such a project given the significant critiques within television and media studies of the utility of this kind of quantitative research (see esp. Ang 1991; Gauntlett 1995; Hill 2005; Lotz 2000; Morley 1980; 1992; 1995). Those critiques highlight a problematic familiar across the arts and humanities as these methods can be seen to fail to account for difference in their emphasis on generalisable models of identity. The data provide no ethnographic clues about how and why people watch archaeology on television. Without the ability to follow groups of television viewers we cannot investigate the ways in which people interact with the flow of programming and advertising across their screens. We cannot explore the multiplicity of individual programmes to discuss viewers' productive watching. None the less, I was keen to be involved in this project precisely because of my own interest in the ways in which heterogenous human behaviours are transformed through these tools and methods of measurement.

The research took place in the summer of 2006 and the first stage focused on the question of definition: what is archaeological television and how does it differ from other forms of television programming? Within the field of television and media studies archaeology is seen to sit within the more general genre of history programming. Ann Gray's current AHRC-funded Televising History project incorporates factual programming that covers archaeological subject matter as 'history' (Bell and Gray 2007). Broadcasters, however, see archaeology very specifically in terms of programmes that either focus solely on the practice of archaeology (eg., *Time Team*, Videotext Communications for Channel 4) or that cover 'recognisably' archaeological subjects such as ancient monuments and civilisations (Laurence Rees, Creative Director of BBC Television History and producer/writer of 2005 *Auschwitz: The Nazis and the 'Final Solution* and Ralph Lee, Head of Specialist Factual, Channel 4, pers. comm.). Archaeologists, however, continue to open out such definitions, informed by some of the recent debates around materiality (eg., Olsen 2003) contemporary and historical archaeologies (e.g., McAtackney, Palus and Piccini 2007; Holtorf and Piccini 2009) and time (e.g., Lucas 2005). Karol Kulik has significantly nuanced the categorisation of archaeological programming in her recent (2006) critical engagement with Bill Nichols' (2001) and Stella Bruzzi's (2000) canonical documentary taxonomies. For example, Kulik describes *Time Team* as a 'backstage' archaeology programme.

For the purposes of the research, however, Don Henson (Council for British Archaeology) and I discussed the need to connect these definitions in order to engage with programmes in which archaeologists, television scholars and industry professionals could recognise something 'archaeological', as this would more closely approximate the views of heterogeneous audiences—from the avid and expert *Time Team* fan to the chance viewer of a series such as BBC's *Timewatch* strand. We agreed on the term 'heritage television' in order to signpost:

> ...any 'factual' programming transmitted on both analogue and digital platforms that concerns material culture, the historic environment and ancient monuments. History programming that focuses on artifacts and sites recovered through archaeological practices is also considered to be heritage television. As such, programmes range from *Antiques Roadshow* through to *Time Team* and *Horizon* (Piccini 2007b).

The study looked at the range of heritage TV (as defined above) over a 12-month period in 2005-2006: 162 programme titles on 25 separate TV channels (compare this with the 201 titles broadcast in 2002 in Kulik 2006: 81). Data for the study was produced by the Broadcasters' Audience Research Board (BARB) and statistically analysed by Television Research Partnership (TRP). BARB is responsible for providing estimates of the number of people watching television. It has set-top boxes in 5,100 homes, which means around 11,500 viewers are monitored as a sample size to represent the viewing behaviour of over 24 million UK households.

Watching Practices

During 2005-2006 some 13,000 programmes were transmitted and 98 percent of all adults saw at least one heritage programme during the year. Of these programmes, the top five made a 60 percent contribution to the amount of viewing in the study and all were antiques-related programming, due to our initial definition of 'heritage programming' (Table 17.1). As we were concerned that viewing figures and patterns for antiques programmes might differ significantly from data for other forms of heritage programmes, we separated the research into two phases: Heritage1 (which included the antique programmes) and Heritage2 (which simply removed the top five programmes). Yet, even when we removed the Heritage1 data a remarkable 96 percent of all TV viewers in the UK watched at least one heritage programme during the year, with 20 percent of viewers having watched 99 or more programmes. People certainly appear to engage with archaeology far more readily and frequently via screens than they do via historic sites or museums. The Department for Culture, Media and Sport's *Taking Part* research (DCMS 2006: 2) indicated that of the UK's population only 69 percent attended at least one type of historic environment site during the past 12 months while 42 percent attended a museum/gallery at least once in the same period.

Viewing profiles provide interesting detail, particularly when figures for the total television and Heritage1 audiences are compared with Heritage2 audience figures (see Piccini 2007b for data). Where both the total television and Heritage1 audiences are biased towards women (55 percent and 56 percent respectively), in Heritage2 we find a more 'gender-balanced' audience of 51 percent women, which is closer to the proportion of women within the total UK population. These percentages can, of course, be used to ground contradictory arguments. I could say that those programmes that are popularly identified as 'archaeological' (rather than our concept of the 'heritage' programme) have a more gender-neutral appeal than other programming. On the other hand, given that the television audience comprises a greater proportion of women viewers than are represented in the UK population I could also argue that archaeology programmes alienate many traditional television viewers. This interpretation would sit nicely within heritage and museum studies scholarship through the 1990s to the present, which has characterised 'heritage' as appealing by and large to middle-aged, middle-class white men. Without contextualising quantitative and qualitative research from the same time period that explored, for example, the scheduling

Table 17.1 Viewing figures contribution of the top 10 heritage programmes for 2005-2006

Rank	Title	Channel	Contribution %
1	*Flog It!*	BBC2	17
2	*Cash in the Attic*	BBC1	14
3	*Bargain Hunt*	BBC1	13
4	*Antiques Roadshow*	BBC1	10
5	*Car Booty*	BBC1	6
6	*Coast*	BBC1	4
7	*Rome*	BBC2	3
8	*Time Team*	Channel 4	2
9	*A Picture of Britain*	BBC1	2
10	*Egypt*	BBC1	2

Note: This list represents only 73 percent of the audience share for all heritage programming, with the top 5 totalling an overwhelming 60 percent.

of individual programmes—perhaps archaeology appears on television precisely when Sunday dinners are still being prepared, in the main, by women (Lader, Short and Gershuny 2006: 38)—and individual attitudes towards archaeological subject matter it is impossible to generate anything more specific out of the raw gender profiles.

What happens if we consider age groupings? At 9 percent of the total audience, fewer younger people (16-24) are watching television than are represented within the total UK population (the Office for National Statistics puts this at around 11 percent). It is, therefore, notable that a much smaller proportion (4 percent) watches any kind of heritage programming, although it is interesting to note that there is little difference between the Heritage1 and Heritage2 figures. In keeping with earlier research (Piccini 1999; Kulik 2006) around three quarters of viewers of Heritage1 are over 45 compared with only 58 percent of the total television audience. For Heritage2, in which antiques programmes have been removed, the percentage of over-45s drops slightly to 72 percent. However, if I compare these percentages with the fact that only some 40 percent of the UK's population is currently over 45 this continues to be a highly significant audience. While older people are over-represented generally within audiences, this is particularly striking in all forms of heritage programming. Consider age alongside gender. Given that over the age of 73 women outnumber men, with the ratio increasing progressively from 1.2 at age 73 to nearly 3 by 90, the figures suggest that women present an even lower proportion of heritage television viewers than might at first be thought. Certainly, many archaeology programmes are repeated on daytime television, which largely excludes school-age children and people in work. In population terms we ought to be seeing many more women viewers.

What of the ethnic makeup of heritage programming audiences? Heritage1 audiences make up the same proportion of viewers as the total television viewing audience at 6 percent, compared with around 9 percent of the total British population. For Heritage2 this drops by half to 3 percent. The very low numbers of Black and other visible ethnic minority viewers presents us with a similar picture to that of age. Heritage2 programmes attract about half the audience that watches television more generally and represents only around a third of the population as a whole. However, I can still interpret this data to suggest that heritage and archaeology speak to a more diverse UK audience than is apparent in other sectors. Benjamin has reported that only 2 percent of UK archaeological undergraduates are Black or ethnic minority (2003), while in the workplace only 0.7 percent of professional archaeologists are (Aitchison and Edwards 2003: 25). This might suggest that while archaeology may be seen to be a worthwhile thing to watch on television, it is not something to pursue with a view to work. Yet, much the same might be said for women: only around one third of professional archaeologists are women (Aitchison and Edwards 2003) compared with around half of the television audience. What is inescapable is that White males are over-represented in archaeology's television audiences, student populations and professional domains.

In terms of social class, the C2DE groups make up the larger group of television watchers overall, at 58 percent. That 60 percent of Heritage1 viewers and only 52 percent of Heritage2 viewers are C2DE might appear to confirm the traditional social 'distinction' arguments of Bourdieu for the arts (1984) and Merriman for archaeology (1991). However, whereas in 1990 60 percent of the population belonged to C2DE, by 2000 this shrank to 40 percent (Geddes 2002: 70). That the C2DE groups dominate all types of heritage programming would appear to suggest that conventional arguments about archaeology's exclusive status are perhaps not all that they appear (eg., Duke, P. and Saitta, D.J. 1998). Yet again, these figures need contextualising via more detailed information about scheduling. If the majority of programmes (including repeats) are shown in the day and the majority of viewers are over 45 years of age, then people classified within the E grouping may be retired pensioners or older people with physical disabilities rather than people relying on state unemployment benefits, for example.

Another stage of the research involved splitting all Heritage1 viewers into three equally sized groups (Heavies, Mediums and Lights) based upon viewing hours. Heavy Heritage Viewers (HHVs) watch the most programming and are also heavy television viewers overall. The average UK adult watches 27 hours of television per week. HHVs for the terrestrial channels watch 38 hours of television per week, while those watching non-terrestrials watch 34 hours per week. HHVs are also:

- 15 percent less likely than the average adult viewer to have access to a computer at home;
- 12 percent less likely to have access to a computer at work;
- 17 percent more likely not to use a computer at all;

- 5 percent more likely to watch period drama, documentaries, news, current affairs and programmes covering classical music and nature programming;
- 5 percent less likely to watch reality TV, US comedy and pop music;
- and 5 percent more likely to be interested in gardening and reading activities (Piccini 2007b).

These 'lifestyle' data point towards the wider set of activities in which viewers are likely to be engaged. Thus, the generalisable trends that the data appear to support indicate that viewers of both Heritage1 and Heritage2 programmes are white, middle-aged and older men from lower socio-economic classes. They appear to shun or not have access to computers, but use a range of other media to engage with 'quality' television and to participate in other activities that are more closely related to people within social classes ABC1. This perhaps supports a view that the bulk of heritage television viewers are people who have entered C2DE through retirement.

How Archaeologies are Made

None of this data will come as a surprise to archaeologists, who have painted broadly similar pictures since the 1990s. In my own research (1999) I described these patterns in terms of *habitus*, a material way of signalling social value and distinction. In short, I described heritage as 'good to think' and television watching as just one activity within a range of others that value 'pastness'. While these baseline data on UK heritage television viewing trends for 2005-2006 provide some good evidence for what archaeologists have argued in common-sense terms over the years, such attempts at mapping the field only get us so far. There is a range of more finely detailed issues to be explored that might begin to indicate the complicated relationships between television viewers and heritage, relationships that are themselves productive of archaeology.

Table 17.1 indicated the popularity of antiques programmes, yet once these are removed, an interesting picture emerges of programmes that proved popular in terms of audience contribution (Table 17.2).

I suggest that these programmes may be categorised either in terms of the power of practised place (*pace* English Heritage 2000) or as ancient civilisation docudrama. *Coast*, *Time Team*, *A Picture of Britain*, *Build a New Life in the Country*, *Tales from the Green Valley* and *Map Man* all emphasise the creative relationships between humans and things in acts of making place. The particularly archaeological combinations of mapping, digging, walking, imaging and shaping has been discussed elsewhere (Webmoor 2005; Witmore 2006). However, this model is a relatively new one and differs significantly from the conventional TV expository documentary format evident from BBC's *Chronicle* series (1966-1991) through to the current programmes within the *Horizon* or *Timewatch* strands in the

Table 17.2 Top 10 programme titles in terms of audience contribution

Rank	Title	Channel	Contribution %
1	*Coast*	BBC1	10
2	*Rome*	BBC2	8
3	*Time Team*	Channel 4	6
4	*A Picture of Britain*	BBC1	6
5	*Egypt*	BBC1	5
6	*Build a New Life in the Country*	Five	4
7	*Tales from the Green Valley*	BBC2	4
8	*Around the World in 80 Treasures*	BBC2	2
9	*Castle in the Country*	BBC2	2
10	*Map Man*	BBC2	2

Source: Data provided by BARB and analysed by TRP.

UK or *Nova* in North America. Similarly, the presence of docudramas within this table also presents a departure from what archaeologists at least see as conventional programming about their subject area.

What is it about dramatic reconstruction and archaeology's material rather than textually interpretative practices that hold our attentions? While critics and archaeologists alike are scathing about docudramas (Caveille and Fox 2005), they have an enviable hold over audiences. Perhaps viewers desire their pasts with identifiable *dramatis personae*. Perhaps the immediacy of spectacle makes heritage something to welcome into living rooms. Perhaps, as has been suggested by some media researchers, viewers trust drama to express some deeper truth than can be found in the conventional documentary (Hill 2005). Yet, what is at stake here is not trust, per se. Instead, performance and place are linked through their shared concerns with performative materialities, with the placed, spatialising and productive entanglements of bodies and other matter. Rather than see dramatic simulation as factual television's abandonment of empirical observation (Dovey 2004: 234), this kind of programming foregrounds Cowie's arguments concerning tensions between documentary's 'discourse of sobriety' and its reliance on desire and pleasure (1999).

Both forms of television archaeologies rely on the realism of material entanglements, whether that realism is expressed through attempts to document archaeological practice or the realism of conventional, character-driven narrative fiction. Both modes aim to efface their representational technologies and techniques to create believable stories. Perhaps more importantly, these modes of archaeological programming differ from the presenter-led expository factual fare in that there is no one obvious figure who is in control of the narrative, who gatekeeps the knowledges. *Time Team* and *Coast* do not share the same format, but they do share a similar focus on collaboration. A group of people enter into an apparently incomprehensible, multiple totality, collectively fragment that multiplicity and then work to suture those fragments into a coherent whole. The dramatisation that we see in drama-docs similarly operates by transforming contradictory human behaviours into identifiable, stable character traits that are further transformed into scenarios that aim to convince us that this does approximate real life. In neither instance do we, as viewers, confront a single author of such visions. Instead, what plays out closely resembles the collective workings-out, the knowings-how that we all engage in from moment to moment.

Of course, the relative popularity of these forms of heritage television is linked as much to their popularity amongst programme commissioners as it is to viewer preference. However, commissioners see that these programmes achieve good viewing figures: they 'work'. And the audience for which they seem to be working is older, male, White and either does unskilled casual labour or is on state benefits of some sort. It is an audience that also values drama and gardening on television and has few opportunities to use computer-based technologies. Yet, this audience also has an intriguingly complex relationship to heritage television and seems to value its ability to put the viewer into a direct, creative relationship with making, even at several removes. The double and triple shapings of heritage programmes perhaps serve as a useful argument against Habermas' derision of media's destruction of the public sphere. Humans and non-humans interact in the shaping of events, in the daily acts of perception that slice lived durations into single images which are brought together to fashion stories; archaeologists group together to transform the disorganised chaos of the traces of such acts into sensible fragments that are stitched together into a coherent narrative; filmmakers transform the multiplicity of archaeological practices into single shots, sound recordings and graphic renderings that are rearranged into seamless sequences. These programmes invite viewers to share in this process and implicitly link these activities to what viewers themselves do. These programmes thus take shape as a new form of public sphere.

Furthermore, for heritage programming to operate as spaces in which viewers may collectively engage with materialities, the relationships between audience and programme involves processes of memory and perception. Following Halbwachs (1992), the multiplicity of the material practices in the programming discussed would appear closer to what he terms 'collective memory' rather than the singular vision of 'history'. Of course, as part of his rejection of Henri Bergson, Halbwachs espoused a Durkeheimian reading of objects as vehicles for social signification. Susan Crane has problematised this by arguing that these memories cannot reside *in* sites (following Nora) or traces (after Derrida) but must reside within individuals (Crane 1997: 1381). I will take up both of these points in the section below in more detail. Here I suggest merely that neither humans nor non-humans should be seen as 'containers' of meaning, that the entanglement of viewers, producers and materials that we see in the viewing patterns discussed above rather points towards a co-creative relationship between things and people in an ongoing production of collective memory.

Co-creating Archaeologies and Putting Things Back into Memory

A second case study in archaeological viewership allows me to explore further questions of memory, perception and materiality. Although there appears to be some evidence that committed heritage audiences do not have access to computers, there is also ample evidence that at least a small number of viewers actively use computers and the internet as part of their television-related practices. In addition to the growing number of academic blogs, wikis and websites devoted to discussion about archaeology's relationship to camera-based technologies, a YouTube search (12 January 2009) using 'archaeology' results in 6,450 videos, there are both official and unofficial fan forum sites for popular television programmes such as *Time Team* and broadcaster websites present on-demand repeats, programme production notes, additional materials and opportunities to comment and even get involved in local archaeology initiatives. In 2005, the BBC's *Coast* series engaged with locative media practice by collaborating with Hewlett-Packard Labs and Gavitec to produce data code technology designed to trigger the delivery of additional information to viewers' mobile phones as they passed by specially tagged signs as they followed in the programme-makers' footsteps along coastal walks. In short, it has become very difficult to sustain any form of sharp boundary between television and computer screens in terms of how people engage with archaeology media.

The permeability of these boundaries is supported by audience studies more generally. Since Henry Jenkins' groundbreaking work on television fan cultures (1991), Matthew Hills' ethnographic approach to the subject (2002) and Jenkins' opus on convergence culture (2006) scholars have come to develop an extended definition of audience and media whereby fan activity across media platforms becomes a television reception context. My interest in this section is in how information from fan forums might inform my emerging argument about the unsettling of the past evident in the focus of contemporary producers and audiences on co-creation and practice.

The *Time Team* Forum is the oldest of Channel 4's forums, having started life in 1997 for what was originally intended to be just a few days to accompany that year's *Time Team Live* (http://community.channel4.com/eve/forums/a/frm/f/8896096411, 27.03.2007). It is continually shifting and growing but at the time of writing the forum consisted of some 661 separately listed topics, 20,473 individual replies and 1,685,912 individual views of posts. 198,642 people are registered to the Channel 4 forum service and can post to any forum area. I am a member of the forum, although I am a 'lurker' and did not announce my research activities on the site. It is best practice in fan studies, as with any ethnographic research, to engage in discussion with forum participants about the researcher's role on the site and to come to some form of collective agreement about how the researcher engages with people and presents that work in a research context. In the context of this research, however, my intention was to provide a snapshot of quantitative data rather than to engage in an ethnography of individual practices. I have anonymised all forum user names for this chapter.

I identified particularly intensive areas of member discussion by tabulating and averaging numbers of individual topic replies and views. The most popular topics for forum visitors to read and reply to deal specifically with other television and radio programmes, activities organised by forum members, news and alternative archaeologies (Table 17.3).

I then organised the forums into a more detailed taxonomy. I based this on my interpretation of the ways in which contributors named their topics and also from the focus of discussion within each topic area. In the Table 17.4 I list only those categories that registered at least.1 percent of replies (n=20,473), views (n=1,685,012) and total number of topics (n=661). Categories not listed in the table are: dinosaurs, druids, games, Ice Age, jobs, language, Mesolithic, Native American, photography, underwater and Vikings. Topic areas that produced the most intensive forum activities were discussions focused on the *Time Team* personalities, the forum itself, other media dealing with archaeology, opportunities for forum users to participate in archaeological practice, local Friends activities (TTFF), details of individual episodes and Romans.

These topics all concern people's productive relationships with archaeology: about how archaeology is made and who makes it. The discussions are all intermedial in that they do not occur within a firmly bounded mediascape but instead draw in and make use of practices beyond the computer screen. Even the focus on Romans was due to the level of discussion around *Time Team*'s 'Big Roman Dig' of 2005, which gave people the opportunity to participate in a large number of professionally managed excavations of Roman sites. One may easily argue that media practices are always-already intertextual. If I think beyond the semiotic textual metaphor here, however, it seems clear that the forum is a space in which participants co-create archaeologies in a similar manner to that of the

Table 17.3 Most popular topics on *Time Team* forum

Topic	Start Date	Replies	Views
Interesting TV and Radio Programmes	20-09-02	531	98,614
North West FF	19-09-02	1,702	85,107
The Trench Fell Inn Returns	01-02-05	1,532	82,966
Archaeology News	21-12-04	488	58,774
More questions for Steve Platt	01-04-04	1,205	56,515
North East Forum Friends	17-09-02	490	52,319
All Ireland Forum Friends	02-01-04	571	47,195
SW/Wessex FF	16-09-02	379	29,068
St George Thread—Reincarnated	23-10-05	423	26,100
Siluria II—TTFF Gwent and Glamorgan	16-08-04	1,657	10,1594

Table 17.4 Forum topics by category

Topic Category	% of Replies	% of Views	% of Total Topics
Arch & You	3	3	5
Alt Arch	3	2	1
Anglo Saxons	.2	.3	1
Astronomy	.1	.1	.3
Banter	.9	1	2
Bronze Age	.1	.2	.5
Celts	.6	.6	.7
Digs	1	2	3
Education	.8	1	3
Egypt	.1	.2	.6
Environmental	.1	.2	.6
Episodes	15	18	20
Finds	1	1	3
Forum	16	12	5
Historical Arch	1	1	3
Intermedia	9	13	9
Iron Age	.2	.3	.7
Mediaeval	1	1	4
Merchandise	1	2	3
Metal Detecting	4	4	2
Museums	.5	.3	.3
Neolithic	.9	1	2
News	2	3	.1
Personalities	5	7	10
Policy	.7	.2	.7
Practice	.3	.5	1
Romans	2	2	5
Royal	1	1	1
Sites	1	2	4
Stonehenge	.2	.3	.7
Technology	.1	.2	.9
Ttff	26	21	2
Tudors	.2	.2	.3
Wales	.6	.7	2

Note: Percentages refer to total individual replies (20,473), views (1,685,012) and topics (661).

programme-makers of *Coast* and *Time Team*. Fragments and artefacts are identified and brought together in never-ending attempts to connect and 'make sense' of them, a direct expression of the tensions between the multiplicity of actuality and the human drive to craft that actuality into order and coherence.

The importance of personality in the forum is signalled through the intimate tone of discussion around key *Time Team* members and might be taken as evidence of the programme as a 'site of contestation and struggle within and between professional practices' (Tulloch 1990: 181, quoted in Hills' and Williams' argument 2005: 347). As username3 writes, 'I even had the pleasure to meet Tony last year in person on his solo theatre tour YES!!-really nice man!!' (29 January 2007).

Discussion of the series, however, tends to focus on where *Time Team* gets it wrong:

> [username5], 14.01.07: Fingers crossed for the new series. Hopefully they have learnt from previous mistakes, and have noted some of the comments made on here, and elsewhere.
>
> [username6], 15.01.07: They seem to have truncated or even dropped at times the 'after the break' teasers and lengthy recapping after the adverts. Perhaps they've realised viewers of this programme have longer attention spans than golfish (*sic*)?

I suggest that from the level of forum topic popularity to individual response the examples above focus on archaeology as something that emerges out of an entangled human and non-human practice. Moreover, these examples signal the pleasures that audiences derive from their engagement with programmes, whether that pleasure is produced by meeting up with other fans for do-it-yourself archaeology, in meeting television presenters or through actively working through programmes to find and highlight mistakes. Although archaeologists may be in a privileged position to communicate with an interested public (Holtorf 2007; Russell 2006), what emerges from the forum is the sense that audiences feel as though they are equal participants in the production of television programmes.

Moreover, these audiences are participating in a virtual public sphere that, through the percolation of past and present, human and non-human, engages with serious questions of memory and perception. As Middleton and Brown argue in their discussion of Halbwachs, 'by retreating into our images of the past, in effect what we do is strengthen the connections between past and present milieus. We produce an expanded version of ourselves as social beings (2005: 36). Yet, I would argue that humans are not in control of this production as such. In archaeological terms people and things co-create encounters between past and present as multiplicity and duration rather than via a relationship between subject and object. Viewers, technologies and things produce the archaeological: following Heidegger, 'the craftsman is not the cause of the chalice in our sense at all, but a co-responsible agent in bringing the chalice into appearance' (Feenberg 2005: 17). Where that presence suggests completion, discussed by Webmoor, after Barthes, in terms of the still photograph as the completion of the image (2005: 63), I would return to Law and Singleton's mutable mobiles (2003) and Henri Bergson's notions of duration as difference to suggest that both the audience and *Time Team* forum data support a more open-ended life of the televisual moving image which itself echoes the duration of the practices that people are watching.

Alongside the symmetrical relationship between things and people (see Webmoor and Witmore 2005), I need finally to consider the processes by which we come to recognise the images, sounds and practices that seem to attract us most powerfully. The work of Henri Bergson on matter and memory (1912) is useful here in that Bergson wishes both to dispel the idea that matter possesses the power to produce representations in us and the idea that matter is reduceable to the representations we have of it. The relationship between matter and human does involve the transformation of matter as image-for-itself into image-for-me, into representation. Representation is always within matter, but only virtually, in terms of its future potential. It is through perception that the virtual is made actual representation, subtracting from the virtual's multiplicity. Representation is always less than the virtual in that it slices a fragment out of the undifferentiated multiplicity of the world (ibid.: 38). Similarly, memory passes into something else by becoming actual in that the progress of memory is towards the material (ibid.: 174) and that materialisation begets memory's obliteration (ibid.: 232).

Critiques of Bergson tend to focus on the apparent bias of the perceiving human subject, which is taken to be represented by his 'cone of perception' (see Olsen 2003: 97). In the first part of this chapter I suggested that this may, itself, be the result of the spatial bias that Bergson seeks to problematise. The audience and fan forum data that I have discussed here appears to support an understanding of archaeology's more radical potential to subvert the 'totalizing quest for meaning' (Minh-ha 1993: 90-107). The interest in archaeological media appears to be emerge from the dynamism of the active interplay of people and things. Paradoxically, television and convergent media facilitate a sense of encounter between audiences and material practices that sits in proximity to archaeology.

To refer to the title of this volume, the 'past' here is certainly 'unquiet'. In terms of our understanding of matter and memory, these forms of archaeological television extend our ability to think beyond armchair cultural critique. However, I find it necessary to return to the study I undertook for English Heritage as a closing cautionary note to suggest further avenues of research. Dynamic entanglement and symmetrical relations between humans and things may all present fabulous potential for regenerating archaeological practices. Yet, this appears to speak most powerfully to a very select audience that is not representative of the UK's population. I do not suggest some kind of moral point here. Nor can I conclude that it is materialising practice alone that appeals to the middle-aged White male—despite the echoes of Irigaray that I cannot help but hear. Further work on fan forums, ethnographies of audiences and university students (currently being carried out by doctoral student Greg Bailey), practice-based documentary research and even traditional narrative and formal analysis are needed in order to develop a more multi-modal approach to working through the complexities of the transformations of archaeology across media.

Academics may continue to criticise television's offerings because it cannot do what research does (Hills 2003). However, the narrative 'content' of television archaeologies is only one aspect and despite academic anxiety about the quality of information on the television there is surprisingly little research to indicate that audiences use television primarily as a knowledge source—although the Glasgow Media Group's extensive research into the relationship between broadcast news and general knowledge might suggest otherwise (eg., Philo and Berry 2004). What does seem clear is that audiences behave as though televisual archaeologies belong to them in a way that museums and degree programmes perhaps do not. It is important that archaeologists continue to call for nuanced, multivocal media narratives that are not necessarily framed by the broadcast media industry (Pearson and Shanks 2001; Piccini 2003-2004; 2007a; 2009; Schofield 2006; Witmore 2006). At the same time, we can begin to trouble any easy marginalisation of broadcast media as not presenting audiences with similar opportunities for active participation in materialising practices. The diverse entanglements of people and the media present the possibility of a space in which archaeological dreams really do come true.

Acknowledgements

This research could not have taken place without Don Henson and Gill Chitty, Council for British Archaeology. My work on the CBA's and British Universities Film and Video Council's Committee for Audio-Visual Education (CAVE) led to this research project for English Heritage. I am therefore extremely grateful to the CBA's vision and tireless work to keep archaeology a vital, accessible and diverse practice. This chapter has benefited enormously from conversations with Karol Kulik, Mike Pitts and students on Bristol's MA in Archaeology for Screen Media. The final editing took place at the University of British Columbia. In keeping with academic convention, I lay individual claim to any mistakes or oversights.

References

Aitchison, K. and Edwards, R. 2003. *Archaeology Labour Market Intelligence: Profiling the Profession 2002-2003*, Bradford: CHNTO/IFA.

Ang, I. 1991 *Desperately Seeking the Audience*, London: Routledge.

Badiou, A. 2007 *Being, Existence and Death*, 4 April 2007, British Library.

Bell, E. and Gray, A. 2007 History on television: charisma, narrative and knowledge, in H. Wheatley (ed.), *Re-viewing Television History*, London: I.B. Tauris.

Benjamin, R. 2003 Black and Asian representation in UK archaeology *The Archaeologist*, 48: 7-8.

Bourdieu, P. 1984 *Distinction: A Social Critique of the Judgement of Taste*, R. Nice (trans), London: Routledge and Kegan Paul.

Bruzzi, S. 2000 *New Documentary: A Critical Introduction*, London, Routledge.

Carman, J. 2002 *Archaeology and Heritage: An Introduction*, London: Continuum.

Caveille, S. and Fox, K. 2005 TV in BA *British Archaeology* 80: 44-5.

Cowie, E. 1999 The spectacle of actuality, in J. Gaines and M. Renov (eds), *Collecting Visible Evidence*, Minneapolis: University of Minnesota Press, 19-45.

Crane, S.A. 1997 Writing the individual back into collective memory *American Historical Review Forum* 102: 1372-85.

DCMS 2006 *Taking Part: The National Survey of Culture, Leisure and Sport.*

Deem, R. 2004 The knowledge worker, the manager-academic and the contemporary UK university: new and old forms of public management? *Financial Accountability and Management* 20(2): 107-28.

Dovey, J. 2004 Its only a game show: *Big Brother* and the theatre of sponaneity, in E. Mathijs and J. Jones (eds) *Big Brother International*, London: Wallflower Press, 234-51.

Duke, P. and Saitta, D.J. 1998 An emancipatory archaeology for the working class *Assemblage* 4 http://www.assemblage.group.shef.ac.uk/4/4duk_sai.html (27.03.2007).

English Heritage 2000 *Power of Place*, London: English Heritage, http://www.english-heritage.org.uk/server/show/nav.1447 (27.03.2007).

—— 2006 *Heritage Counts: The State of Englands Historic Environment*, London: English Heritage, Available at: <http://www.english-heritage.org.uk/hc2006/server/show/nav.9535>. Accessed: 23 March 2007.

Feenberg, A. 2005 *Heidegger and Marcuse: The Catastrophe and Redemption of History*, New York: Routledge.

Gauntlett, D. 1995 *Moving Experiences: Understanding Televisions Influences and Effects*, Eastleigh: John Libbey.

Geddes, D. 2002 The last resort: why the UKs traditional seaside destinations must rethink their long-term strategy *Locum Destination Review* 9: 68-70, http://www.locum-destination.com/LDR9.html (27.03.2007).

Grierson, J. 1926 Flahertys Poetic Moana, *New York Sun*, 8 February.

Halbwachs, M. 1992 *On Collective Memory*, Chicago: Chicago University Press.

Heidegger, M. 1977 *The Question Concerning Technology and Other Essays*, W. Lovitt (trans.), New York: Harper Torchbooks,

Hill, A. 2005 *Reality TV: Audiences and Popular Factual Television*, London: Routledge.

Hills, C. 2003 What is television doing for us? Reflections on some recent British programmes *Antiquity* 77 (295): 206-11.

Hills, M. 2002 *Fan Cultures*, London: Routledge.

Hills, M. and Williams, R. 2005 It's all my interpretation: Reading Spike through the subcultural celebrity of James Marsters *European Journal of Cultural Studies* 8: 345-65.

Holtorf, C. 2006 Can less be more? Heritage in the age of terrorism *Public Archaeology* 5: 101-109.

—— 2007 *Archaeology is a Brand! The Meaning of Archaeology in Contemporary Popular Culture*, Oxford: Archaeopress.

Holtorf, C. and Piccini, A. (eds) 2009 *Contemporary Archaeologies: Excavating Now*, Frankfurt am Main, Berlin, Bern, Bruxelles, New York, Oxford, Wien: Peter Lang.

Irigaray, L. 1999 *The Forgetting of Air in Martin Heidegger*, M.B. Mader (trans.), Austin: University of Texas Press.

Jenkins, H. 1991 *Textual Poachers: Television Fans and Participatory Culture*, New York: Routledge.

—— 2006 *Convergence Culture: Where Old and New Media Collide*, New York: New York University Press.

Koerner, S. 2006 Towards archaeologies of memories of the past and planning futures: engaging the Faustian bargain of 'Crises of Interpretation', in I Russell (ed.) *Images, Representations and Heritage: Moving beyond Modern Approaches to Archaeology*, Springer, 187-220.

Kulik, K. 2006 Archaeology and British television *Public Archaeology* 5 (20): 75-90.

Lader, D., Short, S. and Gershuny, J. 2006 *The Time Use Survey 2005: How we Spend our Time*, London: Office for National Statistics.

Landsberg, A. 2004 *Prosthetic Memory: The Transformation of American Remembrance in the Age of Mass Culture*, New York: Columbia University Press.

Latour, B. 1986 Visualization and cognition: thinking with eyes and hands, *Knowledge and Society: Studies in the Sociology of Culture Past and Present* 6: 1-41.

—— 2005 From Realpolitik to Dingpolitik—An Introduction *Making Things Public-Atmospheres of Democracy at ZKM*, Cambridge, MA: MIT Press.

—— 2006 *We Have Never Been Modern*, Harvard University Press.

Law, J. 2004 *After Method: Mess in Social Science Research*, London: Routledge.

Law, J. and Singleton, V. 2003 Object lessons, Available at: <http://www.lancs.ac.uk/fss/sociology/research/resalph.htm#law>. Accessed: 20 March 2007.

Lotz, A.D. 2000 Assessing qualitative television audience research: incorporating feminist and anthropological theoretical innovation *Communication Theory* 10 (4): 447-67.

Lucas, G. 2005 *The Archaeology of Time*, London: Routledge.

Lyotard, J.-F. 1989 [1975] Beyond representation, in A Benjamin (ed.) *The Lyotard Reader*, Oxford: Blackwell Books, 155-68.

McAtackney, L., Palus, M. and Piccini, A. (eds) 2007 *Contemporary and Historical Archaeology in Theory*, Oxford: British Archaeological Reports.

Macdonald, S. and Fyfe, G. (eds) 1996 *Theorizing Museums: Representing Identity and Diversity in a Changing World*, Oxford: Blackwell.

Matsuda, M.K. 1996 *The Memory of the Modern*, Oxford: Oxford University Press.

Merriman, N. 1991 *Beyond the Glass Case: The Past, the Heritage and the Public in Britain*, Leicester: Leicester Museum Studies Series.

Middleton, D.J. and Brown, S.D. 2005 *The Social Psychology of Experience*, London: Sage.

Minh-ha, T. 1993 The totalizing quest of meaning, in M. Renov (ed.) *Theorizing Documentary*, London: Routledge, 90-107.

Morley, D. 1980 *The Nationwide Audience: Structure and Decoding*, London: BFI.

—— 1992 *Television Audiences and Cultural Studies*, London: Routledge.

—— 1995 Theories of consumption in media studies, in D. Miller (ed.) *Acknowledging Consumption*, London: Routledge, 296-328.

National Readership Survey 2005 Available at: <http://www.businessballs.com/demographicsclassifications.htm>. Accessed 27 March 2007.

Nichols, B. 2001 *Introduction to Documentary*, Bloomington: Indiana University Press.

Office for National Statistics, Population statistics, Available at: <http://www.statistics.gov.uk/>. Accessed 26 March 2007.

Olsen, B. 2003 Material Culture after Text: re-membering things *Norwegian Archaeological Review* 36(2): 87-104.

Pearson, M. and Shanks, M. 2001 *Theatre/Archaeology*, London: Routledge.

Philo, B. and Berry, M. 2004 *Bad News from Israel*, London: Pluto Press.

Piccini, A. 1996 Filming through the mists of time: Celtic constructions and the documentary, *Current Anthropology* 37: S87-S111.

—— 1999 *Celtic Constructs: Heritage Media, Archaeological Knowledge and the Politics of Consumption in 1990s Britain*, unpub PhD thesis, University of Sheffield.

—— 2003-2004 *Guttersnipe*, 14-minute single-channel video and performance presented at CHAT 2003 (University of Bristol), TAG 2003 (University of Wales Lampeter), Pixelache 2004 (Helsinki, Finland), Orange Ashton Court Festival 2004 (Bristol).

—— 2007a Faking it: why the truth is so important for TV archaeology, in T. Clack and M. Brittain (eds) *Archaeology and the Media*, Arizona: Left Coast Press.

—— 2007b *A Survey of Heritage Television Viewing Figures*, CBA Research Bulletin 1, http://www.britarch.ac.uk/publications/bulletin/piccini_toc.html (accessed 01.12.2008).

Russell, I. 2006a Images of the past: archaeologies, modernities, crises and poetics, in I. Russell (ed.) *Images, Representations and Heritage: Moving beyond Modern Approaches to Archaeology*, Springer, 1-38.

—— 2006b Imagining the past: moving beyond modern approaches to archaeology, in I. Russell (ed.) *Images, Representations and Heritage: Moving beyond Modern Approaches to Archaeology*, Springer, 361-66.

Schofield, J. 2006 *Constructing place: when artists and archaeologists meet*, www.diffusion.org.uk. Accessed 10 April 2007.

Travers, T. 2006 *Museums and Galleries in Britain: Economic, Social and Creative Impacts*, London: National Museum Directors Conference and Museums, Libraries and Archives Council.

Webmoor, T. 2005 Mediational techniques and conceptual frameworks in archaeology: a model in mapwork at Teotihuacán, Mexico *Journal of Social Archaeology* 5: 52-84.

Webmoor, T. and Witmore, C. 2005 Symmetrical Archaeology, Metamedia, Stanford University, Available at: <http://traumwerk.stanford.edu:3455/Symmetry/home>. Accessed 10 October 2008.

Witmore, C.L. 2006 Vision, media, noise and the percolation of time: symmetrical approaches to the mediation of the material world *Journal of Material Culture* 11: 267-92.

Chapter 18
Teletubbylandscapes: Children, Archaeology and the Future of the Past

Thomas Kador and Jane Ruffino

This chapter is based on a series of research projects and workshops with primary school students in which the children engaged with objects and spaces from both the present and the past in a variety of ways. We wanted to challenge, on a practical level, the 'top-down' model of education, along with the 'top-down' manner of archaeological interpretation, and so the children were encouraged to define their own priorities of engagement. If archaeology is about human behaviour and relationships, then a top-down model is not only insufficient for widening access, it is insufficient for explaining human relationships themselves. The result for us as archaeologists was a renewed enthusiasm for the value of unexpected interpretations and approaches to explaining the human past. Drawing on our work with the students, we provide a glimpse into their experiences and the ways they chose to engage, express and interpret. The chapter also offers some ideas about how we can truly encourage constructive, multiple interpretations of the past that allow and empower all to participate as active agents in the present.

> What we want to see is the child in pursuit of knowledge, not knowledge in pursuit of the child (G.B. Shaw).

Introduction and Background

The value of inclusive and participatory archaeologies has been demonstrated and widely documented internationally (e.g., Merryman 2004, 5-8; Smardz Frost 2004, 80) However, in Ireland public and community based archaeology remains something that practitioners and enthusiasts feel the need to justify. There are outreach programmes in schools, and there are efforts within museums and heritage organisations to provide better understanding of their holdings, but since these are existing collections and sites, they deal primarily in distilling expert-derived analyses and knowledge. These are, of course, useful activities, but it is equally important to engage the public in the analytical and interpretative processes of archaeology, not just for the sake of archaeological understanding, but because it can provide an intellectual space for people in the present to value their own forms of knowledge.

As archaeologists we place the need to democratise our field within the wider context of democratising knowledge. Among designers, technology experts, businesses, and many others, 'crowdsourcing' has become an important approach to problem-solving, information-gathering, and knowledge creation. It is not to replace the expert, but to value a range of types of expertise, curiosity and perspectives. In our work with children, they are (and should be) considered full members of the community, not merely a minority subset. We build on their existing curiosities and their literacy with the material world, using archaeology as a means to engage creatively with their own landscapes. They may not name this 'phenomenology', but they are perfectly capable of identifying how they perceive and understand places, spaces and objects. Their research helps to inform our research, partly by demonstrating to us the importance of dropping the jargon and figuring out how to say what we really mean.

This Discussion focuses on a series of archaeological workshops and projects for primary school students in disadvantaged areas of Dublin, Ireland which we conducted between 2004 and 2008. Initial contacts with the schools grew out of our involvement with the UCD New Era access programme, which 'aims to encourage and facilitate increased participation in higher education' (UCD New Era nd). Our work with New Era involved engaging with both primary and secondary school students for short workshops and lectures on archaeological topics, as part of New Era's school outreach programme.

Informal conversations with the students and their teachers made it clear that the students were excited about archaeology and wanted to find out more about the subject than the short workshops at UCD allowed. To facilitate this, we began designing a more elaborate programme that gave school students the opportunity and skills to critically engage with the archaeological past in their own neighbourhood. While we have worked with both primary and secondary students, this chapter will focus on our work with children in primary schools.

Motivations

We embarked on these projects out of a belief that interpreting the past should not be a minority exploit reserved for professional, specialist archaeologists but should be open to as broad a section of the community as possible. Since archaeology's research agendas have broadened to include the less-well-represented voices from the past, it seemed relevant, even necessary, to find ways to include the marginalised and non-expert voices in the present. In turn we hoped that a successful opening-up of the field for archaeological interpretation could also contribute in small measure towards a more inclusive present.

Moreover, there are clear additional benefits to working with and critically engaging primary school students in this way. Some of these have been recently documented by Isherwood (2009); they include: a first-hand learning experience outside the classroom, an opportunity to advance the study of local history, wider community involvement, and extending and developing the curriculum. These benefits alongside the goals of UCD New Era of broadening participation in Higher Education were all on our minds when designing and embarking on the projects. We hoped that by experiencing first-hand the excitement of analysis as discovery, it would help the students to cultivate a sustained curiosity about the world around them, even if it did not result in their eventual taking up a place in an archaeology undergraduate course.

From a research point of view, we felt that working with children on archaeological projects, and in particular through allowing them to be at the forefront of interpreting their experiences, might have to potential to contribute positively to our understanding and study of childhood in the past. In contrast to gender, discussions of age in general and childhood in particular appear poorly represented in the archaeological literature (see e.g., Chamberlain 1997). In this respect it was our hope that putting children in the driving seat of archaeological interpretation would go some way towards aiding the exploration of childhood in the past.

Objectives

While the above were some of our motivations that drove us to starting this programme there were a number of more immediate objectives for the work we carried out.

First, there was a very simple idea of exposing primary school students from socially disadvantaged areas to the methods, techniques and theories of archaeology, which represents a field of study generally reserved to university education and thus viewed as beyond these particular children's reach. Overall in Ireland, just over 40 percent of young people go on to third-level education or courses, although the rate of university attendance in some disadvantaged areas remains low (CSO 2009, 51-52). In a longer term sense, and if in line with the aims and objectives of the UCD New Era access programme (see www.ucd.ie/newera), is the hope that critically and creatively engaging children in 'higher learning', for instance through archaeological research problems, can help them advance further in their educational career.

A second objective was to help the students acquire skills for studying and interpreting the past which in turn would also necessitate the mastery of some more generic, transferable skills, such as critical observation and recording, report writing, numeracy, team-working and graphic representation skills. Because archaeology is not a subject on the Irish school curriculum, the students could feel they were all beginning on equal footing. In other words, by not having any universal standard by which they would be judged, the students could engage more fully, with less anxiety than they might have felt in conventional subjects.

Third, by focusing our investigation on archaeological features in the wider locality of the school we aimed to facilitate a critical dialogue on, and greater appreciation of local heritage in the wider communities within which the particular school is located.

Finally, another key objective was to produce a methodological framework for engagements with primary school students. It was important that we be able to assess how well we were meeting our objectives, so that we could improve and refine future approaches. Flexibility was crucial, especially since one of our key goals was to allow children to set their own agendas. We felt it was important to establish a framework that could be adapted easily depending on the number of students involved, resources available, time allotted, and the subject matter to be investigated (including any weather- or transport-related contingency plans).

Practical Issues

The workshops ranged from short sessions of two hours, to a six-week project during which we spent one full day per week with a group of school students. It is not insignificant to note that the enthusiasm of the students, teachers and administrators in the schools was always a key factor in the success of the projects. While community-based archaeology programmes in schools can help in certain areas, their efficacy relies very much on an existing framework of dedication and enthusiasm.

As organisers and facilitators we learned important lessons about how to engage in a primary school setting. There were a number of surprises, not least that what younger students lack in years of formal education, they more than make up for in curiosity. In a very serious sense, working with these students also changed our outlook on practical and theoretical questions.

However, as is to be expected, the projects also brought with them a number of challenges that are certainly worth discussing. Some of these relate directly to intrinsic and well-documented problems with the Irish primary education system such as the dominance of faith based education in Ireland, while others appeared to be more related to social and other issues that particular students and schools were facing. We will return to both the benefits and challenges in the discussion below.

Outcomes

By and large all students engaged critically and creatively with the artefacts, places and spaces we investigated. Some were more reluctant than others, but eventually, they all participated. Moreover, we were impressed with the high standard of their work in documenting, interpreting and creatively expressing their engagements with archaeological subject matters. While one of the difficulties of our methodology was the necessary minimisation of measureable objectives, the work that the students produced usually met and frequently exceeded our expectations.

For example, the six-week project involved students from a primary school in Dublin doing research on the 18th-century hunting lodge known colloquially as The Hellfire Club, located at the top of a nearby hill and clearly visible from the area near their school. We chose this because of its visibility and its proximity, and its mystique. The students researched the history of the building, constructed by William 'Speaker' Connolly in (1725) as a lodge that would sit between his castle at Rathfarnham and his mansion at Castletown House, Co, Kildare. In library research, one student observed that one of the walls—though bare—in the Hellfire Club looked like a picture she found of one of the walls in Castletown House, a possibly deliberate architectural reference. On an even more practical level, the same student noticed that, at eye-height, there was some previously unnoticed 19th-century graffiti (dated to 1848). Indeed, any researcher doing a survey of the building's interior might have found it, but it was a reminder that people of different sizes, shapes, abilities and backgrounds can all add to the kinds of information that is gathered, and bring different skills to gathering it.

There were also a number of unanticipated outcomes. As mentioned above, one of the benefits of the projects to us as archaeologists was that working creatively with primary school students and facilitated them to take charge in developing archaeological interpretations. This, in turn, allowed us to advance and challenge our own attitudes about the role of children and the contribution they make to society in the present, and prompted us to become more aware of the need to understand the meaning of childhood in the past and the roles of children as social actors in past societies.

Furthermore, some of the students' interpretations also served as eye openers to our archaeologically trained and often rather narrowly focused ways of thinking about certain features and thus helped broaden our horizons to equally valid alternative interpretations. For example, one group of students were brought to examine the interior and exterior of Fourknocks passage tomb, Co. Meath. On catching the first glimpse of the mound virtually all the students shouted out at once 'this is the Teletubbyhouse'. This striking similarity—whether the creators of Teletubbys had intended it or not—had never previously occurred to us but clearly was plainly obvious to the children. While we do not wish to suggest that Fourknocks could equally be interpreted as a Teletubbyhouse this event highlighted that we are frequently out of tune with children's perspectives and that to familiarise ourselves with their ways of seeing can enhance our own engagements.

Theoretical and Methodological Considerations

Theoretical underpinnings

Initially we did not start working on the projects with a body of well-formulated theory in mind. We were inspired projects such as Mark Edmond's research at Grandom's Edge and David Austin's work at Strata Florida, Wales, and had outlined some of our motivations, but mostly we enjoyed working with younger students, and felt that they had both a lot to offer to and a lot to benefit from archaeological workshops. After some short workshops as part of the UCD New Era on-campus programmes, opportunities began to arise, mostly from teachers who contacted us looking for extended workshops.

Once we started a programme of more elaborate and detailed work with primary-school students we began to look for theoretical and practical resources to support our practice. Consequently, we embarked on a search for literature on children participating in archaeology, and on the role and position of children both within the discipline of archaeology and in the past. While there is a reasonably large amount of material on practical archaeological work with children of various age groups—including hands-on and how-to guides (e.g., Branigan 2001; Duke 1997; Panchyk 2001), and publications designed as guides for primary school teaches (e.g., Limerick Education Centre 2005)—we found that beyond this, the literature on children 'doing archaeology' was rather sparse. Most of what was available were guides to allowing children to mimic processes and methods used by archaeologists, but either did not encourage them to see their interpretations as valid in their own right, or encouraged them to measure their analyses against those of experts, thus promoting the problematic idea that there are 'right' interpretations that can be verified.

Additionally, the question of childhood appears to receive relatively little discussion in archaeology, with most references to children or juveniles in the archaeological record representing 'pseudo-inclusion', or they are included in statistics, but not in analyses (Scott 1997, 3). Childhood itself might well be a relatively modern and Western concept (Montgomery 2009, 53-55), even this—if it is in fact true—would deserve some discussion and debate. This is additionally notable in Ireland, where the Western, middle-class ideas of childhood are only a few generations old. For example, while the youngest of the children in our programmes were about ten years of age, the eldest were approaching teenagehood. Their age-equivalents even forty years ago would have been on the brink of maturity, or at least the end of their formal schooling, after which they would have taken on increasing adult responsibilities (some of which they may already have had).

While there was some theory for us to draw on and we had some frameworks from which to develop practical guidelines, we also discovered that our workshops with children highlighted areas that were clearly under-researched and under-theorised in archaeological terms. Consequently, we began exploring the possibilities of the children's perspectives we encountered having potential to making an actual contribution towards researching children and childhood in the past. It also prompted us to see the need to reconsider many of the under-challenged presumptions about gender, age, abilities, and even the division between work and leisure, which we as archaeologists continue to project onto past societies. 'Man, the Hunter' has not fully been demythologised, we simply pay slightly more attention to his 'wife' and 'children'. We also still tend to view the past in the context of family units, especially when it comes to children, but young people have rich lives of their own, as individuals and in groups with other children and with adults. Scott notes the importance of children's culture 'transmitted child to child through social contact, [that] exists to the extent of being prime movers in intra- and inter-generational linguistic change and cultural preferences (1997, 3).' It would be remiss to suggest that the mainstreaming of teenage slang into adult parlance is comparable with how past societies worked, but it is surely likely that children and the younger members of the community may have affected change in the more distant past, if in different ways than they have been in the more recent periods.

The fact that most of the research explicitly discussing the role of children in the past appears to be conducted within a gender studies or feminist framework (e.g., contributions in Moore and Scott 1997; Dawson 2000), provides us with an interesting parallel. Gender issues only became prominent in archaeology within a context of feminist archaeology which, in turn, resulted from a greater involvement of women in academic research and public life from the 1960s onwards. It was not until the 1980s, when greater numbers of women—and specifically feminist women—began to access positions of academic authority, that the idea of gender archaeology really came into view. Many of the fundamental outlines of past societies had been laid out without concern for gender or age categories, and focused mainly on the achievements and lives of adult males, or defined achievement as androcentric. While lip service has been paid to women and children in the past, the narratives found in museum displays were generally androcentric ones that, if they included non-adult males, did not do so in ways that challenged the narrative (Sorensen 2000, 33). As Scott (1997, 7) points out, 'Even within much feminist research children retain a residual status, a form of domestic identity through the duties and emotional ties of motherhood.' When we visited sites with the students, we encouraged them to imagine how someone their age in the past, a might have experienced this place and how they may have been perceived by others. At the Hellfire Club, we asked the students to imagine what a child might be doing at a hunting lodge, or at a club reserved for the rakish sons of the lesser gentry, and they began to think about the kinds of jobs that would have been done by children. Instead of imagining themselves as the adult male card-sharps, they wondered what it must have been like for a child servant, hauled up Montpellier Hill to serve the hard-drinking men.

Our workshops served as clear reminders of the necessity to bring those who are marginalised in scholarship into the centre, both in how we include others in the present, and how we read archaeological material. It should also be noted that childhood and youth are temporary categories that also relate to gender, social roles and norms, and identities. As Chamberlain (1997, 249) emphasises, '[C]hildren contribute to the archaeological record whether or not we are competent to recognise them.' In other words, by including children in the research agendas, we are also including adults; we're just starting earlier. In this sense we would like to suggest the possibility that involving children in the process of archaeological interpretation could have a positive effect on our understanding of the role of children in the past. It would be, in part, much like the effect feminist and gender archaeology had on the role of women. But additionally, it might go some way toward destabilising the categorisation of groups, where children and women tend to be lumped together, thus making it more difficult to recognise the agency of either independently.

As adults who inhabit adult worlds and partake in adult discourses, we can sometimes be guilty of forgetting about younger and smaller people. Rather than look for material evidence of their presence without which we continue to assume a place was adults-only, we need to start from the premise that they were present. Witnessing the engagement of children with sites and monuments serves as a key reminder that we need to broaden our intellectual and notional access to past landscapes in past and present. This led us to a series of fresh and unexpected insights about the role of children in the past and to alternative interpretations and narratives associated with some of the places we investigated. The realisation that simply by interacting with a place and set of material culture—without us telling them what it is and what it means—the children could challenge our perceived wisdoms and interpretations led us to re-frame the practical methodology in such a way to optimise these critical engagements.

Methodological issues

As stated above developing a methodology for working with school students on archaeological programmes was one of the key objectives of the overall programme. More specifically, our intention was to devise a flexible methodological framework that could be easily adapted to fit a specific project's need, be it a relatively short workshop or a longer, ongoing project. While an integral part of any of these projects and workshops was a first hand, practical experience of working with actual archaeological objects, landscapes or monuments, it was up to us to provide the physical tools and practical archaeological knowledge from which the students could build their interpretations. They needed to learn to think and view the world like archaeologists.

One element that made a notable difference was to provide the students with authentic archaeological objects for interpretation in the workshops. Being entrusted with an actual archaeological artefact, as opposed to a replica, allowed the children to feel more authentic as archaeologists. Additionally, with archaeological objects, the patina of age invokes a sense of mystery and wonder that can be more easily harnessed and built upon. The challenges that this in turn provokes will be discussed below. What we did not anticipate, however, was that when we brought the children to sites or introduced them to objects, their concepts of 'old' and 'new' differed from our own. At Fourknocks, Co. Meath, the metal door of the fuse box, no more than a few decades old, was considered 'old', as it was badly corroded, while the concrete roof, of the similar date, was 'new'. Again, this provided a reminder that concepts of past and present change, not only with generations, but with the age categories we belong to. The eagerness with which the children embraced multiple functions and meanings of tools, secondary modifications, and unintended uses was equally exciting. The main challenge was to structure a workshop that would provide a framework for the kinds of ideas that came easily to the children, rather than try to direct their interpretations.

The Three Stage Model

To help them acquire the required understanding and skills we divided the process of archaeological investigation into to four central fields of engagement; observation, description, interpretation and communication. Furthermore every workshop or project was divided into three stages, regardless of its duration.

Stage one

The first stage involved acquiring and putting into practice each of the four basic principles (or fields of engagement) of archaeological investigation. Once these principles had been introduced, the students had to master some simple practical skills for investigating spaces and objects. To test their skills we use a technique best described as 'making the familiar strange'. As an ice breaker exercise we introduced the children to personal objects belonging to us and encouraged them to be nosy about us by making inferences about our lives, taking them slowly through each step of observation, description and then interpretation. In the shorter workshops, this step would lead directly to the presentation of archaeological artefacts.

In longer workshops, through practical exercises that can be carried out inside the classroom, in the school gym or outdoors students learned to use measuring tapes, draw a sketch of their subject and appropriately record and document every step along the way. They investigated their own classroom or another familiar part of their school utilising the techniques and terminology they had just been introduced to. We encouraged them to describe the objects around them purely based on their size, shape and appearance without using their actual names. For example a table could be referred to as 'a flat rectangular wooden board suspended on four metal rods, all of the same length and in equal distance from one another'. For the longer projects, the students were also asked to keep a journal in which to record all their observations, descriptions and interpretations. Depending on the duration and scope of the particular project the students could also present their findings to their colleagues at the end of stage one, which represented a good way to let the children really own the legitimacy of their work.

Stage two

Having mastered the first stage, the students then apply the same techniques to older material, with the reverse aim of 'making the strange familiar'. In shorter workshops, this generally involved stone or bronze axes, post-mediaeval pottery, clay pipes, and selected other lithics in order to provide a range, and help the students use their interpretations of one object to build a set of questions to ask of others. In longer workshops, this meant a visit to an archaeological site or feature in the local area. At this point, they would take the same techniques that they have applied to familiar material and apply them to a strange subject—an artefact, building, site or monument—in a strange environment. This formed the second stage and main part of the investigation. In groups of two to four students got to explore a manageable segment of the feature that the whole class investigated. We aimed to focus investigations as much as possible on features located in the locality of the school. Apart from reducing travel time and costs, through the focus on local heritage we hoped to encourage a wider interest and discussion among the students and their peers, their parents and families, their teachers and the wider community.

For this stage of the project we always aimed for maximum parental involvement. This allowed a wider share of the community to partake in the project, but the fact that the students felt adults were interested in what they were doing provided another motivating factor and further validation. More than the other two stages of the projects, the second stage required detailed tailoring for each group or class. Thus the research might entail architectural building survey, archaeological landscape investigation, artefact analysis, library and/or desktop based background research. However, regardless of the particular duties involved, all students get to deal with 'real archaeology' and have to maintain a record of their duties, insights and experiences.

We must also point out here that the programme to date has not involved any excavations. There are several reasons for this. First, the Republic of Ireland has stringent national monuments legislation which decrees that all archaeological work that interferes with the ground or could potentially lead to the discovery and/or removal of artefacts must be licensed by the National Monuments section of the Department of the Environment, Heritage and Local Government (DoEHLG 2004). Therefore—even apart from the financial implications—it would be virtually impossible to conduct an excavation in the context of our programme without having planned the exact nature of the excavation several months in advance. Second, we explicitly set out to challenge the primacy of excavation in the public's perception of archaeological research. Put differently, the projects are aimed at demonstrating that there is much more to archaeology than digging holes in the landscape and that the majority of archaeological research is possible without disturbing what lies underground. We wanted to provide students with the analytical tools to do archaeology at their leisure, if they so chose, after we were gone. This way, too, the students, their teachers and parents come to realise that the landscapes they inhabit are all fundamentally archaeological and are full of evidence of the activities of past generations. All it takes to start uncovering these are some basic skills and a new perspective of looking at their surroundings. Thus after having participated at one of our projects the students will hopefully have gained a greater awareness of and sensitivity towards the local heritage that surrounds them, and we hoped, helped them to value their existing curiosity about the world in general.

Stage three

Throughout the projects we stressed that an archaeological research is only as good as the quality with which it is communicated, as there is no point making significant discoveries without telling anyone about them. Consequently, the third stage and ultimate outcome of every project is the presentation of the discoveries to a wider audience. With the smaller projects this could take the form of some visual displays exhibited on the school grounds. The larger projects might be presented through the publication of a report and an open day at which the students get the opportunity to share their discoveries with fellow students, teachers, parents and other members of their community.

Part of the methodology, was also to attract a wider community interest in the projects, where it was appropriate to do so, and the presentation of the research in the form of an open day often grants an excellent opportunity for this. To facilitate wider community interest and involvement, the projects drew on existing networks, such as the generally wide reaching contacts of teachers, school principals and parents. Utilising these channels has brought local media and political attention to the projects and has thus helped the school to act as a centre of convergence for the wider community.

Problems, Challenges and Resolutions

While we met the very basic objectives of our programmes, there were certainly some challenges that cannot be ignored. Some appeared relate to wider problems with the Irish primary education system. Perhaps the foremost of these is the faith-based patronage of more than 90 percent of all schools in the state (88 percent alone are Catholic, accounting for 92 percent of all Irish primary school students; O'Toole 2009). This contribution is not the appropriate place to discuss the woes of the Irish primary education system, but we do feel it worth highlighting it as one of the areas that has directly affected our work.

The most striking impact of the religious ethos of the schools was in relation to gender. Although there are many Catholic national schools that are co-educational (i.e., boys and girls together) there are also many that are segregated. During our programme we had the opportunity to work with both types of schools, although whether by coincidence, or associated with gender-specific stereotypes and perceptions of what boys and girls are interested in, none of the schools we worked with were girls only. In one boys' national school, we worked with a group of students who enthusiastically participated in the artefact analyses and interpretation. When we introduced our personal objects in the beginning of the class, their perceptions of how modern objects were gendered did not seem alarming. But when we then introduced some neolithic stone axes, the boys were adamant that no female of any age would or could possibly have made or used these. The responses implied women in the past are and should remain invisible and suggested that the boys related to adult males rather than to children in the past. Over the duration of the project it transpired that some of these preconceptions about gender roles were already so entrenched in 10-12 year olds that they directly affected the outcome of the workshops. It made it quite difficult to get them to break out of interpretative habits regarding modern material culture, in order to challenge the biases that were projected onto the past.

In contrast to this, the next school we worked with, a mixed-sex primary school with children of similar ages, we decided to explicitly deal with the gendering of objects. We laid a series of objects on a table, and asked the students to identify who owned which one and why, and then asked them to think of alternative interpretations that would lead to a different answer. One of us brought a screwdriver, and the other a sewing kit. What stumped them was not the gendering of the objects, but that they were the least obviously personal of the assemblage. When asked if they could tell if a woman or a man owned either of these, they said it was not possible, because a woman might need a screwdriver as much as a man would.

We also felt an impact of the religious ethos on the project as part of our work with one of Dublin's two Islamic schools. However, this could have been partly due to our own familiarity with Christian and Western value systems and also due to the fact that none of the (lay) teachers who we worked with were of Muslim background themselves. They were thus overly cautious not to cause offence to any of the students or their parents by engaging them in activities that may not be considered appropriate, without however being entirely familiar with Islamic teaching themselves, despite the fact that both the school and the parents had already agreed to the workshops taking place. In some cases parental involvement was a challenge in its own right, but this was largely an extension of the existing difficulties in encouraging parents to become involved in their children's schools.

One of the biggest challenges is the institutional one to opening up the process of 'academic' research to non-specialists, whether they are children or adults. Apart from a reluctance and traditional attitude to university teaching and research still held by a small number of colleagues, the difficulties are compounded by the reliance of departments on funding that is largely determined by research output, and on the immense pressure academics face to produce publications in peer-reviewed journals. While it is possible to write up projects like ours for academic journals, without specific funding for community-based projects, even an eager department would face difficulty in providing the resources for student-facilitated workshops or programmes. Because of the simplicity of the approach, even for a fairly long-term project, budgets do not need to be very large, and most of the tasks can be completed with resources available in schools. What they require most of all is the time commitment of at least one, but preferably two archaeologists as well as the required time being made available by the responsible teacher and school principal. It is in relation to making time available and compensation for this that the programme has most often run into difficulty.

Most schools would not have the necessary budget to finance one, let alone two archaeologists with at least some teaching experience give up several hours, or even days to work with their pupils. With the exception of access offices like UCD New Era, few universities seem to appreciate the relevance and importance of collaborating with the primary school sector. Third-level institutional funding might be somewhat easier to come by for projects at secondary schools where there is arguably a more straightforward relationship between taking an archaeology workshop and an interest in subsequently taking archaeology at university. However, an interest in archaeology can develop at pre-secondary level, and among students who might feel unenthusiastic about conventional subjects. The idea that there is more 'out there' than what is taught in secondary school may help the struggling student retain some perspective and stay motivated.

However, most universities have a stated general commitment to working for the benefit of their local communities and consequently, it is up to those of us who work within academic archaeology to make them hold true to these commitments. Given the stretched nature of most university and archaeology department budgets, and in particular in the current context of declining exchequer contributions to Higher Education, one way that universities can honour their commitment to the community is to frame workshops and projects like the ones piloted here within a community engagement and service-learning context. This, in turn, would also create some exciting learning environments for final year undergraduate and postgraduate students in archaeology and would give them an early (career) experience of teaching the subject (National Service Learning Cooperative Clearinghouse 1984). Admittedly, this would take some re-drafting and re-designing of existing teaching and learning structures but we are confident that, once implemented, a community engagement, service-learning module in archaeology would deliver immediate returns. One key benefit of such modules would be the improvement of communication skills for archaeology graduates, which are arguably not our greatest strengths. When it comes to communicating archaeological problems and discoveries with a wider public, professional archaeologists seem to find it difficult to connect, and this would help people to develop archaeological vocabularies and an understanding of the purpose of archaeology as it is practiced in the early 21st century.

Bringing a wider range of people into the dialogue about the past will make our discourses more interesting and challenge the received wisdom about heritage priorities in terms of preservation as well as presentation. The existing heritage legislation is detailed, complex, and wide-ranging, but no legislation can take into account the individual sites and spaces that are important to people and to communities. By encouraging people to engage more fully with their landscapes, beginning in childhood, we hope that archaeology can become more inclusive in the past and present, but also that that inclusivity and engagement can be part of a model for community empowerment and dialogue. Archaeologists have always relied on information from 'locals' about archaeological sites and landscapes, but that information should be valued in its own right, and, in tandem with the research of academic archaeologists, used as a springboard for the construction of multiple narratives of the past. Valuing local knowledge and encouraging its development would help professional archaeologists as part of the evaluation process in conservation of heritage, and it would feed into existing community structures, and empower them to find solutions to land-use conflicts that are in the interest of the people who engage with it most intimately. For the future, it will help determine what heritage is protected for subsequent generations, but contributing to land-use discourses in the present also shapes the archaeology of our lives that we leave as a legacy.

Conclusion

It would be unreasonable to suggest that archaeology can make the world a happier or better place (McGuire 2008, 14-15), but making the discipline more inclusive benefits everyone. Archaeology relies on making tangible connections between past and present, relating objects to people, and then people to each other. But apart from archaeologists relating to one another, or sharing the results of their analyses with the public, most of archaeology's current connections and relationships seem to confined to past communities. The nature of the discipline became more interesting, complex, enlightening and exciting as gender and ethnicity became issues for exploration. Likewise the addition of indigenous archaeologies has infused the discipline with new vigour. Children, it seems, remain the 'social baggage' who, to archaeologists, only become important in the past upon reaching maturity. And yet, when we think about biographies of notable individuals, it is their childhood that often fascinates us most. Consequently, we feel it is time to take on the bias towards adult males inherent not only in our narratives of the past but in the ways these narratives are constructed. Actively involving members of minority communities, but most prominently, children in writing the story of the past will provide one of the most effective ways to get there.

Acknowledgements

We would like to acknowledge all the assistance we received from the teachers, school principals and parents at the schools who participated in the programme to date. Additionally we must thank the staff of the UCD New Era Programme for support and partly funding the programme. For additional funding and resources we would like to acknowledge the Michelle Schofel foundation of Pacific-American Securities and the UCD School of Archaeology. Most importantly, we wish to thank all the students for their hard work and enthusiasm and for the enjoyment and critical insights that working with them has granted us.

References

Branigan, K. 2001 *Make it Work! Stone Age People: the hands-on approach to history*, Minnetonka, MN: Two-Can Publishing.

Chamberlain, A.T. 1997 Missing stages of life—towards a perception of children in archaeology, in *Invisible People and Processes: writing gender and childhood into European archaeology*, (ed.) J. Moore and E. Scott Leicester: Leicester University Press, 248-50.

CSO 2009 *Measuring Ireland's Progress 2008*, Dublin: The Stationary Office.

Dawson, T.A. 2000 Why Queer Archaeology? an introduction. *World Archaeology* 32(2), 161-5.

DoEHLG 2004 National Monuments (Amendment) Act 2004, (ed.) Department of the Environment Heritage and Local Government Dublin: Government Publications Office.

Duke, K. 1997 *Archaeologists Dig for Clues,* New York: Harper Collins.

Isherwood, R. 2009 *Involving Schools in Community Archaeology Projects*, Council for British Archaeology.

Limerick Education Centre 2005 *Archaeology in the Classroom: It's about time!* Limerick: Limerick Education Centre.

McGuire, R.H. 2008 *Archaeology as Political Action*, Berkeley: University of California Press.

Merryman, N. 2004 Diversity and dissonance in public archaeology, in *Public Archaeology*, (ed.) N. Merryman London: Routledge, 1-17.

Montgomery, H. 2009 *An Introduction to Childhood: Anthropological perspectives on children's lives,* New York: Wiley-Blackwell.

Moore, J. and E. Scott (eds) 1997 *Invisible People and Processes: Writing gender and childhood into European prehistory*, Leicester: Leicester University Press.

National Service Learning Cooperative Clearinghouse, N.S.L.C. 1984 *Defining Service-Learning*, California: National Service Learning Cooperative Clearinghouse (NSLC).

O'Toole, F. 2009 Lessons in the power of the church, in *The Irish Times,* Saturday 6 June 2009.

Panchyk, R. 2001 *Archaeology for Kids: Uncovering the mysteries of our past*, Chicago: Chicago Review Press.

Scott, E. 1997 On the incompleteness of archaeological narratives, in *Invisible People and Processes: Writing gender and childhood into European prehistory*, (ed.) J. Moore and E. Scott. Leicester: Leicester University Press, 1-14.
Smardz Frost, K.E. 2004 Archaeology and public education in North America: view from the beginning of the millennium, in *Public Archaeology*, (ed.) N. Merryman. London: Routledge, 59-84.
Sorensen, M.L.S. 2000 *Gender Archaeology*, Cambridge: Polity Press.

Chapter 19
Serial Closure: Generative Reflexivity and Restoring Confidence in/of Anthropologists[1]

Caroline Gatt

In the past decades public engagement has been marked by ambivalence within the circles of academic anthropology. Prominent anthropologists such as Nancy Scheper-Hughes (1995, 2009), Julian MacClancy (1996), Thomas Hylland Eriksen (2006) and Michael Herzfeld (2009) are increasingly calling for a more publicly engaged anthropology. However, a benchmark report produced in 2006 by the UK Economic and Social Research Council noted that the 'invisibility' of anthropology in public engagement was not mainly due to a lack of anthropologists actually participating in public life. Rather anthropology was invisible because anthropologists did not readily identify themselves as 'anthropologists', preferring the title of 'expert'.

In this chapter I review the argument that aspects of post-modernism and the older ontology of mono-naturalism/multi-culturalism create a two-way lack of confidence: on the one hand, certain public spheres lack confidence in anthropology; on the other, anthropologists' lack of confidence in the value of the knowledge they produce. This lack of confidence in and of anthropologists prejudices engagement with public or ethical matters. Uncertainty about methods partially explains why alternative theoretical paradigms to post-modernism have not been already widely adopted in anthropological research in order to counter the hegemony of mononaturalism/multiculturalism in certain public

1 This chapter is based on two papers, presented at a panel at a ASA 2009 convened by Katy Fox and Caroline Gatt and a CRASSH workshop organised by Lee Wilson and James Leach. I would like to thank the participants at both workshops for their feedback on this work. The fieldwork from which I draw these ideas consisted in six months participant observation with FoE Brazil, five months with the FoE International Secretariat in Amsterdam, six months with FoE Malta as well as the three years between 2003 and 2006 during which I engaged with FoE Malta as an activist, nine international meetings between 2003 and 2007 and continuous email participant observation throughout the period from February 2003 to December 2007. I gratefully acknowledge the University of Aberdeen's Sixth Century Studentship award for funding my doctoral research project. I would also like to thank the FoE activists in Malta, Brazil and Amsterdam, Tim Ingold and Arnar Arnasson, Stephanie Koerner, the participants in the University of Aberdeen's writing-up seminar, and Richard Muscat for their comments on earlier versions of this chapter.

realms. Drawing on doctoral research with Friends of the Earth International, an international federation of environmental NGOs, I propose a strategy of 'serial closure' as an approach to re-store anthropological confidence without prejudice to the complexity or novelty of its matters of concern. 'Closure' refers to aims to focus on the contextually contingent, 'serial' summarises something of the form that arises through the necessary partiality of any such 'closure' as well as through our responsiveness to another 'closure'—no less partial, but different.

Friends of the Earth

Friends of the Earth International was founded in 1971. On the current website run by the International Secretariat, FoEI is described as the 'World's largest grassroots environmental network, uniting 77 national member groups and some 5,000 local activist groups' (www.foei.org). One of FoEI's distinguishing aspects is that it is broadly focused on questions of environmental justice rather than more narrowly on such issues as conservation. When the secretariat refers to FoEI as a 'grassroots' federation it expresses an ideology of decentralisation. Underlying this ideology is the struggle for legitimacy for the policies that the federation promotes. Decentralisation is understood as the best tool currently available to ensure wider representation in FoEI policies, and to avoid a situation in which only the voices of a core, central group would be represented. In FoEI's spheres of operation, representativeness is considered a powerful index of political legitimacy.[2]

The actual discourse and practice of decentralisation in FoEI deserve further analysis, which I cannot undertake here. However, the ideological aim of decentralisation has produced a federation embracing diverse environmentalist ideologies. For instance there are nature conservation groups—such as Friends of the Earth Austria—as well as human rights advocacy groups—such as Friends of the Earth Philippines and each member group has its own unique mission statement. This diversity of member groups and types of environmentalism is taken as proof of the federation's decentralisation. Diversity in FoEI is celebrated and conceptualised as 'inclusivity'.

A number of subtle but significant details on their website, on documents and in speeches given by members of the executive committee or of the secretariat portray the federation as a clearly defined 'thing'. For instance when speaking English most members refer to Friends of the Earth International as FoEI (pronounced *fō-ī*) or as 'the federation' as if it were a stable thing. And it is almost ubiquitous that activists use the first person plural when talking to each other, they talk about 'who *we* are' or 'our goals', the website has a section 'about *us*'—as though the we/our/us were uncontentious. Of course in practice it is not simply that there is no 'FoEI' or no 'us'. In fact there is a core of people who have worked together in

2 Representativeness was a central index for legitimacy in other Maltese ENGOs (Gatt 2001) and in development NGOs (Lister 2003).

what they call FoEI for at least a decade. There are also some activists today who were part of the inception of FoEI in 1971. Others have associated with FoEI for much less although they identify their life trajectories with the history of FoEI. However, it is the tensions the rhetoric of 'unity in diversity' creates in practice in FoEI that have allowed me to explore different understandings of collaboration, from which I draw the idea of serial closure.

Two-way Lack of Confidence

Latour argues that mononaturalism/multiculturalism causes certain audiences to lose confidence in anthropology—and is often used as a means to silence the claims of 'others' (2003: 28-39, 130). He defines mononaturalism as an appeal to a single fundamental 'nature', upon which plural cultures are added on top. This division between one nature and many cultures is enshrined in the separation of the sciences into the natural and the social (see Figure 19.1). Mononaturalism has a silencing effect in public debate: within this logic science claims exclusive access to nature and thus assumes the power to settle differences of opinion.

Within their discipline, however, anthropologists have also lost confidence in the value of their contributions to public debate. Thomas Hylland Eriksen (2006) and Julian MacClancy (1996) list a variety of reasons for anthropologists' varying interest and confidence in public engagement. Among these are: changing funding patterns for academic anthropology, requiring more or less public engagement depending on the institutional support available; issues of representation in the post-colonial period which were heightened by the fact that those whom anthropologists previously 'represented' increasingly claimed and had access to the means of making their own voices heard; and the erosion, by the post-modern deconstruction of the notion of truth, of anthropologists' confidence in the worth of the knowledge they produce for public debate.

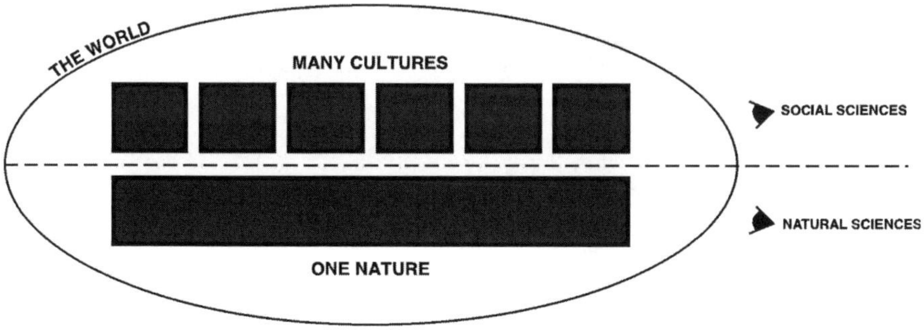

Figure 19.1 A view of the world according to 'mononaturalism/ multiculturalism'

Source: Diagram by Richard A. Muscat.

For this chapter I focus only on the latter, which Eriksen (2006) and MacClancy (1996: 2) argue has caused the authority of ethnographic accounts to be questioned. Edelmen (2002) further argues that the post-modern linguistic turn has distracted attention from political strategies, other than rhetorical ones. Others have even complained that academia has been engulfed by a hopelessness caused by post-modernism (Riles 2000; Miyazaki 2004). My argument here is not with post-modernism. It is only the aspect of post-modernism as Haraway describes it (1991: 153,155), where plurality takes leave of ethics and becomes an extreme form of relativism, that is relevant to the question of anthropologists' confidence for public engagement. I choose to focus on it because of the close affiliation between this sort of post-modern plurality and the multiculturalism that Latour shows creates a lack of confidence *in* anthropology.

Granted that in debates outside of academia anthropologists are considered experts only on multiple cultures whereas psychiatrists, for instance, are considered experts on 'Human Nature' (Eriksen 2006: 29), it is understandable that anthropologists' critiques are dismissed. Beyond this, however, Eriksen (ibid.: 30) argues that complexifying critiques will not displace the currently hegemonic basis for public engagement; only alternative paradigms will.

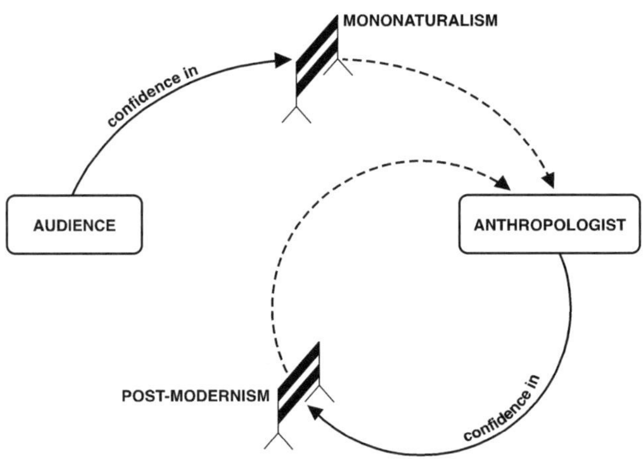

Figure 19.2 A two-way lack of confidence: For public engagement—anthropologists' lack of confidence and a lack of confidence in anthropology

Source: Diagram by Richard A. Muscat.

Ingold Latour Haraway

There are alternatives that can achieve this displacement and they can be found in the works of Tim Ingold, Donna Haraway and Bruno Latour.[3] They offer a way out of the hopelessness of post-modernism and the paralysis of mononaturalism. I do not argue here that together they offer a coherent theoretical paradigm. Rather the resonances between them offer a synergy that is helpful in trying to understand the constitutive forces of particular situations without the trade-off of having to leave politics, ethics and an engaged approach out of the analysis. Their work converges on at least four fundamental points.

First, all three theorists claim that human beings perceive the world directly. This is in contrast to the constructivist view of perception according to which humans receive inchoate stimuli from the outside world through the senses, which their minds then organise according to the patterns of a learned culture. The world is not a blank canvas upon which we inscribe our conceptual constructions (Ingold 1992, 1993; Latour 2003; Haraway 1991).[4] Ingold (1993) shows that without direct perception, no learning would be possible at all. Moreover, things often resist our conceptualisations (Ratto nd). Perception and knowledge can only be partial and situated because they are constituted relationally (Ingold 1992, 1993; Haraway 1991; Latour 2003, 2005). In other words, perception is not a process of reception, where our senses receive ready-made stimuli from a ready-made world. Rather, things emerge from multifarious relations in an environment, and once they emerge they are instantaneously involved in the ongoing coming-into-being of the world. Second, all things, are constructed relationally, as the outcomes co-constitutive relations amongst multifarious aspects (Ingold 1993, 2000; Latour 2003, 2005; Haraway 1991, 2003). Latour has often been misunderstood as supporting the notion of multinaturalism—which although multiple in its negation of grand single truths, reinforces a relativism that is unproductive for establishing due process for politics (Latour 2003: 29). To overcome this misunderstanding Latour proposes the idea of a single arena, oriented towards engagement, in which the single collective is being continually made (ibid.: 30). Third, because of this co-constitution it is important that research accounts for diverse relations, not only among persons (Haraway 1991: 198; Latour 2005, 2003; Ingold 2000b). In fact all three theorists comment specifically on the false division between human history and natural history (Latour 2003: 123; Ingold 1999: 252, Haraway 2003). Limiting the scope of social relations, in Durkheimian fashion, to a domain of human society abstracted from other aspects of the world, only leads to the tautology that the social is powerful because it is social in the first place (Latour 2005: 65).

3 There are others working along these lines, for instance Susan Oyama's (2000) *The Ontogeny of Information* in the area of developmental biology.

4 Haraway's (1991: 190) arguments about perception relate more specifically to the necessity to see from different positions. However, she also criticises arguments that propose 'some abstract version of existence' (ibid.: 153).

Fourth, all relations are processual; Ingold, Haraway and Latour put forward an ontology of dynamic becoming (Latour 2003; Ingold 2000, 2007; Haraway 1991, 2003). For Latour (2003) the world is a process, a progressive composition of the collective as it is in Ingold's (2006: 9) words, 'a world-in-formation'.

There are also clear divergences in their work. For instance Haraway's (2003) use of the notions of naturecultures and hybrids, in which earlier nature:culture, human:nonhuman divisions linger, does not sit easily with the idea, in both Ingold and Latour, that everything is the result of mutual constitution. On the other hand Latour does not address Ingold's (2006) critique of the notion of agency, namely that it attributes vitality to objects that are ready-made and rendered lifeless through their excision from the flows of materials in making. In the *Politics of Nature* Latour qualifies his notion of 'things'[5] at length. However, unqualified in this way in *Reassembling the Social* (2005: Third Source of Uncertainty: Even objects have agency) the notion of objects coupled with his use of agency risks perpetuating the very ontology that splits the world into domains of mind and nature, human and nonhuman, subject and object which, in *Politics of Nature*, he argues against. Haraway's (1991: 190) argument for the necessity to see from a number of different situated knowledges offers a practical approach to exploring questions of power and inequality, which Ingold, with his ecological focus (Ingold 2000, see also Ingold and Lee 2007), has not set out explicitly to address. Yet, these very foci are what give strength to the synergy, because these scholars do not replicate each other's work. With different foci they complement each other. Ingold's work elaborates in great detail an ontology that explains the processual and mutual constitution of our world including ourselves. Whereas Durkheimian[6] or Foucauldian[7] accounts are largely tautological because they cannot explain the emergence of power and social relations, Latour (2005) and other ANT scholars (see Ludden, D. 1985 and Heitzman, E. 1987) have elaborated how particular

5 The resonances between Latour (2003: 54) and Ingold (2000) come across also in the fact that both for instance define 'things' not as ready-made object but as relations and not incidentally both refer to the etymology of the word 'thing' lying in public process and discussion. In fact in a footnote Latour (2005, 65) also refers to James Gibson's notion of affordances to describe the relational quality of things; affordances being a notion that Tim Ingold (1992) introduced into anthropology.

6 Biernacki and Jordan (2002) argue that Durkheim's notion of society being *sui generis*, that is that each societies follow their own internal rules and structures not causally linked to other aspects of the world such as ecology (Hatch 1973), allows no room for an *explanation* of how society, and therefore social relations are more powerful than for instance ecological ones, since society is insulated from the rest of the world and yet the most powerful aspect of it. Therefore Durkheim argues that social relations are the most formational aspects of human life because they are social—this is the tautology.

7 Although Foucault's work has created extensive attention to and description of the practices of the imposition of power and governmentality (cf. Inda 2005 and Burchell, Gordon and Miller 1991), his definition of power as omnipresent, since everyone is both subject to power and also exercises power (Kusch 1991: 138) does not explain *what* power is.

social relations come into being and exert power in what they call sociotechnical systems (Pfaffenberger 1992). Finally, Haraway's (1991, 2003) work offers concrete examples of research in this vein of ecological phenomenology that is inherently attentive to political and ethical issues.

If mononaturalism/multiculturalism and post-modernism have led to a lack of confidence in partial accounts, Ingold, Haraway and Latour propose a partiality that yearns to relate: to build relationships and to tell stories. Haraway talks of her accounts as 'non-Euclidean knots of partial connections' (2003: 98). The key is that they offer an alternative to the logic that sees culture (or genes) as cause and behaviour as effect.

Challenge of Methods

Theoretically, then these alternatives could do away with the two-way lack of confidence of and in anthropologists. The question that needs to be asked is why have we not seen many examples of anthropologists engaging publicly using such approaches? One reason is the continuing uncertainty about the methods they entail. Due to a lack of clear methods there is also a lack of established forms of evidence that go with the theories; and evidence, although a highly contested notion (Engelke 2008), is a crucial ingredient for public engagement (Eriksen 2006). The result is that even those anthropologists working in particularly engaged situations, such as in aboriginal land claims around the world, find it difficult to employ these theoretical paradigms (Robert Wishart Pers. comm).[8]

Csordas's 1994 volume *Embodiment and Experience* exemplifies this problem of methods and evidence. Csordas (1994) and his contributors attempt to collapse the mind-body dichotomy (which is an integral part of mononaturalism), but conceptually they end up reinforcing it. Their central method and source of evidence in their analysis of the body is discourse—talk and ideas about the body (ibid.).[9] As Ingold (2000) shows, an analysis of the body that ignores the

8 See James Weiner (2007) for a parallel issue where the juxtaposition of tradition and history in native title cases has meant that anthropologists are often left out of evidence because they have a sophisticated approach that does not understand tradition as something that is ready made and history as something that is processual but rather the two work in concert. The result is that the courts will go to historians for the evidence instead because they do not problematise the dichotomy. Thanks to Rob Wishart for making this point.

9 Csordas (1994: 11), referring to Heideggar, makes a very valid argument that language does not only represent or refer to experience, but being part of that experience 'discloses' something of our 'being-in-the-world'. However, by restricting analysis to what people say is as good as denying that any other form of analysis is valid in the endeavour to understand their experience. Furthermore, in the preface Csordas (1994: xi) specifically says that the volume offers essays 'as a step towards an anthropology that is not merely about the body, but from the body', yet the focus on discourse analysis contradicts this. On the difference between talk *about* the body and talk *from* the body, see Brenda Farnell 2000

organism has the effect of reinforcing the division of the sciences into natural and social. And as I described above, according to Latour's logic of mononaturalism/multiculturalism this division between the sciences reinforces public lack of confidence in anthropology.

Although diverse methods may be being explored it is not necessary to wait for these to become established, mainly because we cannot depend on a single method or recipe of methods. What is needed is a starting point from which to develop, change or adapt methods. In my doctoral fieldwork in Malta, Brazil and the Netherlands with Friends of the Earth International, I set out to address this challenge of methods: How could I pay attention to the co-constitutive relations between different aspects of the various environments I moved through? It turned out that experimenting with different methods in response to different situations was a good way of working. An essential aspect of the strategy I propose, *serial closure*, lies in this emphasis on responsiveness. Serial closure, therefore, is a series of strategic conclusions. In my fieldwork each 'closure' was a decision on my part to stop and focus on one aspect. But the process was serial because the aim of closure was not to cut further responsiveness, but to be open to other aspects calling for attention and to readapt my focus to explore those subsequently. But before I go on to explain serial closure and give examples of it in practice, it is necessary to mine the currently hegemonic ontology/epistemology, resonant with mononaturalism/multiculturalism, which remains present in other forms of enquiry, in order to reveal the different premises from which serial closure arises.

Different Epistemologies[10]

Inherent in mononaturalism is the idea that knowledge is a representation of an external 'objective' nature (Latour 2003). This ontology/epistemology imagines the world to be made up of 'individual entities and events, each of which is linked through an external contact—whether of spatial contiguity or temporal succession—[and that connection] leaves its basic nature unaffected' (Ingold 2007). This is a particulate understanding (ibid.) in which knowledge is assembled from data that specify *discrete portions* of the world (see Figure 19.3). These data can then be integrated into models of smaller or larger scale. This particulate view of the world also underpins the reductionist approach in the 'natural' sciences. When this view is transferred over into 'social' science, 'society as a whole' is divided into sub-groups and sub-sub-groups, and so on, down to the level of the individual. Thus it sees society as an abstract whole comprised of concrete parts.

Getting out of the *habitus*: An alternative model of dynamically embodied social action. *Journal of the Royal Anthropological Institute* (N.S.) 6: 397-418.

10 Since an epistemology is dependent on a particular ontology in the following two sections I switch between them. Using 'ontology' when I refer to an understanding of how the world is, and 'epistemology' in reference to ways of knowing.

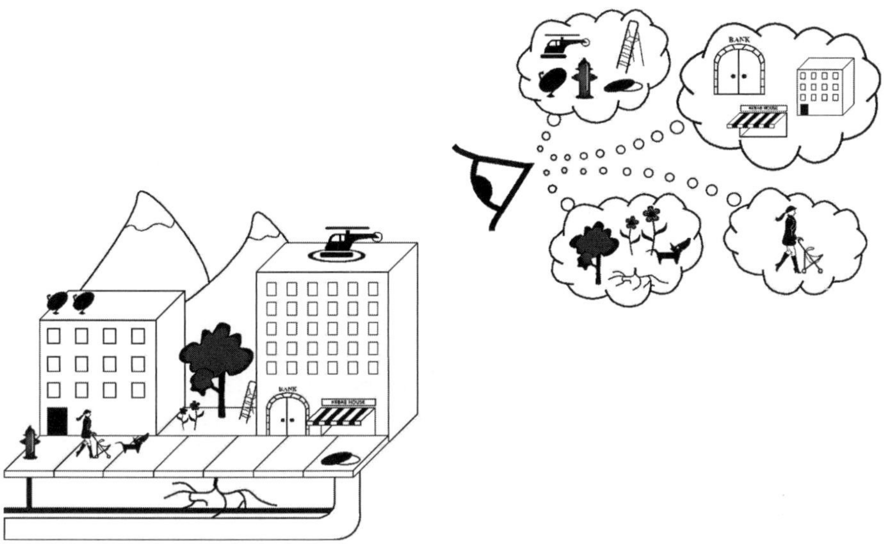

Figure 19.3 A particulate ontology and epistemology

Source: Diagram by Richard A. Muscat.

Within this logic of a particulate world, collaboration is a matter of assembling discrete bits of knowledge; bits that in multi-disciplinary collaboration can be slotted back together to create a fuller model or representation of 'society' or the 'world' (Leach 2008). However, in the same way that reassembling dismembered parts of an organism cannot bring it back to life, so one cannot recover the process of social life by piecing together isolate knowledge of its particulars, of individuals and events (ibid., Ingold 2007).

As I have already shown Ingold, Latour and Haraway offer an alternative ontology, where 'any particular phenomenon…enfolds within its constitution the totality of relations of which, in their unfolding, it is the momentary outcome' (Ingold 2007: 18). Consequently, rather than representing an unchanging world, knowledge in a relational/processual ontology is emergent from any engagement with persons, things or experience. Being part of the world, knowledge is also relational and processual. Furthermore, once such knowledge emerges, it also contributes to the processes in which it becomes enmeshed.

Anthropologists value knowledge that emerges from spending time *with* people and things in places (Ingold 2007). Many examples spring to mind of anthropologists whose theories resonate with the lifeways of the people among whom they carried out fieldwork (see also Tsing 2002: 468). As I mentioned above, my own approach of serial closure also arises largely from FoEI activists' own practices.

Figure 19.4 A relational/processual ontology and epistemology

Source: Diagram by Richard A. Muscat.

The Dialogic Emergence of Serial Closure

Images do count; they are not mere tokens, and not because they are prototypes of something away, above, beneath; they count because they allow us to mover to another image, exactly as frail and modest as the former one—but different (Latour 2002: 32).

In *Politics of Nature* Latour (2003) proposes a 'new constitution', in which matters of concern would shuttle between two houses: the house charged with the power to take into account and the house charged with the power to put in order. The former house is where new matters of concern are constantly being taken into account (ibid.: 109). The latter house is where, through due process matters of concern are instituted as 'provisional essences' (ibid.: 112). Being provisional, essences may once again become matters of concern and instituted as essences again and so on, in a 'feedback loop' (ibid.: 125). I call this process of shuttling back and forth serial closure.

Serial closure can be found in practice in FoEI activists' own current decision-making practices. In FoEI it can take a number of years for activists to reach consensus about a policy or mission statement. Once a decision is reached it is temporarily fixed, and people will make an effort to avoid that discussion being re-opened, at least for a period of time, in order to allow discussion of other issues.

Yet at the BGM of 2006, held at Abuja, the newly instituted 'Mission Vision and Values' were written in white letters on a black cloth hung at the back of the plenary room. This document had taken three years to be negotiated, 'word for word' as Meena Ramaan, then chairwoman of FoEI, described the process. In her opening speech Meena urged us to turn and look at the cloth that embodied, and made tangible and visible, the hard work of three general meetings, the intervening regional meetings and many hours of work of the Executive Committee, the International Secretariat and a number of other teams in FoEI. After the toils 'we' went through to create them, she said, they were not up for discussion anymore. And in fact, rather than contesting the Vision, Mission and Values, the sentiments that were expressed during that first session of the 2006 BGM were of 'strong ownership of *the* values', 'solidarity in the network', 'the common values that we come together with and the strength that comes with the diversity' even 'a feeling of love'. Reading this last comment, Meena 'oooohed' loudly and the plenary erupted into joyous laughter.

Yet, during the same introductory day, it was explained to us that our task in the meeting was to agree on a strategy. Where 'Mission, vision and values' was the task of the 2005 Penang meeting, the task of the 2006 Abuja meeting was to determine 'FoEI's strategy'. Incorporated into the ten-year strategy plan to be 'fixed' at Abuja was a feedback system, allowing the plan to be adapted and changed along the way. Opening and closing debate is incorporated into the long-term planning processes of FoEI—their closures are not end points. This is 'a process without beginning or end, punctuated but not originated or terminated' (Ingold 2009: 9). As for Ingold, Latour and Haraway, so also for FoEI activists, decisions and policies, like research foci and all other things, involve serial closures.

My previous discussion aimed to show how scholarship, and anthropology in particular, is a collaborative process (Ingold 2007, Tedlock and Mannheim 1995, Gatt nd). Some aspects of that collaboration can be of immediate value to the people with whom anthropologists do fieldwork. Life history interviews, for example, can have a therapeutic value (Gatt nd). Yet these things cannot be planned in advance. It is necessary to make space for moments of serendipity. However I propose that serial closure can be used to create things of value from the collaborative learning process of fieldwork in a more deliberate way.

In my fieldwork I explicitly set out to explore methods that responded to particular situations. Most anthropologists do not limit their field methods and theoretical insights *a priori*. Judith Okely (1996) advises ethnographers to be *disponible*, to open their attention as widely as possible. Paul Clough (1999) describes the process of fieldwork as a funnel. In the beginning one's attention is cast wide, and during the process this attention comes to be focused more and more keenly on a narrower range of themes. I would go further, to argue that in practice, whether casting one's attention wide, or gradually narrowing, it is still necessary at any one time to focus on some things more than others—always with a certain degree of peripheral alertness to what you may not be focusing on. Since each time the focus of one's attention falls on one aspect or another, one comes to some sort of conceptual closure in that there is 'something' to be focused on. This focus is never meant to put an end to questioning. Rather, it is a means to gauge how things are constituted in relation to others. This, in turn, is what makes it possible for one's focus to be readapted.

Reclaiming Evidence

So on the one hand, serial closure describes what anthropologists ideally do in fieldwork. But once made explicit, it can also be used to make valuable things for the people with whom we work as we proceed with our fieldwork, or to engage in debates outside of academia. The key issue here is to reclaim the notion of evidence from positivism; with serial closure the confidence in knowledge, and the traces of that knowledge—and I define evidence as traces of knowledge—would no longer be weighted towards generalisable truths or unchanging 'facts'. To the contrary, within the relational-processual understanding of serial closure, the value of evidence would come from being able to show how that knowledge emerges from particular situations, or in other words, that methods, analytic process and provisional conclusions are *responsive* to those particular situations.

Serial Closure in Practice

A couple of first attempts of serial closure in practice should ground this so far rather abstract discussion. The following are two examples of serial closure, one from within research and the other as employed in engagement. These instances arose from the collaborative learning process of my fieldwork and were then woven into the daily work of the respective FoE groups.

In the office of FoE Brazil in Porto Alegre, I noticed that in their regular meetings the activists would often adopt similar postures. At other times people would sit in ways that made continuous forms. When I asked direct questions, such as 'is how you sit at meetings important to how the group works?', the answers I was given were about how circles work well, how proximity can help or hinder group dynamics, but they could not articulate the specifics about mimicking postures or affinity through sitting. This is because these aspects are not explicit. However the activists' relation to each other in proximity drew my attention to how they had similar relations when not sitting close together.

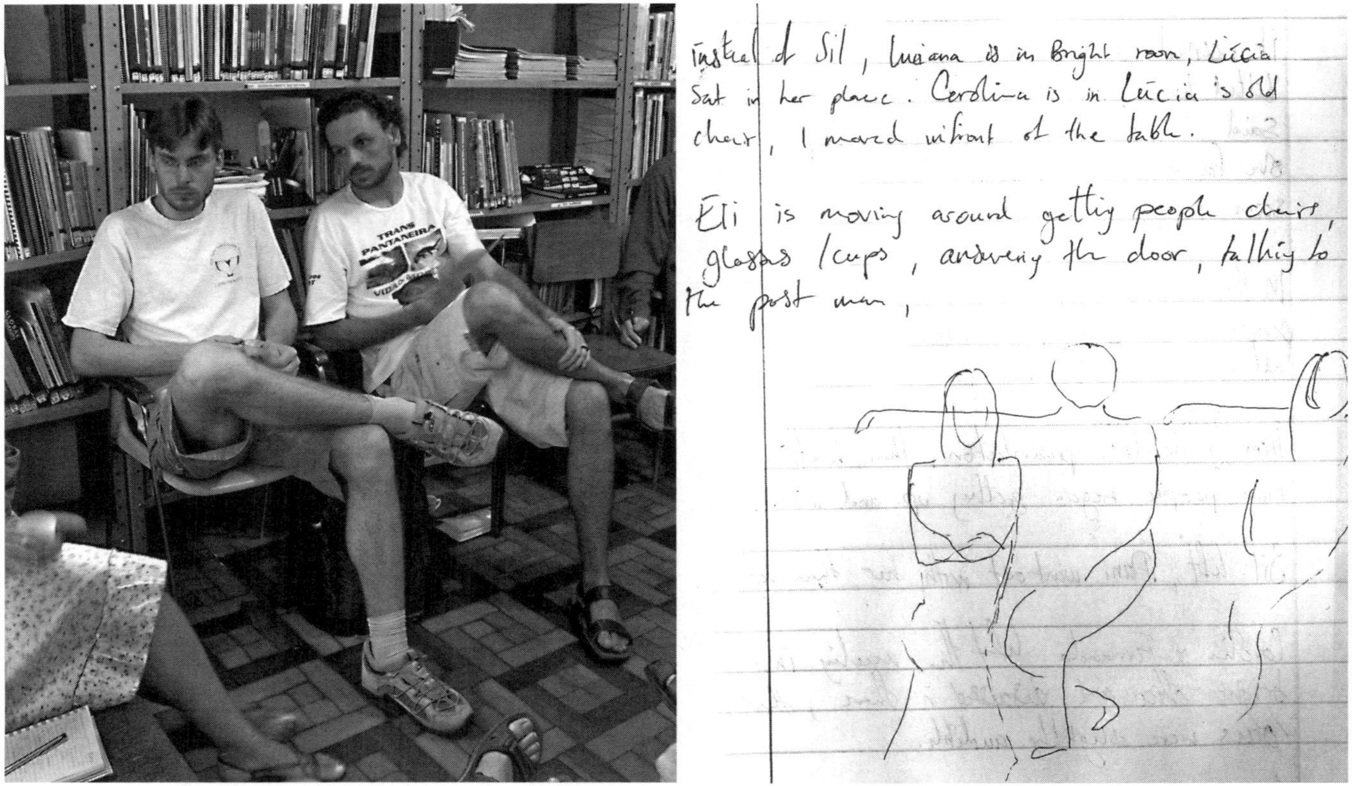

Figure 19.5 Sketch and photo of mimicked postures

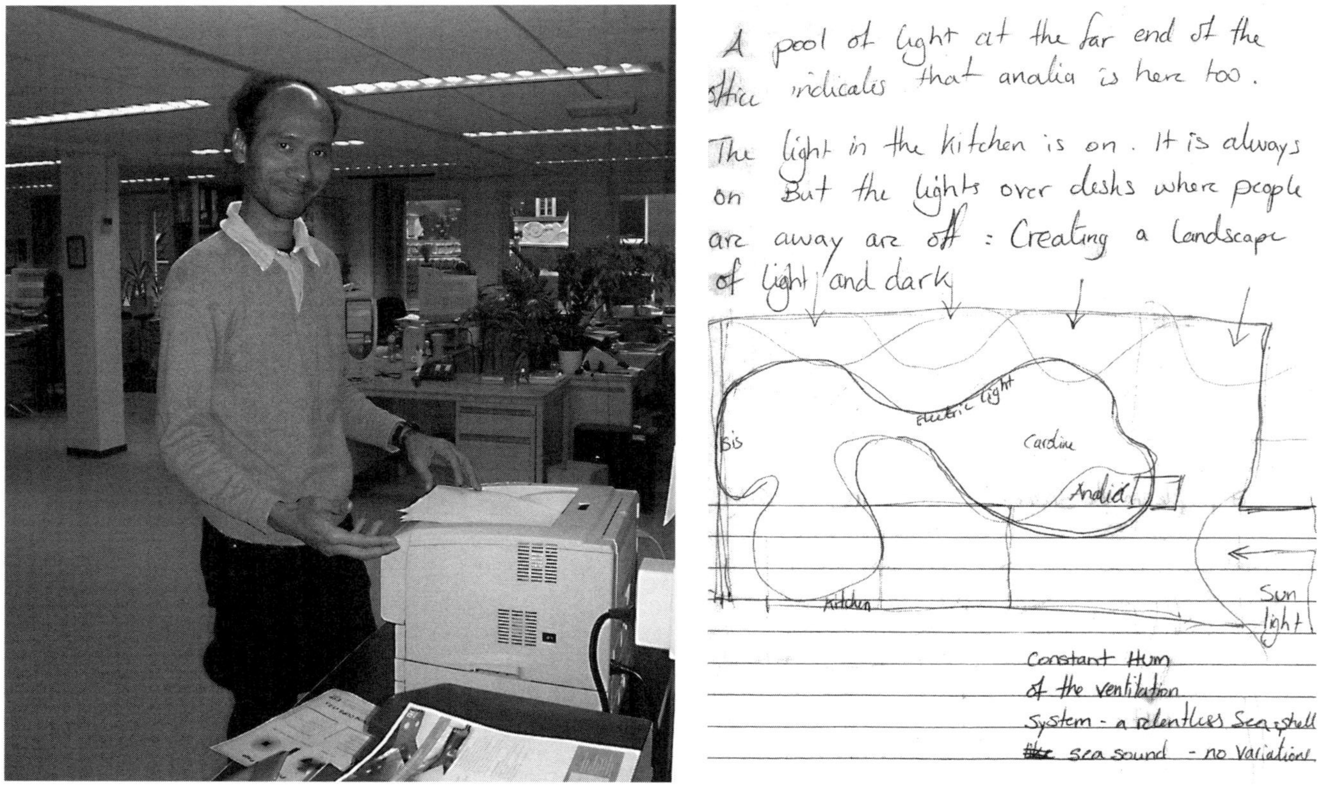

Figure 19.6 Pools of light

In the office of the FoEI Secretariat in Amsterdam I noticed that the electric lights were also part of the office dynamics. The sketch in Figure 19.6 shows the different pools of light the electric lights make. The activists knew who was in the office because the light over their desk was either on or off. Had I paid attention only to who had conversations with whom it would have seemed that half of the office had very little to do with the other half. But by paying attention to the activists' reaction to the lights it became clear that they participated in the life of the office not only by talking to each other.

Working with FoE Brazil, I was asked to draw up a statue and procedural regulations. The documents I produced arose from their need and desire for a guide that could incorporate enough structure to facilitate continuity and enough flexibility to incorporate their principles of diversity and inclusion. Both structure and flexibility were essential especially because they have a lot of staff and volunteers turn over and although they want some continuity, they also want to include what new members have to offer. These documents were not adopted unchanged; that was not the goal. What they provided was a concrete starting point, grounded in their concerns and experiences. With FoE Malta I also worked on the statute together with three other FoE Malta activists. In this case the statute was adopted as we proposed it in 2007. But it has changed over these two years because it is being used as a guide and needs constant updating.

Conclusion

To return to the two forms of epistemology and consequently understandings of evidence and authority that I started with, FoEI's decision-making processes stand in stark contrast to the way mononaturalism/multiculturalism silences debate. This contrast is mirrored in anthropology by the difference between choosing methods *a priori* and following the procedure of serial closure. In serial closure the confidence for public engagement comes from showing that methods clearly respond to particular situations, following a relational/processual epistemology. While it is true that processual approaches may not have immediate potency in public debate, it is also true that such approaches are already beginning to gain currency. In fact, the international bodies that fund FoEI find most value in the 'processes' of the international federation. Furthermore, Robert Bocock (1974) has argued that marginal movements often offer the sources of change that are then picked up by the mainstream.

Serial closure promotes a processual form of critical thinking rather than depending on generalisable truths or unchanging facts. Not only can this emphasis on process help to redevelop confidence in partial knowledge in debates outside the discipline, but it can also be used to promote more critical thinking within such debate. Furthermore, serial closure is a tool that allows that responsiveness to be made deliberately useful to working in groups.

I do not claim that serial closure is the ideal, nor is it the only possible approach. Rather the aim of this chapter is to highlight the continuity between theoretical paradigms, their corresponding methods and evidence and the effects of these on engagement and debates outside of academia. Through this discussion I hope to have offered a starting point to address the issues of uncertainty, ethics and engagement in practice.

References

Bocock, R. 1974 *Ritual in Industrial Society: A Sociological Analysis of Ritualism in Modern England*. London: Allen and Unwin.

Clough, P. 1999 The relevance of religion and culture to commercial accumulation: Fieldwork on Muslim Hausa and Agricultural Trade in Nigeria, in Barbara Hariss-White (ed.) *Agricultural Marketsfrom Theory to Practice*. Great Britain: Macmillan Press Ltd.

Csordas, T. 1994 Introduction: The body as representation and being-in-the-world in Csordas, T. (ed.) *Embodiment and Experience: The Existential Grounds for Culture*. UK: Cambridge University Press.

Edelman, M. 2002 Toward an anthropology of some new internationalisms: small farmers in global resistance movements. *Focaal* 40: 103-122.

Engelke, M. 2008 Objects of Evidence, *Journal of the Royal Anthropological Institute* (N.S.), S1-S21.

Eriksen, T.H. 2006 *Engaging Anthropology: The Case for a Public Presence*. Oxford: Berg.

Friends of the Earth 2008 Friends of the Earth International Home page, Available at: <http://www.foei.org>. Accessed: 21 January 2008.

Gatt, C. 2001 *Environmentalism in the Maltese Context: The Case of Nature Trust*, Unpublished Ba (Hons) Dissertation, Mediterranean institute, University of Malta.

—— 2009 Serial Closure: Doing Anthropology *With* Friends of the Earth, paper presented at a workshop on *Creativity and Innovation in Groups*, CRASSH, University of Cambridge, 8 May 2009.

Graham Burchell, C.G. and Miller, P. (eds) 1991 *The Foucault Effect: Studies in Governmentality*. Chicago: University of Chicago Press, 87-104.

Gupta, A. an Ferguson, J. 1997 *Anthropological Locations: Boundaries and Grounds of a Field Science*, Berkeley: University of California Press.

Haraway, D. 1991 *Simians, Cyborgs and Women: The Reinvention of Nature*. UK: Free Association Books Ltd.

—— 2003 *The Companion Species Manifesto: Dogs, People, and Significant Otherness*. Chicago: Prickly Paradigm Press.

Heitzman, E. 1987. Temple urbanism in medieval south India. *J. Asian Stud.* 46: 791-826.

Herzfeld, M. 2009 Whose rights to which past? Archaeologists, anthropologists, and the ethics of heritage in the global hierarchy of value Keynote speech presented to the 2009 Conference of the Association of Social Anthropologists of the UK and Commonwealth, on *Anthropological and Archaeological Imaginations: Past, Present and Future*, University of Bristol, 6-9 April.

Inda, X. (ed.) 2005 *Anthropology of Modernities*. USA; UK; Australia: Blackwell Publishing.

Ingold, T. 1993 The Art of Translation in a Continuous World, in G. Pálsson (ed.) *Beyond Boundaries: Understanding, Translation and Anthropological Discourse*. Oxford: Berg.

—— 1999 Human Nature and Science, *Interdisciplinary Science Review*, Vol. 24, No. 4 pp. 250-54.

—— 2000 Concluding Comment, in Hornborg, A. and Palsson, G. (eds) *Negotiating Nature: Culture, Power and Environmental Argument*. Lund: Lund University Press.

—— 2005 Movement, Knowledge and Description, in Parkin, D. and Ulijaszek, S.J. (eds) *Holistic Anthropology: Emergence and Convergence*. New York; Oxford: Berghahn Books.

—— 2006 Rethinking the Animate, Re-Animating Thought, *Ethnos*, Vol. 71 (1): 9-20.

—— 2007 Anthropology is *not* Ethnography, Radcliffe-Brown Lecture at the British Academy.

—— 2009 No More Ancient; No More Human: The Future Past Of Archaeology And Anthropology, Keynote speech presented to the 2009 Conference of the Association of Social Anthropologists of the UK and Commonwealth, on *Anthropological and Archaeological Imaginations: Past, Present and Future*, University of Bristol, 6-9 April.

International Benchmarking Review of UK Social Anthropology 2006 Available at: <http://www.esrcsocietytoday.ac.uk/ESRCInfoCentre/Support/Evaluation/ibr/IBR_Social_Anthropology.aspx>.

Latour, B. 2002 What is an Iconoclash? Is there a world beyond the image wars, in Bruno Latour and Peter Weibel (eds) *Iconoclash. Beyond the Image Wars in Science, Religion and Art*, 14-37. London: MIT Press.

—— 2003 *Politics of Nature: How to Bring Science back into Politics*. USA: Harvard University Press.

—— 2005 *Reassembling the Social*. Oxford University Press.

Leach, J. 2008 Gaps in Knowledge? Ethnographic practice, addition and collaborative endeavour, paper presented at the 2008 Conference of the European Association of Social Anthropologists, on *Experiencing diversity and mutuality* Ljubljana, 26-30 August.

Lee, J. and Ingold, T. 2006 Fieldwork on Foot: Perceiving, Routing, Socializing in Simon Coleman and Peter Collins (eds) *Locating the Field: Space, Place and Context in Anthropology*. UK, USA: Berg.

Lister, S. 2003 NGO Legitimacy: Technical Issue of Social Construct? *Critique of Anthropology*, Vol. 23 (2), p. 175-92.

Ludden, D. 1985. *Peasant Society in South India.* Princeton: Princeton University Press.

MacClancy, J. 1996 Introduction in *Popularizing Anthropology* MacClancy, J. and McDonaugh, C. (eds). London; New York: Routledge.

Miyazaki, H. 2004 *The Method of Hope: Anthropology, Philosophy, and Fijian Knowledge*. Stanford: University Press.

Okely, J. 1996 *Own or other Culture*. London: Routledge.

Ratto, M. 2009 Tim Ingolds Shoe, paper presented at SAnECH Seminars, University of Aberdeen, 26 February 2009.

Riles, A. 2000 *The Network Inside Out*. Ann Arbor: The university of Michigan Press.

Scheper-Hughes, N. 1995 The Primacy of the Ethical, *Current Anthropology*, Vol. 36 (3): 409-440.

Tedlock, D. and Mannheim, B. (eds) 1995 *The Dialogic Emergence of Culture*. Urbana: University of Illinois Press.

Weiner, J. 2007 History, oral history and memoriation in native title, in Smith, B.R. and Morphy, F. (eds), *The Social Effects of Native Title: Recognition, Translation, Coexistence*. Australia: ANU E Press.

Chapter 20
Vu jà dé: Cultural Heritage, Urban Planning and Hospitality

Monica Popa

> That is to say, whether looking at things in this way, we can see something we were not able to see before or to see something already known in a new light.
>
> Corsico 1994

Introduction

Cities are subjects to urban trends with major *future* implications in their effort to satisfy the needs and wants of *present* (potential) urban customers, having as ultimate goal the promotion of urban community prosperity. One may wonder: Can the cultural heritage—as that part of the past, employed for contemporary aims (Ashworth and Graham 2005)—be put in good use in order to contribute to a harmonious development of the ever-changing city? This chapter develops a 'city hospitality' approach to this and related questions.

If 'déjà vu' is the feeling that you have had an experience before, even though it is brand new, then 'vu jà dé' is what happens when you feel and act as if an experience (or an object) is brand new even if you have had it (or seen it) hundreds of times (Sutton 2001). In such a view, 'vu jà dé' can and has given rise to major 'paradigm shifts—as people rethink old assumptions and expectations' (Schmitt 1999). Sutton (2001) attributes the notion to Jeff Miller, who argued that 'the same old stuff seems new because it enables you to keep learning small lessons from every race (he refers to sailing) and it keeps you excited about the sport' noting that: 'Jeff's funny comment made me realise that innovative people and companies (organisations) have this same ability. They can keep looking at the same thing but keep changing which aspects they think about and which they ignore. People can learn 'vu jà dé' mentality; they don't have to be born with it'.

Cultural heritage as long been employed by urban planners—but mainly in the restoration of buildings and areas to enhance their economic value and/or to enhance their value in urban tourism sectors. Can there be other new ways of integrating heritage into urban development, for instance amongst means to address from perspectives offered by the notion of 'hospitality' some of the most difficult challenges facing urban planners under circumstances summarised by expressions like 'globalisation' and 'risk society'?

Changing Orientations Towards Urban Development Planning

In order to explore these questions, it helps to provide an overview of some key aspects of urban planning. Cities in general and city governments, city managers and other (semi) public institutions, have as their ultimate goal promotion of the prosperity of the urban community (Van den Berg and Braun 1999). Two major elements influence the decision-making in urban planning when it comes to promotion of urban community's prosperity. These are: the needs and wants of the urban customers; and the developments in economics, technology, demography and politics are reshaping the environment for towns and cities in Europe. Among others Van den Berg et al. 1990; and Ashworth and Voogd 1990 distinguish three general categories of customers: (potential) residents, (potential) companies and (potential) visitors. The needs and wants these customers are direct related to the following elements:

- For the (potential) residents the welfare depends on three elements: (1) the quality of the home and environment, (2) the availability and diversity of jobs and the ensuing income differences, and (3) the supply of such facilities as shops, schools, health care, culture leisure activities, social welfare and religion. The residential *attractiveness* of a location is relevant as long as it is related not only to the relevant environment but also to the level of facilities to which people aspire.
- For (potential) companies the decision to migrate or to invest inclusive locate in a region is determined by their access to location factors such as: raw materials and auxiliaries, labour, services, customers, and transport facilities for them all. Nowadays the location attractiveness of an urban region, for companies in order to relocate, depends more and more of: its environmental status, its quantity and diversity in manpower, its quality of gatherer of information, personal contact and agglomeration economies.

A city's attractiveness to *visitors*, *as* urban tourist destination depends on the following basic conditions: the city's ability to supply a range of highly competitive—where originality is perceived as a major strength—urban primary (e.g., natural, historical and cultural characteristics, attractions created for the specific purpose of drawing visitors, and events) and secondary tourist products (e.g., hotels and restaurants). Complementary tourist products do not themselves draw visitors, but contribute to the attractiveness of a city's primary tourist products.

The developments that are reshaping the environment for towns and cities in Europe (Bramezza 1996) making them to compete and complement each other at regional, national and international levels are identified as six according to Braun (2008). The first fundamental development is globalisation and internationalisation, which according to Camagni (2001) is the growing planetary interdependence of societies and territories. A second relevant development concerns the rapid changes in information and telecommunication technology, which seems to have 'opened up the globe'. A third factor is the political developments such as the integration processes within EU and its expansion, the NAFTA and the GATT negotiations. Increasingly, the global shift of political and economic power to South East Asia and in particular China has been labelled as another geopolitical change with has an effect on the economy and jobs in cities in the Western world (World Bank 2005; Van Winden 2005). A fourth fundamental change is the terrorist threat. According to Savitch and Ardashev (2001) cities have become the central venues for terror. Hence the urban safety in general is a top priority for many cities (see also Van der Berg, Pol, Mingardo and Speller 2005). The changes in terms of transport and infrastructure as well as changes in the transportation industry are considered as the fifth fundamental development. The advancement of the high-speed train network (see also Pol 2002) that and the introduction of low-cost airlines are urban phenomena's with substantial impact on urban tourism and air traffic among cities. Finally the prominent position of the (mass)media. The printed media has since long been important for cities, but we refer in particular to the tremendous development of the audiovisual media with big media conglomerates. Goldhaber (1998) argues that we are drowning in information and messages, and that there is great competition for our attention in and from media.

One might conclude that in such a competitive environment, the old 'certainties' no longer exist. The death of distance will at least equalise the location advantages of all places in the world allowing almost infinite decentralisation from higher-cost or less-efficient locations to lower-cost, more efficient ones, and thus transferring activity to wherever people happen to be (Hall 1998). Consequently the cities will compete based on such competitive advantages as the quality of living environment (expressed as quality of life for inhabitants, environmental status for companies and city image for—but not only—visitors), the level of cultural opportunities and access to knowledge. These examples of location factors come more and more up front on the 'list of needs and wants' of the urban customer. In this amalgam of trends with major *future* implications for city development, there is no easy answer for the satisfaction of the *present* needs and wants of the urban customers. Hence one may wonder: Does *the past* speak to the city? And in what way? *Can the cultural heritage*—as that part of the past, employed for contemporary aims (Ashworth and Graham 2005)—*be put in good use in order to contribute to a harmonious development of the ever-changing city?*

Integrating Heritage

In this respect I would like to succinctly (1) define the cultural heritage; (2) define its functions; and (3) present two particular cases of cultural heritage intervention in the present European urban environment. An item of heritage could be defined as (1) tangible e.g., cultural site itself, whether single location or scattered; a cultural object or objects, of one or many types, housed at a location or locations

(Snowball 2008); or (2) intangible e.g., mythologies, beliefs, folklores and the products of creative imaginations (Ashworth and Larkham 1994). The worth attributed to these artefacts rests less in their intrinsic merit, and more in the fact that heritage is that part of the past which we select in the present for contemporary purposes, whether they be economic or cultural (including political and social factors) and choose to bequeath to a future (Ashworth and Graham 2005). Thus a conscious selection of these artefacts from an entrepreneurial city is desirable. The two functions of heritage are the economic and the cultural. There is a direct relation between these and the place identity: they employ and reinforce the place identity through creation of a diversity and community feeling. As Ashworth and Larkham (1994) put it: 'after all, it is diversity—the desire to see different places and things—which fuels tourism, particularly heritage-related tourism'. And Hall (1998) noticed it: the cultural places with 'a unique fizz and buzz (…) will prove more magnetic' for 'a communal experience' (The Jerde Partnership International 1999).

In this view, heritage places are places of consumptions and are arranged and managed to encourage consumption, heritage being the most important single resource for international tourism. Tourism fuels regional or urban regeneration and employment creation (G.J. Ashworth and B. Graham 2005). Challenges in this sense for urban planning might be on one hand the effect of 'museum'-ification and heritage destruction on the other. The cases of Venice and large parts of Amsterdam city centre are typical examples for the former challenge, with relocation of the businesses and residents. For the latter the challenge is to respond to the tourist needs without destroying the urban heritage (Ashworth and Larkham 1994)—as in the case of many Eastern European cities including Prague and Bucharest where the creation of secondary tourist products are made at the cost of tourist attraction itself. One may conclude that tourism and heritage coexist when tourism does not dominate the local economy, even though the number of tourists may be large (Ashworth and Larkham 1994). When the cultural function is directly related to the creation and/or maintenance of diversity and communality can become a powerful tool for the city's hospitality towards its (potential) inhabitants. As such heritage is (1) a condition for what Relph (1976) coined as overcoming the loss of individual place-identity and (2) a condition of pluralist, multicultural societies, based on inclusiveness (Tunbridge and Aswhorth 1990).

Examples from Leinefelde and Rotterdam

An interesting example of overcoming the loss of individual place-identity is the one present by Kil (2009) in 'The Marvel of Leinefelde'. The town is a typical industrial new town in Germany facing the shrinking process of the ex-industrial cities of Eastern Europe. The housing offer was 'typical abstract, undecorated cubes'. The place was lacking identity, becoming a space unappealing even to its occupants. The past seemed unable to bequeath its heritage to the present. A solution was called upon including 'a good portion of anarchy, individual spirit as well as a number of bottom-up initiatives' (Kil 2009) taken by the city Mayer. Because of this 'new functions filled up empty links between single housing blocks, lifts were added to buildings which only had stairs before. Former housing blocks were transformed into offices, little shops or sun studios. When such enterprises succeeded, they changed the original shape or the buildings by adding spontaneous constructions (Kil 2009). The conclusion was: 'you shouldn't through (the past) away but convert it and take it into the project,'[1] having a retrofitted city as result.

In their plans, cities are trying to attract and maintain different segments of population e.g., young families with children, double income households without kids, well to-do-seniors, and students encouraging them to contribute to the social-economic and cultural urban identity. What about the 'undocumented people' (people without a residence permit) whom no plans for active participation were designated? Can these (temporary) citizens positively contribute to the urban cultural identity? The following example is one of creating space (even though at small scale and temporary) for pluralist, multicultural societies. In 2006 and 2007 the Mamre Project Rotterdam and the Rotterdam District Theatre offered a surprising alternative for a group of 'undocumented people': a unique opportunity to be heard. The theatre project called 'Status' offered the actors the possibility to share their experiences in a creative way, and spectators could go on a journey through colourful lives and cultures, without leaving the city of Rotterdam. A real encounter between fellow citizens who didn't know much about each other took place thanks to the way the play was designed and because producers, actors and visitors shared a meal during the performance (Bohmer 2009). One may argue that the above examples are congruent with Ashworth and Larkham (1994) words: what may be mor*e* appropriate, however, is the fostering of national, regional and even local diversity. By no means of a greater bureaucracy, standardisation and sameness, nor the subsuming of individual place-identities but rather to be widespread acceptance of existing diversity and plurality in parallel with the development of linking themes of Europe-wide relevance.

Hospitality

A conclusion can be drawn therefore. From the city development point of view a planning strategy for redevelopment projects should include: (1) an integrated vision of urban development, (2) the ability to develop strategic networks, (3) leadership and (4) knowledge regarding the dynamics urban systems and its managerial implications (Van den Berg and Braun 1999). The cultural heritage whenever applied to city development should be anchored in hospitality—as in space creation for identity and diversity. As Nouwen (1975) put it: 'Hospitality,

1 Interview with W. Kil at the New Towns for the 21st Century; the Planned vs. The Unplanned City, Almere, The Netherlands, June 2009 conference.

therefore, means the creation of a free space where the stranger can enter and become a friend instead of an enemy. Hospitality means (…) to offer them space where change can take place, (…) to open a wide spectrum of options for choice and commitment. Hospitality is not a subtle invitation to adopt the life style of the host, but the gift of a chance for the guest to find his own.'

References

Ashworth, G.J. and Graham, B. 2005 *Senses of Place: Senses of Time*, Burlington: Ashgate Publishing Company.

Ashworth, G.J. and Larkham, J. 1994 *Building a New Heritage: Tourism, Culture and Identity in the New Europe*, London: Routledge.

Ashworth, G.J. and Tunbridge, J.E. 1990 *The Touristic-Historic City*, London: Belhaven.

Ashworth, G.J. and Voogd, H. 1990 *Selling the City: Marketing Approaches in Public Sector Urban Planning*, London: Belhaven Press.

Berg, L. van den and Braun, E. 1999 *Urban Studies*, Vol. 36, Nos 5-6, 987-99 *Urban Competitiveness, Marketing and the Need for Organising Capacity*, Rotterdam: Euricur.

Berg, L. van den, Klaassen, L.H. and Meer, J. van der 1990 *MarketingMetropolitan Regions*, Rotterdam: European Institute for Comparative Urban Research.

Berg, L. van den, Mingardo, G., Pol, M.J. and Speller, C.J.M. 2005 *The Safe City; Safety and Urban Development in European Cities*, Aldershot: Ashgate.

Bohmer, M. 2009 Status: *A Stage for People without Papers*, Rotterdam: Wijk and Wijk Publishers/4u-im.

Bramezza, I. 1996 *The competitiveness of the European city and the role of urban management in improving the city's performance*, Amsterdam: Thesis Publishers.

Braun, E. 2008 *City Marketing: Towards an Integrated Approach, Erasmus Research Institute of Management* (ERIM), Rotterdam: Erasmus University.

Camagni, R. 2001 The economic role and spatial contradictions of global cityregions: the functional, cognitive and evolutionary context. In A. Scott, (ed.) *Global City-regions: Trends, Theory, Policies*, New York: Oxford University Press, 96-118.

Goldhaber, M. 1998 The attention economy and the net. Available at: <http://www.firstmonday.org/issues/issue2_4/goldhaber/>.

Hall, 1998 *Cities in Civilization*, New York: Pantheon Books.

Kil, W. 2009 Normalization of New Towns (Some East German Experiences). In *New Towns for the 21st Century* Conference Reader, Almere: International New Town Institute.

Nouwen, H. 1975 *Reaching Out: The Three Movements of the Spiritual Life*, New York: Doubleday.

Pol, 2002 *A Renaissance of Stations, Railways and Cities,* Ph.D. Erasmus University Rotterdam, Delft: TRAIL Thesis Series.

Relph, E.C. 1976 *Place and Placelessness*, London: Pion.

Savitch, H.V. and Ardashev, G. 2001 Does terror have an urban future? *Urban Studies*, 38(13), 2515-33.

Snowball, J.D. 2008 *Measuring the Value of Culture: Methods and Examples in Cultural Economics*, Berlin: Springer.

Sutton, R. 2001 *Weird Ideas That Work 11½ Ways to Promote, Manage and Sustain Innovation*, New York: Free Press.

The Jerde Partnership International 1999 *You are Here*, Phaidon Press Limited, London.

Winden, W. van 2003 *Essays on urban ICT Policies*, PhD Erasmus University Rotterdam, Rotterdam: Thela Thesis.

World Bank, Global Economic Prospects 2005 *Trade, Regionalism, and Development*, Washington: The World Bank.

Chapter 21
Creating Localities against the Grain of the Myth of the Eternal Return

Alia Ullah

This chapter explores the usefulness of Edwin Ardener's approach for illuminating some of the variety of ways in which 'rooted cosmopolitan' localities are created and transformed in light of research carried out in the neighbourhoods of Rusholme and Longsight in Manchester, and participation as artist and curator in an exhibition in Manchester's Whitworth Art Gallery in 2006, entitled 'Who are you? Where are you really from?'

Introduction

'Cosmopolitanism—derived from the Greek conjunction of 'world' (cosmos) and 'city' and/or society (polis)—has since antiquity referred to the notion of a 'citizen of the world'—a member in a universal circle of belonging that involved the transcendence of the particular and blindly supposedly 'given' ties of kinship and country' (Cheah 2006: 487). Writing in their introduction to an edited volume based on papers and discussions in the 2006 meeting of the Association of Social Anthropologists in April 2006, entitled 'Cosmopolitanism and Anthropology,' the editors note that:

> The year 1990 was a watershed one for the new cosmopolitan scholarship. The fall of the Berlin Wall, signaled the end of the Cold War, coinciding with an awareness of a speeding up economic globalisation, the spectacular rise of extraterrestrial media during the 1992 first Gulf War and increasing consciousness of the perils of ecological disasters...Against 'globalisation', a term implying free movement of capital and the global (mainly Western) spread of ideas and practices,...'new cosmopolitan' [and 'rooted cosmopolitanism' concerns] empathy, toleration and respect for other cultures and values. [It] is about reaching out across cultural differences through dialogue, aesthetic appreciation and respect; of living together with difference. It is also about the cosmopolitan right to abode and hospitality in strange lands and alongside that, the urgent need to devise ways to living together in peace in the international community (Werbner 2008: 2).

Important contributions to envisaging 'cosmopolitanism' along such lines have been made by efforts to revisit Immanuel Kant's (1784) formulation, as well as analogous themes in the works of such influential contributors to 20th century scholarship on political implications of aesthetics and 'mechanical reproduction' and 'mass communication' as Walter Benjamin, Theodore Adorno, Juergen Habermas and quite a number of others. (for instance, Habermas 1999). Kant developed his notion of 'cosmopolitan' pedagogical and political ideas and practices in an essay entitled 'Towards Perpetual Peace. A philosophical sketch' (1970 [1795]). The essay's most immediate motivation was the signing of the Treaty of Basel by Prussia and revolutionary France. For Kant the event was not a 'peace treaty', but simply a cease fire amongst the most powerful around an agreement on who to exclude from negotiations and partition. Kant's (1970: 93-130) essay objects to the 'senseless plunder' by the supposedly 'civilized' and 'the injustice which they show to lands and peoples they visit (which is equivalent to conquering them is carried to terrible lengths.' He went on to argue for democratising 'cosmopolitan' perpetual peace could be realised if new forms of 'publicity' would create a global public sphere in which people can regard themselves and others as free and equal 'citizens of the world'. But although he invoked the idea that this would begin at local levels and become more and then become globally widespread, he did not specify what sorts of contexts would be most likely to initiate the process. Instead a clash develops in his essay between his hope for public spheres—which would seem to be the contexts where the task might be most tractable (see Koerner's introduction to the second section of this volume). Instead, much of his 'philosophical sketch' argued for the relevance to the task of state's giving philosophical experts the authority needed to 'speak freely and publicly on universal maxims of warfare and peacemaking' for others (Kant 1970: 115).

In the views of Benjamin (1992 [1936]), Habermas (1999) and others, such a view is both undemocratic and unnecessary. For Benjamin, new technologies of mass reproduction of art, photography and cinema had undermined the 'aura' of artworks and created new democratic possibilities for publicity. For Benjamin—as for many of those arguing that the 'new cosmopolitanism is about reaching out across cultural differences through dialogue, aesthetic appreciation and respect; of living together with difference' (Werbner 2008) the arts were of crucial importance for creating and sustaining diversity amongst such possibilities. The challenges such efforts face are considerable. Many dynamics of 'globalisation and multi-culturalism' actually arises out of deepening inequalities, which form the contexts of the sorts of 'clashes' that Koerner attempts to summarise in the introduction to this volume. Writing on these inequalities, Richard Farden says:

> I think it is a matter of varied conjuncture of unequally empowered processes and imaginings that are localized conceptually as figures against the ground of global space that is both traversed and simultaneously classified, possessed and defended exhaustively…[I]t is a matter…of the complexity of agency recognised in particular places…If we give ourselves some elbow room with respect to

> the present these preoccupations with space and locality…and with how things occur in spaces and how they seem, appear symptomatic of the same condition (Fardon 1995: 2).

The above mentioned edited volume concludes with a conversation between Pnina Werbner and Stuart Hall entitled 'Cosmopolitanism, Globalisation and Diaspora' (Werbner 2008: 345-57)' that begins with Werbner asking Hall: '…so how do you see cosmopolitanism today in the world with all its apparently endemic, terrible conflicts, intractable…?' For Hall, how one envisages cosmopolitanism relates to ones experience of globalisation:

> Once our perspective becomes planetary, and there is a possibility of global citizenship, then cosmopolitanism as a utopia becomes potentially more possible. Of course, the actual form that globalisation takes—this interconnectedness—has taken, is exactly opposite. It connects disjunctive histories, the very early, the developed, the developing and the underdeveloped, the colonials and the colonialisers, the pre- and the post-colonial,…of global or transnational inequalities and conflicts etc.… (Hall 2008: 345-46).

For Werbner, as for Hall and a number of contributors to the *Anthropology and the New Cosmopolitanism* (2008) volume, the most pressing (and perhaps the most tractable) problem is not that of 'globalising democracy' but 'democratising globalisation', and the latter is likely to hinge upon new appreciation of 'rooted cosmopolitanism'. Or as Werbner (2008: 2) puts it: 'Against the slur that cosmopolitans are rootless, with no commitments to place or nation, the new post-1990s cosmopolitanism attempts to theorise the complex ways in which cosmopolitans juggle particular and transcendent loyalties—morally, and inevitably politically.' In such views, Richard Werbner (2008: 175) says: 'You can be cosmopolitan—celebrating the variety of human cultures; rooted—loyal to one local society (or a few) that you count as home; liberal—convinced of the value of the individual; and patriotic—celebrating the institutions of the state (or states) within which you live'. Werbner and Hall's conversation—and especially the concept of 'rooted cosmopolitanism' builds upon a very considerable history of anthropological efforts to explore processes summarised by the expression 'globalisation' through the 'prism of the local' (e.g., Tonkin et al. 1989; Appadurai 1981, 1991, 1995, 2001; Miller 1995; Inda and Rosado 2002). Over a decade ago, Arjun Appadurai asked:

> What is the place of locality in schemes about global cultural flow? Does anthropology retain any special rhetorical privilege in a world where locality seems to have lost its ontological moorings? Can the mutually constitutive relationship between anthropology and locality survive in a dramatically delocalized world?…[And:] How does *locality*, as an aspect of social life, relate to *neighbourhoods* as substantive social forms? Is the relationship of locality

> to neighbourhoods substantially altered by recent history, and especially by the global crisis of the nation-state? A simpler way to characterize these multiple goals is through this question: what can locality mean in a world where spatial localization, quotidian interaction and social scale are not always isomorphic? (Appadurai 1995: 204).

To the best of my knowledge, few authors have been more concerned to develop strongly reflective and deeply historically oriented approaches to such issues than Edwin Ardener (1989). In their pursuit of means to address problems posed by approaches to questions of 'How did the past lead to the present. And, how does the present create the past' centring on such dichotomies as those of real-historically contingent, reason-tradition, nature-culture, and so on, the editors of *History and Ethnicity* (Tonkin et al. 1989) included as the first chapter Ardener's (1989: 22-33) essay, 'The Construction of History: 'vestiges of creation''. In that paper, amongst other things, Ardener developed a framework, which suggests that questions Appadurai (1995) poses might be addressed by exploring how 'localities' are created through:

1. retrospective interpretations of events and historical processes,
2. means whereby events are remembered,
3. expressions of experiences of relationships between events,
4. personal relationships,
5. narratives and other forms of expressing historical experiences of (1)—(4), including those of researchers.

This chapter explores the usefulness of Ardener's approach for illuminating some of the variety of ways in which rooted cosmopolitan localities are created and transformed in light of research carried out in the neighbourhoods of Rusholme and Longsight in Manchester, and participation as artist and curator in an exhibition in Manchester's Whitworth Art Gallery in 2006, entitled 'Who are you? Where are you really from?'

Social Implications of Interpretations of Events and Historical Processes

> But then Asians happened…Not that a huge horde of them swamped the place overnight. Took decades. Little by little, house by house, the Asians moved in and the whites moved out. The trouble with Asians, especially pakis is they're different. Different clothes, different language, food, skin, and, of course, we got a different God. That's why the whites move out…After that, the only ones who'll move in are more pakis because whites don't want to know, not once the place has become polluted. And on and on it goes until you get these little enclaves, some would say ghettoes, sprawling up all over town. And then when the young punks start kicking up a fuss for whatever reason,

> in comes some smart ****** who tells the world that a place like Bradford suffers from self segregation. No ****ing shit Einstein. The whole world is segregated in a million different ways so why should Bradford be any different? (Alam 2002: 300-301).

Today, together with 'integration', 'fundamentalism' and 'extremism', self segregation iis the focus of much academic literature relating to multiethnic communities in many parts of the world, as well as of considerable public controversy. M.Y. Alam's novel '*Kilo*' 2002 provides a unique take on the multicultural city of Bradford and the experiences of British born Pakistanis. Les Back asked 'about the way Bradford has been written about as a segregated city' and Alam replied that 'What pissed me off about the whole discourse of segregation and fragmentation is that it was just simplistic beyond belief, it's intellectual junk food—it might taste ****ing good but too much of that shit is bad for you. Even now some Home Office report has come out saying some of Bradford's Muslim communities are self segregated, they're in self seclusion zones...Why are they segregated? They're segregated because it's a money thing' (Back 2004). Alam's concern with the issues of self-segregation amongst Pakistani Muslims in a multicultural British city is not unique. He is amongst many British Pakistani Muslims who feel they have become unwillingly embroiled in conflict over segregation vs. integration dichotomies.

There is a large literature of retrospective interpretations of the history of Pakistani Muslim 'ethnic groups' in the UK. The literature represents tendencies to divide between two cultures as though British and Asian were timeless facts. *The Second Generation Punjabi or English?* (Thompson 1974); *The Half-way Generation* (Taylor 1976); *Culture Conflict and Young Asians* (Ballard 1976); *In Search of Identity* (Meadows 1976). The notion that British Pakistani's are stuck between two cultures suggests that cultural identity is homogenous and static, but identity is heterogeneous.

Following World War II and the break up of the British Empire, South Asian migration to the UK increased through the 1950s and 1960s from Commonwealth of Nations countries such as India, Pakistan and Bangladesh, alongside immigrants from former Caribbean colonies. South Asians who came to Britain during the 1950s were 'part of this broader movement of labour migrations in Europe. Almost all the jobs available to them were those which the white workers did not want' and mainly included 'unskilled jobs involving unsociable hours of work, poor working conditions and low wages' (Brah 1996: 21).

There is much disagreement over how much of the migrants life ways was voluntary or not so. Ballard and Vaughn tend to stress voluntary choices made by migrants for instance about 'self segregation'. But considerable evidence questions how 'voluntary' the migrants choice was to move into concentrated inner city areas of Asians. Brah (2005) points out immigrants 'responsible for inner city areas' (Brah 2005: 37) became the object of resentment. 'This resentment was reflected in the negative constructions of the 'immigrant.' As was the case during the British

Raj, it was the Asian cultural practices which first came under attack. According to the stereotype, the Asian was an undesirable who 'smelled like curry', was 'dirty', wore 'funny clothes', lived 'packed like sardines in a room', practiced 'strange religions,' and so forth' (Brah 2005: 37).

During the 1960s and 1970s British Asians felt the brunt of racial discrimination and hatred. There are many examples of imposed segregation upon British Asian communities during this period. In the early 1960s the 'bussing' of Asian children to schools outside the areas they lived occurred (Brah 2005: 39). Politicians such as Enoch Powell made speeches which 'evoked images of the Asian as the archetypal 'alien.' The practice of 'Paki bashing' (an epithet naming the violence penetrated against South Asians in this period) reached its peak during the late 1960s' (Brah 2005: 41). In 1978 Margaret Thatcher pledged that if elected, her party would 'finally see an end to immigration' (Brah 2005: 50). Brah looks at how the Asian in Britain experienced the decade of the 1970s and the dawn of the 1980s. She suggests they 'witnessed a further entrenchment of institutionalised racism, particularly in the form of immigration laws and the British Nationality Act' (Brah 2005: 51).

Life history change goes unremarked. Young Asian people are in this relation uninteresting except when they present a problem, with consequences for attention focusing on problematic males. Alexander (2006) suggests that both the gang and the fundamentalist are viewed as male. 'This is linked in turn back to the wider representation of Asian Muslim cultures as inherently patriarchal and misogynist. Young women appear only as the victims of culturally rooted masculine control and oppression (Macey 1999), an image which belies the ongoing struggle by British Asian young women for their own space' (Alexander 2006: 267-68).

One of the key areas of controversy in academic and wider public discussion of globalisation and multiculturalism is—perhaps needless to mention—connections between religion and politics. One of the questions posed is whether politics is making people more religious. This huge question lies beyond the scope of my present aims but it does bear mentioning. But two questions bear mentioning. One is that of the extent to which characterisations of British Pakistani communities in terms of their religion may relate to emphasis in mainstream British society on the importance to democracy of secular politics. Second—and that relating more directly to topics of this chapter are problematic associations of characterisations of communities in terms of religion (and not just Islam) with images of their culture as variously centring on what Eliade (1963) might have called *the Myth of the Eternal Return*. For the religious historian Mircea Eliade, the 'myth of the eternal return' was a theory about a belief that the practice of certain customs and rituals made it possible to return to the time of origins, of myth—to become contemporary with the events described in one's myths.

Where the expression is used in literatures on multicultural communities, it is often associated with discussions of resistance to assimilation. Alison Shaw (1994) stresses that if we use this expression it should be—contrary to Eliade's conception—to stress the contextual circumstances and dynamics of the variety of forms that cultural customs can and have taken. For instance she notes that far from developing in a realm of 'eternal return' much commitment to Muslim identity relates to its providing young Pakistani's with a powerful and ideologically effective justification for the maintenance of family and *biradari* as solidarity' (Shaw 1994: 57). One of the advantages of the notion of 'rooted cosmopolitanism' (Werbner 2008) and the approaches developed by Appadurai and Ardner to 'vcreating localities' is that they help us avoid the impacts such stereotypes have on the variety of ways in which multicultural communities create localities (Figure 21.1).

Figure 21.1 The so-called 'Curry Mile', Rushholme, Manchester
Source: Photograph by the author.

Means whereby Events are Remembered 'Curry Mile' and 'Little Pakistan'

The idea of 'grassroots globalisation' activities may be useful for understanding the creation of and dynamics of multicultural localities in Manchester (e.g., Appadurai 2003: 9). According to Werbner (2002), the emergence of 'postcolonial Pakistani Muslim communities' has created 'new diasporas' in Britain around contrasting social, cultural and national ideals. For Werbner 'Pakistani diasporas' are on the margins of contrasts between the ideals of 'the South Asian, with its aesthetic of fun and laughter, of vivid colours and fragrances, of music and dance;

the Islamic with its utopian vision of moral order; and the Pakistani, with its roots in the soil, in family, community and national loyalties' (Werbner 2002: 251). Contextualising these ideals may help replace notions of diasporas with those of grassroots globalisation and/or rooted cosmopolitanism.

Rusholme and Longsight population dynamics

In the 1991 census from the Rusholme Ward Profile, 18.4 percent of the Rusholme population came from the Indian subcontinent. The vast majority of this percentage fell into the Pakistani category (Rusholme Ward Profile 1991). In 2001 the number of Pakistani's in the area rises to 5.9 percent of the population of ethnic minority groups in the Rusholme area, (Rusholme Ward Profile 2001) this increases in 2004 to 12.7 percent of Pakistani's in the area, the majority from an ethnic minority background (Rusholme Ward Profile 2004).

In the 1991 census from the Longsight Ward Profile the ethnic minority group with the largest population in the area came from the Indian sub-continent. Again, a vast majority o fell into the Pakistani ethnic category (Longsight Ward Profile 1991). In 2001 the number of Pakistani's in the area rose to 5.9 percent of the population in Longsight from a Pakistani background (Longsight Ward Profile 2001). The number of Pakistani's increased to 29.7 percent of the population in 2004 (Longsight Ward Profile 2004).

The percentage of Muslims in Longsight was 40.5 percent in 2004, the majority religion in the area. In Rusholme the percentage of Muslims is 25.4 percent, the second majority after Christianity (Religion Census 2004 Wards). All this suggests something of the influences of Pakistani Muslim communities on the landscape.

Re-thinking notions of self segregation; 'Curry Mile?' and 'Little Pakistan?'

Nasser (2006) has examined the changing dynamics of 'postcolonial spaces' with attention to British Asian landscapes. He says the growing importance of tourism as a generator of income has been a significant factor in this trend as 'essentialised, romantic notions of South Asian culture' has increasingly become a commodity for consumption. Manchester's 'Curry Mile' is an example of tourist led development projects (Nasser 2006 389-90). The name that the Manchester City Council gave to Wilmslow Road of 'Curry Mile' shows something of the cultural factors that impacted development of this predominantly Pakistani Muslim space in Manchester. By presenting Wilmslow road as a place for tourists to have a British Asian experience, the place is perceived as stagnant and is viewed as a way in which one can experience the exotic 'Other.' The 'Curry Mile' does provide a unique experience with distinct sight, smells and sounds. The restaurants and shops on Wilmslow road in the evening are lit up, and resemble the city of Lahore at night, all the way down the road the aroma of South Asian cooking is nostalgic of being in Pakistan, and the sounds from South Asian songs played in cars and shops as well as people openly speaking Urdu and Punjabi reminds one that the 'Curry Mile' is a space in which South Asians especially British Pakistani Muslims can congregate and remember their 'homeland', as well as enjoying being in Manchester. I argue that the 'Curry Mile' is not a tourist area, but a space in which diverse 'rooted cosmopolitans' often come into conflict. Manchester City Council has tried to resolve this issue by presenting the area as a tourist attraction to enable people from outside the local community to feel comfortable entering the area. Visitors have started to refer to Longsight as 'Little Pakistan'. But few inhabitants identify with this expression.

Neighbourhood heterogeneity

The 'Curry Mile' does not only have restaurants and takeaways, it is a place which provides the needs of British Pakistani Muslims. There are many halal grocers which cater for the Islamic needs of the community alongside South Asian fabric, clothing and jewellery stores that supply British Asian women with traditional attire from different areas of the Indian sub-continent, mainly the Punjab amongst Pakistani sweet shops (Figure 21.1). There are beauty salons exclusively for women, complying to the Islamic rule that men and women should not mix. There are music shops that supply the latest bollywood, bhangra beat and Asian garage music, one being the renowned Paan Rhythm Centre. Situated on the 'Curry Mile' are the two only Islamic bookstores in Manchester; 'Rolex Bookshop,' and the other 'The Islamic World Bookstore,' both shops cater exclusively for the needs of Islamic communities in Manchester, selling a wide range of goods for an Islamic lifestyle (see Figure 21.2). These different South Asian and Islamic facilities are interspersed amongst accountants, income tax consultancies, immigration advisory bureaus, insurance firms, real estates, travel agencies and newsagents. 'This suggests that the changing character of these commercial areas is the result of spatial and functional rather than morphological transformations. The active use of the pavement as an extension to the shop space for the arrangement of displayed goods and produce reflects a different conceptualization of public space by the incoming community originating from the traditional market economy' (Baumann 1996: 382).

The most attractive feature of 'Little Pakistan' is Longsight market which is located on Dickenson road and one of the busiest in the Northwest (Figure 21.3). The Longsight Market is renowned throughout the UK by the British Pakistani community as it is highly acclaimed for its extensive number of Pakistani stalls, which sell fabric, jewellery, shoes, accessories and household goods, there is also a new Islamic Book stall. These South Asian and Islamic stalls are interspersed amongst others. The market is reminiscent of traditional shopping areas in South Asia, where customers are able to barter prices, and a lot of the stall holders are able to speak Urdu and Punjabi, so first generation British Asians enjoy visiting the market for the traditional shopping methods. Second and third generation Pakistani's are not as used to bartering as it is not a cultural practice in Britain, some enjoy shopping with the older generation as they get a better bargain, but others feel embarrassed by these foreign cultural practices.

Figure 21.2 South Asian fabric, clothing and jewellery shop, Manchester
Source: photograph by the author.

Outsiders dread driving through the Longsight however, insiders like nothing more. Young men virtually all of Pakistani parentage, get together to cruise down the 'Curry Mile' and through 'Little Pakistan.' This requires the right make of car, and a sound system that can blast music at deafening volumes. The cruisers' satisfaction grows with the number of heads that are turned and with the whispers that identify one or the other as a 'big guy' in Manchester. Along the stretch, drivers pass the Pakistani Consulate which is situated on Dickenson road, between the busy areas of Rusholme and Longsight. It acts as a marker uniting both areas, and is visited by numbers of people from around the UK acquiring a Visa to visit Pakistan. Whilst they are waiting for their Visa to be made, they are able to experience both the 'Curry Mile' and 'Little Pakistan' which will send them down a path of nostalgia in preparation for their travels back 'Home.'

Nasser (2006: 384) suggests that 'the conscious act of remodelling elevations with ornamental features and decorative motifs has been a major development in the metamorphoses of the British urban landscape as a means of redefining the presence of the Other'. The development of postcolonial Pakistani architecture is both evident in Rusholme and Longsight. This is an example of the reversal of the exchanges during the colonialist period when British hegemony was expressed in the South Asian landscape through the architecture. The overt display of Pakistani Islamic symbolism in the landscape is a marker of the presence of British Pakistani Muslim communities in Manchester 're-defining their identity in no uncertain terms' (Nasser 2006: 389).

Figure 21.3 Open air market, Longsight, Manchester
Source: photograph by the author.

The main shopping centre in Longsight is near the corner of Stockport road and Dickenson road, it contains the only two Pakistani community centres in Longsight as well as the only Islamic bank which was built in 2005. 'Little Pakistan' is quite similar to the 'Curry Mile' as they both contain similar facilities that cater exclusively for the Pakistani Muslim community. They are presented differently as the 'Curry Mile' is viewed as a tourist attraction, whereas 'Little Pakistan' is viewed as an area that the Pakistani Muslim community in Manchester can congregate in, and displays more of a community spirit with extra facilities such as a number of mosques that develop the ideologies of the community.

Expressions of Experiences of Relationships between Events

Over the past decade, Eid celebrations in Rusholme have become focus of local news, as well as renowned for its extravagance throughout British Asian communities. Eid-ul-fitr marks the end of Ramadan; the holy month in the Islamic calendar where Muslims fast for thirty days. Ramadan is a period of reflection for Muslims where time is spent with family and friends and extra Taraveeah prayer is established where the Qur'an is recited every night.

Much controversy has stemmed around Eid-ul-fitr celebrations on Wilmslow road. The 'Curry Mile' is transformed on the night of this holy day with what would seem like one big street party where the colourful displays are visible from far off. Uncertainty and conflict surrounds celebrations of Eid-ul-fitr on Wilmslow road, and police presence is visible around the 'Curry Mile' as well as surrounding areas. In 2004 police in Manchester pleaded for a 'trouble free Eid-ul-fitr celebration', and the Greater Manchester Police (GMP) joined forces with the Pakistani Consulate, Manchester Council for Community Relations and the cities Muslim community to 'try to ensure this years celebrations go off peacefully' (BBC News). There are distinct uncertain boundaries created on the 'Curry Mile' during Eid-ul-fitr celebrations. The concern on Wilmslow road on Eid-ul-fitr stems from British Pakistani youths 'cruising' down the mile-long stretch, driving recklessly in high-performance cars. Due to the consumption of an excessive amount of alcohol (which is forbidden in Islam) and the predominantly male rowdy atmosphere, sometimes this can lead to violence. In 2004 the GMP issued a 'zero tolerance' (BBC News) policy which has applied during Eid-ul-fitr celebrations on Wilmslow road annually since. British Pakistani women are not normally permitted to join in Eid celebrations on Wilmslow road, however the last few years has seen a turn where young British Pakistani women have started to compete with their male counterparts regarding their make of car, reckless driving and anti-social behaviour.

British Pakistani's use Eid-ul-fitr as a marker of identity. The 'Curry Mile' is a platform to do this where nationalist allegiances are made with the Pakistani flag waived out of car windows and draped over people's celebratory traditional Pakistani costumes.

'Chaand Raat'

'Chaand Raat' means 'new moon' and is celebrated to mark the arrival of Eid-ul-fitr in the Indian sub-continent. This is because the sighting of the 'new moon' after Ramadan means that the month of fasting is over. Chaand Raat is a highly gendered occasion where distinct boundaries are made between men and women whom traditionally congregate separately in order to celebrate this occasion. Men usually gather at the mosque, where they check the visibility of the new moon, and women will usually have all female parties where they will prepare for Eid-ul-fitr the following morning, putting henna on their hands and preparing sweet dishes for the next day's celebrations. Chaand Raat is also the last chance to go shopping for any final additions forgotten for the Eid celebrations. In Pakistan markets and bazaars are open till the early hours of the morning, and the streets are filled with last minute shoppers, some people are just out for the experience and this is the same case in Longsight.

Chaand Raat is celebrated by people in and around Manchester in Longsight market which has been hosting Eid celebrations since 2002 and has proved to be a pioneering location. This is due to the fact that Longsight market is situated in 'Little Pakistan' and a five minute drive away from the 'Curry Mile.' Chaand Raat is described by Manchester City Council as 'a fun-filled family evening' which 'includes a fun fair and Bhangra drummers. Specialists selling Asian food, crafts, jewellery and fashion will join regular traders from Longsight Market. Asian Sound Radio, based in Cheetham Hill, usually broadcast at the event live with a stage for performances and free prizes. The colourful celebration normally includes a road show where different DJs perform and play songs' (Manchester City Council).

Over the past two years the celebrations of Chaand Raat have turned from a family event to a predominantly male occasion similar to celebrations of Eid-ul-fitr on Wilmslow road. The last year's Chaand Raat celebration highlighted change of celebratory style even more, with the first Pakistani television channel in Manchester DM Digital, broadcasting the event live to Europe, South Asia and the Middle East. This attracted a wider male audience.

British Pakistani Muslims utilise this event as a marker of identity using various forms of media to portray their affiliations. The event was posted onto youtube.com as well as discussed in many chat forums by British Pakistani Muslim youths. As it was broadcasted internationally to Pakistani Muslim communities this local event has become a concern for Pakistani communities globally. Here is evidence of Appadurai's (1995) idea that neighbourhoods represent anxieties for the nation state as the techniques of nationhood are weak or contested and need to be policed as thoroughly as borders. Also his theory that the mass media plays a major role in the production of neighbourhoods is evident and claims that diasporas are changing due to new forms of electronic media. Werbner's (2002) theory of conflicting diasporas is visible in this performance with Pakistani nationalist affiliation conflicting with the South Asian cultural festival Chaand Raat as well as Islamic values in the British landscape.

Clothing and Plurality of Personal Relationships

Clothing plays important roles in expressing custom and conflict. There is much literature on clothing, gender and ethnicity relationships with widespread assumptions that these are determined by measures of modernity. In a number of Islamic societies modest dress is used to enhance the reputation of a woman's father, brothers or husband because it can be seen that she is under the control of the men folk of her household and as part of a male strategy for achieving upward social mobility, Chapkis (1988) suggests that women's bodies are often repository for 'tradition', when women wear traditional dress; it can be seen as an attempt to preserve or re-create a real or imagined past. In both Rusholme and Longsight, British Pakistani women adopt different types of traditional dress in order to display their cultural affiliation and the level to which their family is considered as 'modern.' The more western the attire the more 'modern' a family is considered, and the level of modesty an outfit displays, the more 'religious' a family is considered. In both these areas, the clothing worn by British Pakistani Muslim women is a conscious process and viewed as a marker of her family values and ideologies.

Ballard (1994) and Saifullah Khan (1977) wrote that Sikh and Pakistani men who migrated to Britain in the 1950s usually adopted western dress, but women who travelled to join their husbands in the 1990s wore the same costume as in their homeland, making only minor modifications to allow for differences in climate. Wilson (1985) believes that women who adhere to re-adopt 'traditional' dress can symbolise authenticity in the face of imperialism, but at the price of being excluded from modernity which is negotiated by men. Crawford (1984) believes that women who adopt western styles are attempting to find a definition of themselves in terms of the modern western world.

When reading the literature on dress in context of British Pakistani Muslim women, one sometimes gets the impression that they cannot win. If they wear concealing outfits or are veiled they are seen as views of tradition, if on the other hand they adopt western dress, they are described as trapped in an image of powerlessness. It is important to look not only at dress, but also at the meanings surrounding it and the environment in which it is worn.

The Pakistani National dress worn by women is *Shalwar Kameez.* This consists of a long tunic (*Kameez*) teamed with a wide legged trouser (*Shalwar*) that skims in at the bottom accompanied by a *duppata*, which is a less stringent alternative to the *burqa.* Modern versions of this National dress have evolved into less modest versions. *Shalwar* have become more low cut so that the hips are visible and are worn with a shorter length of *Kameez* which has high splits and may have a low-cut neckline and backline as well as being sleeveless or having cropped sleeves. British Pakistani Muslim women have adopted a more modern version of the *Shalwar Kameez*, but some prefer the more modest traditional version.

British Pakistani Muslim young women have also adopted the traditional styles of the Indian sub-continent ranging from *saris, lenghas* and *ghararas.* They display their heritage as not only deriving from Pakistan, but also South Asia defining themselves not only as Pakistani but South Asian British Pakistani women use dress as a marker of movement from one cultural world into the other. Second and third generation British Pakistani Muslim women split between western and traditional cultural clothing depending on the occasion, if the occasion is British they will wear British clothing and if the occasion is Pakistani or Islamic they will adopt traditional cultural clothing. Longsight and Rusholme provide occasions for British Pakistan Muslim women to wear cultural attire.

Islamic, national and South Asian dress play important roles as identity markers for British Pakistani Muslim women. Young women are trying to find a distinct identity, some are moving toward a display of Islamic identity adopting the *hijaab* (headscarf) and even the *niqaab* (face cover). This has sparked controversy within the wider British society where it has been claimed that the wearing of Islamic dress is a move away from cultural assimilation and integration into British society which has created conflict between and within communities.

However, with the emergence of a second and third generation British Pakistani Muslim woman, cultural and traditional South Asian and Pakistani clothing have become modified as a marker of multiple identities and a way to integrate into western society. Expecting women to restrict themselves to specific ethnic dress is not realistic. Wardrobes usually contain both 'western' and 'ethnic' dress, allowing them to adapt with ease to communicate effectively with others and establish their desired image as any given situation demands. A young woman has created a new form of hybrid dress called 'jeans-kameez.' This is where a pair of jeans is worn with a *kameez*, and a more modest version of the outfit is worn with a *duppata*. Here is an example of a world fashion item (jeans) teamed with a form of national and cultural dress (kameez), creating a new identity that is recognisable in both western and non-western atmospheres so is able to be worn on any occasion. This style of dress was adapted as a means of creating an identity that would define women as South Asian, Muslim and western, as well as being perceived as an integrative style of dress.

Narratives and other Forms of Expressing Historical Experiences—Including those of Researchers

One of the questions that has motivated much of my research is that of the contributions that the arts—especially arts which help express the diversity of community's cultural histories—can make (to borrow Tim Darvill's 2007 terms) 'creat[ing] and maint[aining] different kinds of knowledge…and contexts for enriching quality of life' (Darvill 2007: 436). Much of my exploration of the ways in which 'rooted cosmopolitan' (Werbner 2008) localities are created and transformed in Rusholme and Longsight in Manchester has centred on more informal contexts. I gained a different perspectives by participating as artist and curator in an exhibition in Manchester's Whitworth Art Gallery in 2006, entitled 'Who are you? Where are you really from?'

Curated by Hammad Nasser, the exhibition at the Whitworth Art Gallery touched upon several themes of the present volume. Dudrah examines the questions; '*Who are you? Where are you really from?*' suggesting 'these two questions, and especially the latter have been frequently asked of people who visibly stand out due to their skin colour or perceived cultural differences' the artworks featured in this exhibition 'encompass an engagement with ideas of migration, identities, belonging and social transformations through the lens of Muslim cultural references that circulate between and beyond tradition—late-modern' dichotomies (Dudrah 2006). Participating artists included Faiza Butt, Shezad Dawood, Ayaz Jokhio, Naeem Rana and Yara el Sherbini. This exhibition was part of the Festival of Muslim Cultures (2006-2007) and was accompanied by a symposium and set of performances which focused on the British Muslim landscape, some of the papers highlighted various issues regarding integration in Britain.

My sister, Ruqqia Badran, and I responded to the exhibition by creating a series of textile art pieces that were then exhibited in the Whitworth Art Gallery from February 2007 until July 2007 (Figure 21.4). We come from Longsight and felt that it would be useful to try to express our experiences of what Werbner and other call 'rooted cosmopolitanism' as third generation British Pakistani Muslim women. Our art pieces are textile wall hangings made out of raw silk as well as embroidered and beaded in an Islamic style reflecting our multiple identities as Muslims, South Asian, Pakistani, British and female. We felt that much emphasis in the exhibition, 'Who are you? Where are you really from?' tried to represent global impressions of culture clash, but not their personal and highly context dependant historical experiences, social loyalties and matters of moral concern. We tried to express such experiences and concerns, but also with much attention to producing artworks that are aesthetically pleasing. We aimed to not only address issues of integration but also to make explicit the difficult challenges facing British Pakistani Muslim communities whose belonging in British society often conflict with diverse other facets of their identities.

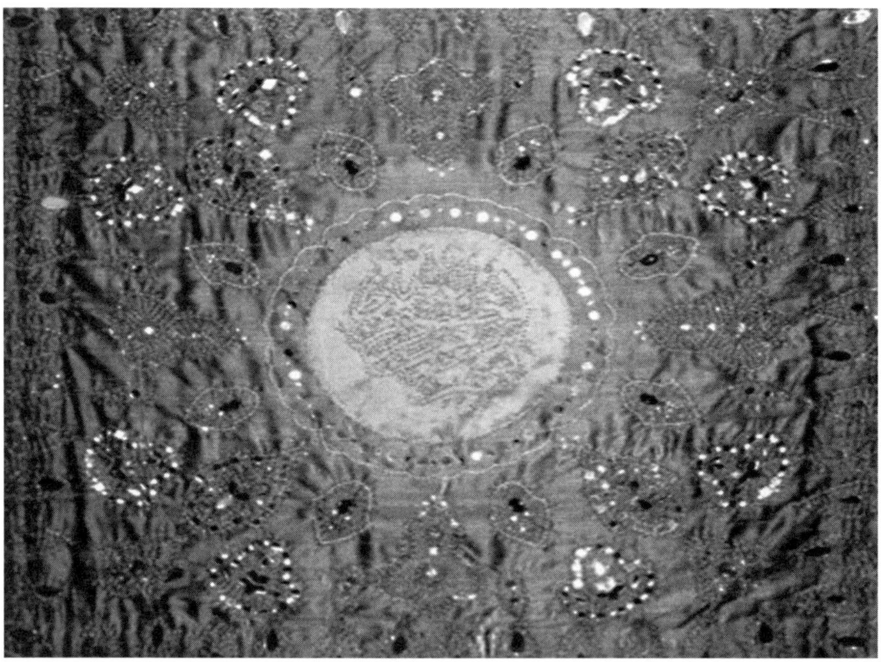

Figure 21.4 Embroidered raw silk wall hanging, Ruqqia Badran and Aliah Ullah

Source: Photograph by the author.

Especially importantly, the Whitworth Art Gallery is situated directly at the beginning of Rusholme and nearby Longsight. Not many people from the local area came. But it did succeed in illustrating the idea suggested by Ardener's (1989) essay, that artistic expressions of discrepant experiences—as is the case for other forms (to borrow James Wescoat's 2007 terms) of 'heritage conservation and conflict conciliation': (a) encompass diverse historical layers, and (b) operate on multiple geographical scales, illuminate diverse people—place connections. Thus, such artistic expressions can help engage in struggles to sustain diversity of human life-ways, experience, spiritualities—without making claims about settling differences amongst interlocutors'. It also drew the attention of quite a number of people in the area to the relevance of insights that, precisely for this reason, the arts cannot provide timeless solutions. Every generation must address 'life quality issues' (Darvill 2007) anew.

References

Alam, M.Y. 1998 *Kilo*. Glasshoughton: Route.

Alexander, C. 2006 Imagining the Politics of BrAsian Youth, in N. Ali, V.S. Kalra, and S. Sayyid (eds) *A Postcolonial People: South Asians in Britain*. London: Hurst and Co.

Anwar, M. 1979 *The Myth of Return: Pakistanis in Britain*. London: Heinemann.

Appadurai, A. 1981 The Past as a Scarce Resource. *Man* (NS) 16(2): 201-19.

—— 1991 Global Ethnoscapes: notes and queries for a transnational anthropology, in R. Fox (ed.) *Recapturing Anthropology: working in the present*, 191-210. Santa Fe, NM: School of American Research Press.

—— 1995 The Production of Locality, in R. Fardon (ed.), *Counterworks: managing the Diversity of Knowledge*. London: Routledge.

—— (ed.) 2001 *Globalization*. Duke University Press.

Ardener, E. 1989 The Construction of History: 'vestiges of creation', in E. Tonkin, M. Chapman and M. McDonald (eds), *History and Ethnicity*. London: Routledge.

Assinder, N. 2007 http://news.bbc.co.uk/1/hi/uk_politics/6270825.stm (last updated April 2007).

Back, L 2004. www.rout-online.com (last updated April 2007).

Ballard, R. 1994 *Desh Pardesh: the South Asian presence in Britain*. London: Hurst.

Baumann, G. 1996 *Contesting Culture: discourses of identity in multi-ethnic London*. Cambridge: Cambridge University Press.

BBC News http://news.bbc.co.uk/1/hi/england/manchester/4008165.stm (last updated Friday 12 November 2004).

Benjamin, W. 1992 [1936] The Work of art in the age of Mechanical Reproduction, in H. Arendt (ed.), *Illuminations*: works of Walter Benjamin, 211-44. London: Fontana Press.

Brah, A. 1996 *Cartographies of Diaspora* London: Routledge.

—— 2006 The 'Asian' in Britain, in N. Ali, V.S. Kalra, S. Sayyid (eds), *A Postcolonial People: South Asians in Britain*. London: Hurst and Co.

Dudrah, R 2006 *Who are you? Where are you really from?* Whitworth Art Gallery exhibition, Oxford Road, Manchester.

Eliade, M. 1963 *Myth and Reality*, translated by W.R. Trask. New York: Harper and Row *Faith and Identity in Contemporary Visual Culture A Symposium.* 10-11 November 2006 Whitworth Art Gallery, Oxford Road, Manchester.

Habermas, J. 1999 *The Inclusion of the Other: studies in political theory*, edited by C. Cronin and P. De Grieff. Cambridge Mass.: MIT Press.

Hall, S. and Werbner, P. 2008 Cosmopolitanism, Globalisation and Diaspora, a Conversation, in P. Werbner (ed.), *Anthropology and the New Cosmopolitanism*, 345-57. Oxford: Berg.

Inda, J. and Rosaldo, R. (eds) 2002 *The Anthropology of Globalisation: a reader.* Oxford: Blackwell Publishing.

Kant, I. 1970 [1795] Perpetual Peace: a philosophical sketch, in H. Reiss (ed.), *Kant's Political Writings*, 93-130. Cambridge: Cambridge University Press.

Khan, Saifullah, V. 1977 The Pakistanis: Mirpuri villagers at home and in Bradford, in J.L. Watson (ed.) *Between Two Cultures: migrants and minorities in Britain,* 57-89. Oxford: Blackwell.

Macey, M. 1999 Class Gender and Religious Influences on Changing Patterns of Pakistani Muslim Male Violence in Bradford. *Ethnic and Racial Studies* 22(5): 845-66.

Manchester Census 1991 Longsight Ward Profile, Planning Studies Group, Manchester City Council, Crow Copyright.

—— 1991 Rusholme Ward Profile, Planning Studies Group, Manchester City Council, Crow Copyright.

—— 2004 Longsight Ward Profile, Planning Studies Group, Manchester City Council, Crow Copyright.

—— 2004 Rusholme Ward Profile Planning Studies Group, Manchester City Council, Crow Copyright.

Manchester City Council, Advertisement for Chaand Raat in Longsight. www.manchester.gov.uk.

Manchester Religion Census 2004 Wards, Planning Studies Group, Manchester City Council, Crow Copyright.

Miller, D. (ed.) 1995 *Worlds Apart: modernity through the prism of the local.* London: Routledge.

Nasser, H. 2006 *Who are you? Where are you really from?* The Whitworth Art Gallery exhibition, Oxford Road, Manchester.

Shaw, A. 1994 The Pakistani Community in Oxford, in R. Ballard (ed.), *Desh Pardesh: the South Asian presence in Britain.* London: Hurst.

Tonkin, E., MacDonald, M. and Chapman, M. (eds) 1989 *History and Ethnicity.* London: Routledge.

Ullah, A. and Ullah, R. (curators and artists) February 2007-June 2007 *Contemporary Islamic Textiles*, exhibition at The Whitworth Art Gallery, Oxford Road, Manchester.

Werbner, P. 1990 *The Migration Process: capital, gifts and offerings among British Pakistanis*. Oxford: Berg.

—— 2002 *Imagined Diasporas Among Manchester Muslims*. Cumbria: Long House Publishing.

—— 2008 Introduction: Towards a New Cosmopolitan anthropology, in P. Werbner (ed.), *Anthropology and the New Cosmopolitanism*, 1-32. Oxford: Berg.

Werbner, R. 2008 Responding to Rooted Cosmopolitans: patriots, ethnics and the public good in Botswana, in P. Werbner (ed.), *Anthropology and the New Cosmopolitanism*, 173-96. Oxford: Berg.

Chapter 22

Blowing in the Wind: Cultural Heritage Management in a Risk Society

Timothy Darvill

Introduction

Archaeology has long been recognized as something much more than an academic subject concerned with piecing together the past. Certainly, the product of archaeological work is what might loosely be termed 'knowledge', but as I have argued elsewhere (Darvill 2007) this cover-term embraces a many different faces of knowledge which reflect the inherent complexity of the subject, the emergent novelty of new applications, and the increasing need to move away simple binary dichotomies relating to discipline boundaries such as those with history or anthropology. For some, the emergence of an independent and more rounded and distinct 'archaeology of our world' which conflates established binary categorical structures of thinking—for example: past and present; us and them; and people and place—poses a crisis of representation and for some at least a retreat into a 'two-cultures' mentality. This is played out very clearly, and ritualistically, in the identification of 'research-focused' and 'development-prompted' investigations, and in the institutionalized distinctions between 'academic' and 'commercial' sectors. Recognizing a plurality to endeavours associated with producing archaeological knowledge, and accepting an inherent diversity within the forms of knowledge produced, provide early steps along the route towards a more culturally constructed discipline with a worthwhile role in today's world.

In this chapter I would like to explore just one strand of archaeological endeavour, that known as Archaeological Resource Management (ARM), and the associated knowledge-set that can be characterized as 'Strategic Knowledge' (Darvill 2007: 450-51). After a brief review of the changing theoretical basis of ARM from the positivism of the seventies through to the instrumentalism of the naughties attention shifts briefly to the values ascribed to archaeological remains, and then to the way that proposed changes to the legislation and related government guidance for England are shaping a new era of heritage awareness. In this, it is argued, we can see new articulations to the use of heritage in a risk society in which the past in playing a stronger than ever role in shaping the future.

Changing Philosophies of Archaeological Resource Management

Looking back with the kind of twenty-twenty vision that retrospection allows, it seems as if it was once all so easy. During the early years of our discipline, the late 1960s and early 1970s, Archaeological Resource Management as it is generally known in Britain (also known as Cultural Heritage Management or Cultural Resource Management in North America) had clear objectives, approaches, and methods that were for the most part grounded in prevailing positivist philosophies (McGimsey 1972; McGimsey and Davis 1977; Schiffer and Gumerman 1977). Many of the overarching ideals of the time, and also some of the contradictions, are well expressed in the words of the World Heritage Convention, adopted by the General Conference of UNESCO on 16 November 1972 at its meeting in Paris. The preamble notes that in the face of increased pollution of the environment the convention seeks to protect those items 'pertaining to the cultural or natural heritage which are of outstanding universal value from the point of view of history, art, science, or aesthetics' (UNESCO 1985: 75).

Yet amid all the definitions and protocols in the Convention there is little that explains exactly why such an agreement is needed, although one of the recitals adopts a typically universalizing, and many would now say patronizing, argument by stating that:

> the existing international conventions, recommendations and resolutions concerning cultural and natural property demonstrate the importance, for all peoples of the world, of safeguarding this unique and irreplaceable property, to whatever people it may belong (UNESCO 1985: 80).

And in available commentaries developing the ideals of the Conventions, great play is made on the need to protect and conserve the heritage for the benefit of future generations (Feilden and Jokilehto 1993: 14).

Even as the ink was drying on the pages of the World Heritage Convention new challenges were being mounted to the youthful ideals of Cultural Resource Management. Who created these documents demonstrating the importance of heritage to the entire world? And who exactly were the future generations spoken of? People woke up to the stark realization that the control of heritage was political. Interpretation of heritage was political. And the stakes were high. As George Orwell presciently anticipated 'who controls the past controls the future, who controls the present controls the past' (Orwell 1949: 213).

Of all the challenges to cultural resource management during the 1980s cultural relativism was probably the most devastating in its impact on traditional thinking about heritage and one that still provides valuable critique (Smith 2006). It demonstrated that what we nowadays call the Western Gaze served to limit options and preserve an essentially imperialist view of heritage in which there was just one view on how it should be looked after and what it all meant. David Lowenthal memorably referred to the past as a foreign country in his book of the same name.

He argued forcefully that the past had by this time ceased to be a sanction for inherited power or privilege, but rather had become a focus for personal and national identity and a bulwark against massive and distressing change. But it remained, he noted, as potent a force as ever in human affairs (Lowenthal 1985).

Meanwhile, on the other side of the world the Australian National Committee of ICOMOS met in the historic South Australian mining town of Burra to create a charter for decision-makers. It tried to deal with all kinds of place with cultural significance, including natural indigenous and historic places with cultural values which embraced the views of those with a deep and inspirational sense of connection to community and landscape, to the past, and to lived experiences (AICOMOS 1999: 1). Opening up the heritage world in this way to alternative narratives, while caring for different kinds of heritage within the context of other cultural norms, philosophical frameworks and belief systems, has served to enrich approaches to heritage management across the world by promoting diversity and localization. The idea of cultural memory was widely adopted as a means of connecting the past with the present in the context of specific communities and it is therefore perhaps no surprise that when the Council of Europe revised the *European Convention on the Protection of the Archaeological Heritage* at a meeting in Valletta, Malta, in January 1992 the aim of the convention was defined as:

> To protect the archaeological heritage as a source of the European collective memory and as an instrument for historical and scientific study (CoE 1992: 4).

Another strand of thinking that surfaced during the 1990s was the idea of sustainability, a perspective that again originated in the western world and which has been evangelically promoted across the planet and linked closely with the 'green debate' (Macinnes and Wickham-Jones 1992). In the heritage context it effectively means making good use of resources for the needs of today without compromising the ability of future generations do the same (EH 1997; HMG 1994). Elsewhere, in the field of environmental economics, we find the development of concepts such as renewable, non-renewable, and transferable assets with trade-offs and exchanges between them offered as a way of mitigating damage in one place through enhancing the environment or heritage of another (CAG 1997).

Both cultural relativism and sustainability were in a sense middle-range theories that mediated high-level political philosophy with low-level solutions to the day-to-day problems of actually dealing with the heritage in terms of land-use, visitor management, education and public information programmes, access, and infrastructure. However, as the naughties unfolded there was a new challenge to heritage thinking, at least within the centre-left political systems now widespread in Europe and North America. The old high-level political philosophy of monetarism so influential through the 1970s, 1980s and 1990s, is being overtaken by an approach known as instrumentalism. This term was coined by John Dewey to describe an extremely broad pragmatic attitude towards ideas or concepts in general, but has come to be applied to the promotion of actions or activities not because they are useful or interesting in their own right but because they are tools or instruments in the attainment of wider ambitions in the realm of human experience. Such experiences are not simply a sensory state of 'happiness' but an aesthetic dimension of life in which the individual citizen is optimizing their individual potential as a member of a global society in an environment that is stable, just, secure, and sustainable (Vickery 2007: 9). Thus, for example, the value of expenditure on hospitals and health care is not judged on the basis of how many people are successfully treated for their ailments, as monetarism might, but rather what contribution new investment in these fields makes to widening access, promoting social mobility, and sustaining economic growth.

Similar questions are being asked on state funding for education, and of course the heritage sector too. Sensing the shifting political agenda at the start of the 21st century in England, English Heritage facilitated a debate on the future of the historic environment, its role in people's lives, and its contribution to the cultural and economic well-being of the nation (EH 2000) which has had wide-ranging repercussions (DCMS 2001; 2006; 2007). Moreover. the potential of culturally-led regeneration seems to have struck a cord with those in government. In a paper entitled *Better Places to Live* Tessa Jowell, at the time the British Secretary of State at the Department for Culture, Media and Sport, set out some challenges to the heritage sector working in the brave new instrumentalist world (Jowell 2005). Among them she identified the need to increase diversity in both audiences and the workforce; the need to capture and present evidence of the value of heritage; to contribute to the national debate on identity and 'Britishness'; to create public engagement; and to widen the sense of ownership of the historic and built environment. They are meaty issues, and in many cases still matters of debate.

On a more practical or operational note this shift in political philosophy also raises questions about heritage legislation and its administration. Most European governments, and indeed most inter-governmental bodies such as UNESCO and the Council of Europe, traditionally operate within the legal framework of statutes, laws, conventions, charters, recommendations, and precedent that was last updated back in the middle or late 20th century under quite different prevailing philosophies. This is especially true in the heritage sector where many countries make a virtue of the antiquity of their antiquities legislation. Reinterpreting and adapting legislation to accommodate ideas of cultural relativism and sustainability has usually been possible, even if the original intentions of the legislation are left far behind and implementation leaves organizations creaking at the seams. But instrumentalism requires a whole new approach not just to the content of legislation but to its very architecture. As Jonathan Vickery has said of these changes in the UK:

> since 1997 and the accession of New Labour to Government, policy statements have become both politically charged and placed under a scrutiny of self-imposed audit, monitoring and assessment; the demand for results that has characterized New Labour's style of governance has generated policy documents that provide a conceptual framework that is interpreted literally on a local level of implementation (2007: 9).

Moreover, it is a way of thinking that harmonizes with two principles deeply embedded in the thinking of the European Union: the democratization and subsidiarity of decision-making. Indeed, item 1 of Article 128 under Title IX (Culture) of the Maastricht *Treaty on European Union* states that:

> The Community shall contribute to the flowering of the cultures of the Member States, while respecting their national and regional diversity and at the same time bringing the common cultural heritage to the fore (CEC 1992: 48-9).

Democratization, subsidiarity, and the orientation of heritage management towards people and the greater well-being of individuals through improvements to their quality of life are already beginning to manifest themselves. Take the idea of cultural landscapes. In 1995 the Council of Europe published a *Recommendation on the Integrated Conservation of Cultural Landscape Areas as Part of Landscape Policies* in which the central idea of 'landscape' was seen as the:

> formal expression of the numerous relationships existing in a given period between the individual or a society and a topographically defined territory, the appearance of which is the result of the action, over time, of natural and human factors and of a combination of both (CoE 1995: 4).

Just five years later this overtly processual definition was replaced in the *European Landscape Convention* by a more humanistic one in which

> landscape means an area as perceived by people, whose character is the result of the action and interaction of natural and/or human factors (CoE 2000: 5).

Likewise, the very definitions of what constitutes heritage are opening up as new and diverse themes are brought to the table. At one level the old Cartesian dualism between culture and nature is no longer sustainable philosophically or worthwhile practically (Habgood 2002). As culturally sensitized social actors the very act of defining something as natural is a cultural activity while the entity itself that we categorize and label as 'natural' is bounded by culturally determined rules, norms, and systems of codification. At another level, foregrounding the experience of being in the world as a way of recognizing 'inheritance' as the very essence of 'heritage' means that the widest possible range of sensory perceptions can be used to chart the territory. No longer can history, art, science or aesthetics alone be the defining criteria of international cultural heritage. In 2003 UNESCO published a *Convention for the Safeguarding of the Intangible Cultural Heritage.* By intangible heritage it means

> the practices, representations, expressions, knowledge, skills—as well as the instruments, objects, artefacts and cultural spaces associated therewith—that communities, groups and, in some cases, individuals recognize as part of their cultural heritage.

Such things, it goes on to explain, are manifest in such domains as

> oral traditions and expressions, language, performing arts, social practices, rituals, festive events, traditional craftsmanship, and knowledge and practices concerning nature and the universe (UNESCO 2003: 2).

Two years later the Council of Europe moved things on further with a *Framework Convention on the Value of Cultural Heritage for Society* in which cultural heritage was recognized as

> a group of resources inherited from the past which people identify, independently of ownership, as a reflection and expression of their constantly evolving values, beliefs, knowledge and traditions. It includes all aspects of the environment resulting from the interaction between people and places through time (CoE 2005: 2).

And more, the Common Heritage of Europe was seen as

> all forms of cultural heritage in Europe which together constitute a shared source of remembrance, understanding, identity, cohesion and creativity, and the ideals, principles and values, derived from the experience gained through progress and past conflicts, which foster the development of a peaceful and stable society, founded on respect for human rights, democracy and the rule of law (CoE 2005: 2).

In just two short steps then heritage, including all the raw materials that archaeology traditionally draws upon, was changed from a purely physical resource to become the cutting edge of a new tool in the attainment of human happiness. But of course behind these public articulations of policy there are fundamental issues pertaining to the social and personal values that give meaning and purpose to heritage resources.

Value, Importance and Conservation Principles

Many attempts have been made to articulate and understand the way in which society values its heritage (Mathers et al. 2005; Samuels 2008 with earlier refs). In most the idea of 'value' is conceived in its sociological sense, that is a set of standards against which things are compared and placed on a gradient between extremes such as desirable or undesirable, appropriate or inappropriate, worthwhile or pointless. I have discussed elsewhere how such value systems emerge (Darvill 1995), but here it is important to recognize that they are constantly being renegotiated and changed, and that their formulation and acceptance is a consensual matter. Changes in people's values are not independent of innovations in cognitive orientation created by shifting perspectives on the social world. There is no rational basis for values, and shifts in outlook deriving from diverse inputs of knowledge have a mobile relationship to changes in value orientations (Giddens 1990: 44).

In a ground-breaking review of archaeological value systems, William Lipe (1984) considered four types of value: Economic, Aesthetic, Associative, and Informational (Table 22.1). These he linked to a range of interests within society, recognizing that archaeological value is embedded in wider issues, although retaining the notion that cultural materials from the past function as resources in the present.

Later, in 1993, I presented a rather different way of treating heritage resources (Darvill 1993; 1995) which looked at value in terms of three interrelated value sets (Figure 22.2). The idea of use values perpetuated the principle that heritage included resources for immediate consumption. In thinking about the range of possible uses I tried to reflect the wide range of consumers that can be identified, and the expectations they have in using the heritage for such things as: research; creative art; education; recreation and tourism; symbolic representation; the legitimization of action; social solidarity; and monetary and economic gain. In defining option values, attention turns to the deferred use the heritage not only the conservation ethic of preservation but also the realization that not all possible uses are currently definable. Other ways of using heritage resources will undoubtedly come along, and our descendants deserve the opportunity to exploit such opportunities. Finally, existence values are related to emotional attachments to things that cannot be directly experienced: the sense of well-being stimulated by knowing that everything is in order; the sense of stability, familiarity, local distinctiveness, attachment, and the physicality of identity within an historic environment. Each of these value sets embrace general socially determined values and relate, in different degrees, to scales of emotion: the pleasures of consumption; the thrill of revelation; the contentment of preservation; the peace of mind in conservation; and so on.

Table 22.1 Summary of Lipe's (1987) value system for heritage resources

Value	Context
Economic	Economic potential; Market factors; Costs of development against preservation
Aesthetic	Aesthetic standards; Stylistic tradition; Human psychology
Associative/Symbolic	Traditional knowledge; Historical documents; Oral tradition; Folklore and mythology
Informational	Archaeological research; History; Art history; Architectural history; Folklore studies, etc.

Table 22.2 Summary of Darvill's (1991) value system for heritage resources

Value system	Application
Use value	Research; Creative arts; Education; Recreation and tourism; Symbolic and representational; Legitimation of action; Social solidarity; Monetary and economic gain
Option value	Social stability; Mystery and enigma
Existence value	Cultural identity; Resistance to change

In a third approach, again constructed in an archaeological context, Martin Carver has argued that all the values listed by Lipe and Darvill ultimately derive from realizing the results of research (1996: 46-7). Instead he offers a rather different view based on the competing interests that reside in the use of a defined piece of land. These he defines as: market values, community values, and human values. He sets archaeology squarely in the last of these three groups, competing with the other two for attention. Although this dialectic model tries to recognize the broader social context of heritage in defining what is valuable, it ultimately fails because it falls back on the myopic position that archaeology is somehow more special than other things (1996: 48). Thus instead of trying to situate archaeological value within a broader range of social values, as others have done, Carver prefers to ring-fence archaeological interests and set them apart.

Inevitably, the question of how heritage values work in the wider world is difficult to see from within the profession. It is interesting and instructive, therefore to consider attempts to look at questions of heritage value from other perspectives. Recent work, for example, has introduced the economist's view, and the property market's view (Allison et al. 1996). Here, formal valuation methods have been applied to particular issues. The Hedonic Pricing Method (HPM) looks at changes in the value of property in order to estimate the value of environmental goods. The Travel Cost Method (TCM) is based on the premise that visitors reveal the value they put on a resource by the amount of time or cost they are willing to spend in order to see elements of it. In contrast, the Contingent Valuation Method (CVM)

Table 22.3 Summary of Carver's (1996) value system for heritage resources

Value	Context
Market values	Capital/Estate value; Production value; Commercial value; Residential value
Community values	Amenity value; Political value; Minority interest value
Human values	Environmental value; Archaeological value

relies on questionnaires to ask people to declare the value they put on different kinds of resources or the existence of some feature. These, and other indicators of value, operate mainly at the level of case-study, although potentially provide insights of at a far more general scale.

As instrumentalist views emerged in the late 1990s news ways of examining and documenting value emerged. In a review of the so-called 'Heritage Dividend', English Heritage (1999; 2002) concluded that giving grants for conservation attracted matching funding from the private and commercial sectors, and that this was prompted by desires to enhance cultural distinctiveness, pride of place for local communities, establish local cultural identity, and contribute to the sense of place. At the same time such work created new jobs and secured the future of others, created new floorspace in formerly redundant historic buildings, and this in turn provided opportunities for new jobs and economic revival.

On a wider scale, Robert Hewison and John Holden of the think-tank Demos, suggested that the heritage generates three kinds of cultural value (Hewison and Holden 2006: 14-15) that can be conceived as a pulling in three directions at once (Figure 21.1). First is the value of heritage itself, its intrinsic value in terms of an individual's experience of heritage intellectually, emotionally, and spiritually. They argue that it is these values that people refer to when they say things like 'This tells me who I am', or, quite simply, 'this is beautiful'. A second value of heritage is the purely instrumental, referring explicitly to the ancillary effects of heritage as it is used to achieve a social or economic purpose. They cite urban

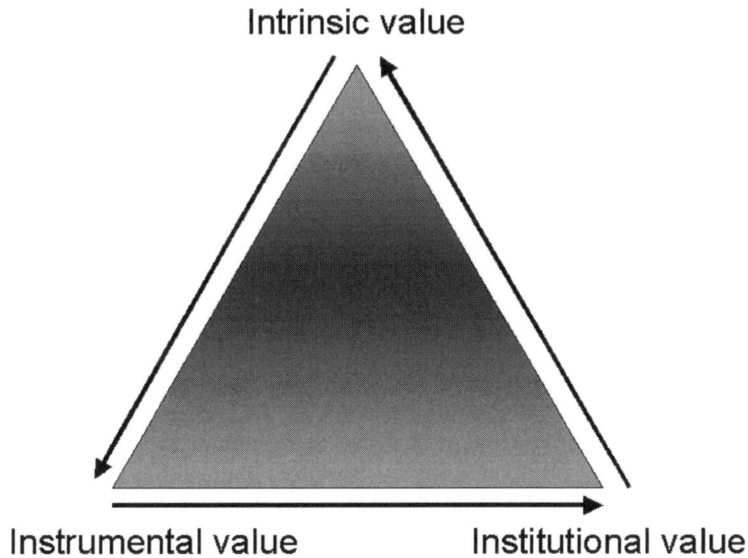

Figure 22.1 Ternary view of competing heritage values
Source: Based on Hewison and Holden 2006.

regeneration as one obvious example where the instrumental objective is making the world a safer place by, in this case perhaps, cutting crime. Their third heritage value is institutional value. This relates to the processes and techniques that organizations adopt in creating value for the public. Thus institutional value is generated or destroyed by how organizations engage with their public, it flows from their working practices and attitudes and is rooted in notions of public good. Through its concern for the public an institution can achieve such public good as creating trust and mutual respect between citizens, enhancing the public realm and providing a context for sociability and the enjoyment of shared experiences. In this way, heritage organizations should be considered not just as repositories of objects, or sites of experience, or ways of generating cultural meaning, but as creators of value in their own right.

Implicit to these essentially aesthetic notions of heritage value is that they recognize both an emotionally-based 'human' side to our activities as well as a structured process-based 'professional' side. English Heritage have usefully proposed what might be termed the 'heritage cycle' which focuses attention on the four successive phases of understanding, valuing, caring-for, and enjoying aspects of the historic environment (Figure 22.2). Alongside this we may see a 'management cycle' which promotes a sequence of expert review, validation and enhancement, informed decision-making, and the execution and monitoring of activity relating to the heritage (Figure 22.3). Indeed this is a cycle that already exists in relation to informed decision-making within the spatial planning systems of most European countries, and is seen at a wider scale both in Europe and North America through Environmental Impact Assessment regulations (Darvill 2004: 418).

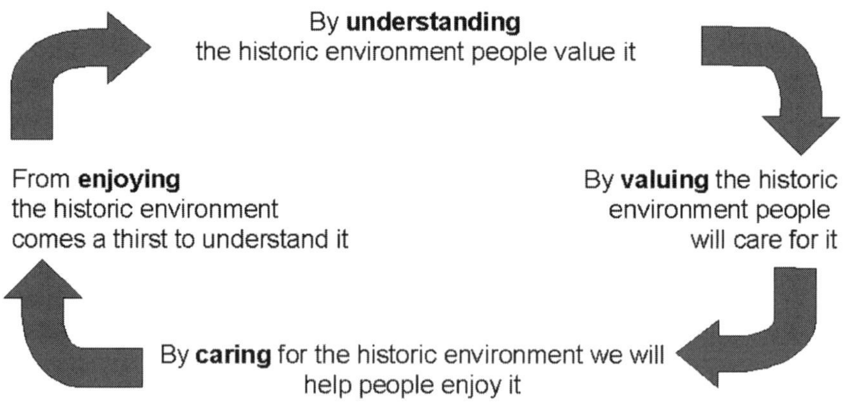

Figure 22.2 The 'Heritage Cycle'
Source: Based on EH 2006.

Underpinning the use of such systems and the application to decision-making through any of the value systems currently available are general conservation principles. These may relate to the scope and basic philosophy of conservation efforts (Darvill 2004: 420), or the processes of achieving it (EH 2008), but either way must nowadays align with the agenda of instrumentalism if they are to be successful.

As a starting point the following can be considered:

- The heritage is a shared resource.
- It is essential to understand and sustain those aspects of the heritage considered valuable.
- Everyone can make a contribution.
- Understanding the values of places and the significance they have for different communities is vital.
- Places should be managed to sustain their significance.
- Decisions about change must be reasonable and transparent.
- It is essential to document and learn from decisions.
- Maintain diversity in the content and representation of the heritage.

Core to these principles is an increasing democracy to the definition and valuing of heritage in general and heritage resources in particular, themes that resonate with the latest cycle of making heritage work within an instrumentalist context.

Figure 22.3 The 'Management Cycle'

Source: Based on Darvill 1999, figure 17.5.

Heritage Realigned: Assets and Risk Reduction

In broad terms, in England today, archaeological remains in particular and heritage resources in general are subject to two parallel sets of legislative frameworks and control. First is a wide raft of measures concerned with specific heritage resources as protected areas, sites, objects, materials, or traditions. Second, are measures relating to spatial planning and environmental assessment. In both cases there are international, European, and national conventions, charters, directives, recommendations, acts, instruments, and guidance, some of which has already been explored in earlier sections. In practice, implementing of all these arrangements in England at least falls to national agencies and government departments, regional agencies, and local authorities. Needless to say, there is a degree of duplication in the work of different organizations, and the multiplicity of designations inevitably causes confusion in the public eye. Since about 2006 a start has been made on tidying up heritage protection, orchestrated by English Heritage for England (Taylor 2009) with other initiatives happening in Scotland (Gilmour and Taylor 2009) that are not considered further here. Culturally and professionally there is a great of support for these initiatives which will undoubtedly simply administration and clarify obligations. In the background, however, are the more subtle drift in thinking and the embedding of new political philosophies which increasingly foregrounds heritage as an instrument of political control while at the same time placing its future within a culture focused around risk management.

In the field of heritage legislation, the Government presented to Parliament a *Draft Heritage Protection Bill* in April 2008. This related to England and Wales, and for the first time proposed the establishment in law of what was termed a 'heritage asset' which effectively combined ancient monuments, Scheduled Monuments, historic buildings, Listed Buildings, World Heritage Sites, Historic Wrecks, historic parks and gardens, battlefields, and many other items that previously enjoyed separate and specialist treatment. Heritage Assets would be listed on a Register, after which any proposed works affecting them would be controlled through a Heritage Asset Consent scheme. As previously mentioned, Heritage Assets could be taken into State Guardianship under certain circumstances; their conservation, management, investigation, and enhancement supported by grants, loans, and partnership agreements; and particular restrictions imposed on the use of metal detectors within the geographical limits of a defined Heritage Asset. As it turned out, the Bill was not picked up for debate in the 2009-10 session of Parliament, and some believe it has been sidelined for the foreseeable future even though English Heritage and other parts of government are committed to implementing elements of it without legislation. In this connection it is notable that the language of the Bill is already finding widespread usage. In a recent consultative review of World Heritage Sites (DCMS 2008) for example, we read that: 'World Heritage is not just about the identification of future sites. It is also about the protection, conservation, presentation and transmission to future generations of our heritage assets' (DCMS 2008: 37). It is also evident in the second strand of legislative support: spatial planning.

Town and Country Planning in England is structured through a series of core enabling acts that provide the framework for Strategic Planning and Development Control. Environmental Impact legalization forms part of this, with Strategic Environmental Assessment (SEA) relating to broad plans and Environmental Impact Assessment (EIA) relating to specific proposals and schemes. Both require the production of detailed information about likely and anticipated impacts and the placement of that information in the public domain ahead of any decisions being taken about the implementation of plans or proposals (and any associated mitigation). The same principles cascade down to more routine development proposals and while the rule of engagement for these are currently set out in *PPG15: Planning and the Historic Environment* (DoE 1995) and *PPG 16: Archaeology and Planning* (DoE 1990), July 2009 saw the publication of a consultation on the replacement document in the form of *PPS15: Planning for the Historic Environment* (DCMS 2009), together with a set of supporting guidance notes produced by English Heritage (English Heritage 2009). Much that is familiar from the two earlier PPGs remains in place, but the idea of 'artistic' interest is added to the more traditional factors of historic, archaeological and architectural in terms of defining the significance of heritage assets. It notes that the Government's objectives are that heritage assets 'should be conserved, enhanced and enjoyed for the quality if life they bring to this and future generations' ensuring that the principles of sustainable development are applied, that conservation and enhancement is carried out in a manner appropriate to the significance of the assets, and that these assets are to contribute to our knowledge and understanding of our past (DCMS 2009, 14). Elsewhere there is an emphasis on the use of heritage assets in 'place-making' (e.g., DCMS 2009, HE9.4), an implicit assumption that the public and well as specialists have a role to play in determining the significance of particular heritage assets, and that the view of the local community should be sought.

Clearly the top-down approach to the definition and control of Heritage Assets provides an interesting counterpoint to the increasingly bottom-up approach implicit to balancing Heritage Assets within the planning system. The possibility of tensions and unintended consequences is already a matter being debated, with Nigel Hewitson, for example, recognizing that all protestors opposing a proposed development have to do is 'persuade the local authority that a particular building, site or landscape affected by the proposed development has the requisite [heritage] interest' (Hewitson 2009). Taking a broader perspective, it is noticeable that through the heritage legislation, whether existing or as proposed, the State has a valuable tool to protect its interests and control risk, while, on the other side, the general public are being handed increasing powers to control what it sees as risk to its familiar structures, places and landscapes and at the same time minimize personal or corporate gain through the development of property in a way that contributes little or nothing to social well-being and the quality of life. Where all that leaves experts and professionals working in the heritage sector remains to be seen. Many ideas are now blowing about in the winds of change, but one potentially fruitful approach is to expand the idea of multiple faces of knowledge and to think in terms of property development and change as an opportunity to create a variety of new kinds of knowledge from the Heritage Assets represented there rather than a constraint on change or the mitigation of impact. In this way, archaeologists can create new alliances, enhance trust in their work, and participate in novel kinds of active engagement with the wider world and the risks inherent to its future.

Acknowledgements

Following the 'Unquiet Past' session at the 12th Annual Meeting of the EAA in Krakow, Poland, in September 2006, aspects of this chapter were developed as a keynote address to the *Rethinking Cultural Heritage: Challenges and Dilemmas* conference held in the Maritime Museum, Vittoriosa, Malta, on 15 September 2008, and are included in the conference proceedings with the equally Dylanesque title 'The times they are a-changin'.

References

AICOMOS (Australia International Council on Museums and Sites) 1999 *The Burra Charter for Places of Cultural Significance*. Burwod: Australia ICOMOS.

Allison, G., Ball, S., Cheshire, P., Evans, A. and Stabler, M. 1996 *The value of conservation? A literature review of the economic and social value of the cultural built heritage*. London: English Heritage.

CAG Consultants 1997 *What matters and why. Environmental Capital: A new Approach*. Cheltenham: Countryside Commission.

Carver, M. 1996 On archaeological value. *Antiquity*, 70, 4-56.

CEC (Council of the European Communities/Commission of the European Communities) 1992 *Treaty on European Union*. Luxembourg: Office for Official Publication of the European Communites.

CoE (Council of Europe) 1991 *European Convention on the Protection of the Archaeological Heritage (revised)*. Valletta 16.1.1992. Strasbourg: Council of Europe European Treaty Series 143.

—— 1995 *Recommendation No R(95) 9 of the Committee of Ministers to Member States on the integrated conservation of cultural landscape areas as part of landscape policies*. Strasbourg: Council of Europe.

—— 2000 *European Landscape Convention*. Florence 20 October 2000. Strasbourg: Council of Europe European Treaty Series 176.

—— 2005 *Council of Europe framework Convention on the Value of Cultural Heritage for Society*. Faro, 27 October 2005. Strasbourg: Council of Europe Treaty Series 199. Available at: <http://conventions.coe.int/Treaty/EN/Treaties/html/199.htm>. Accessed: 9 November 2006.

Darvill, T. 1993 *Valuing Britain's archaeological resource* (Bournemouth University Inaugural Lecture). Bournemouth: Bournemouth University.
—— 1995 Value systems in archaeology. In M.A. Copper, A. Firth, J. Carman and D. Wheatley (eds) *Managing archaeology*. London: Routledge. 40-50.
—— 1999 Reeling in the years: The past in the present. In J. Hunter and I. Ralston (eds), *Archaeology in Britain*. London: Routledge 297-315.
—— 2004 Public archaeology: A European perspective. In J. Bintliffe (ed.), *A companion for archaeology*. Oxford: Blackwell. 409-34.
—— 2007 Research frameworks for World Heritage Sites and the conceptualization of archaeological knowledge. *World Archaeology*, 39(3), 436-57.
DCMS (Department for Culture, Media and Sport) 2001 *The historic environment: a force for our future*. London: DCMS.
—— 2006 *Government response to the Culture, Media and Sport Committee Report on Protecting and Preserving our Heritage*. London: The Stationery Office Cm 6947.
—— 2007 *Heritage protection for the 21st century*. London: DCMS.
—— 2008 *World Heritage for the Nation: identifying, protecting, and promoting our World Heritage. A consultation paper*. London: DCMS.
—— 2009 *Consultation paper on a new Planning Policy Statement 15: Planning for the Historic Environment*. London: DCMS. Available at: <http://www.communities.gov.uk/publications/planningandbuilding/consultationhistoricpps>. Accessed: 30 August 2009.
English Heritage 1997 *Sustaining the historic environment: new perspectives on the future*. London: English Heritage.
—— 1999 *The Heritage Dividend: measuring the results of English Heritage regeneration*. London: English Heritage.
—— 2000 *The Power of Place*. London: English Heritage.
—— 2002 *The Heritage Dividend 2002: measuring the results of heritage regeneration 1999-2002*. London: English Heritage.
—— 2008 *Conservation principles for the sustainable management of the historic environment*. London: English Heritage.
—— 2009 *PPS Planning for the Historic Environment: Historic Environment Planning Practice Guide*. London: English Heritage. Available at: <http://www.english-heritage.org.uk/server/show/nav.21136>. Accessed: 30 August 2009.
Feilden, B.M. and Jokilehto, J. 1993 *Management guidelines for World Cultural Heritage Sites*. Rome: ICCROM/UNESCO/ICOMOS.
Giddens, A. 1990 *The consequences of modernity*. Cambridge. Polity Press.
Gilmour, S. and Taylor, A. 2009 Heritage protection in Scotland. *The Archaeologist*, 72, 16-17.
Habgood, J. 2002 *The concept of nature*. London: Darton, Longman and Todd.
Hewison, R. and Holden, J. 2007 Public value as a framework for analysing the value of heritage: the ideas. In K. Clark (ed.), *Capturing the Public Value of Heritage*. Proceedings of the London Conference 25-26 January 2006. London: English Heritage. 14-18.

Hewitson, N. 2009 Heritage asset definition fogs protection motives. *Planning*, 1832 (21:08:09), 9.

HMG (Her Majesty's Government) 1994 *Sustainable development*. The UK Strategy (Command paper 2426). London: HMSO.

Jowell, T. 2005 *Better places to live: Government, identity and the value of historic and built environment*. London: Department for Culture Media and Sport.

Lipe, W.D. 1984 Value and meaning in cultural resources. In H. Cleere (ed.), *Approaches to the archaeological heritage*. Cambridge: Cambridge University Press. 1-11.

Lowenthal, D. 1985 *The past is a foreign country*. Cambridge: Cambridge University Press.

Macinnes, L. and Wickham-Jones, C.R. (eds) 1992 *All natural things. Archaeology and the Green debate* (Oxbow Monograph 21). Oxford: Oxbow Books.

Mathers, C., Darvill, T. and Little, B. (eds) 2005 *Heritage of value, Archaeology of renown*. Gainesville: University Press of Florida.

McGimsey, C.R. 1972 *Public archaeology*. New York and London: Seminar Press.

McGimsey, C.R. and Davis, H.A. (eds) 1977 *The management of archaeological resources. The Airlie House Report*. Washington: Society for American Archaeology.

Orwell, G. 1949 *Nineteen Eighty-four*. London: Secker and Warburg.

Samuels, K.L. 2008 Value and significance in archaeology. *Archaeological Dialogues*, 15.1, 71-97.

Schiffer, M.B. and Gumerman, G.J. (eds) 1977 *Conservation archaeology. A guide for cultural resource management studies*. New York and London: Academic Press Studies in Archaeology.

Smith, L. 2006 *Uses of heritage*. London: Routledge.

Taylor, A. 2009 Heritage protection in England and Wales. *The Archaeologist*, 72, 14-15.

UNESCO (United Nations Educational, Scientific, and Cultural Organization) 1985 *Conventions and Recommendations of UNESCO concerning the protection of cultural heritage*. Paris: UNESCO.

—— 2003 *Convention for the safeguarding of the intangible cultural heritage*. Paris: UNESCO.

Vickery, J. 2007 *The emergence of culture-led regeneration: a policy concept and its discontents*. Warwick: University of Warwick Centre for Cultural Policy Studies Research Papers 9.

Interface 3
After the 'Myth of a Clean Slate': Towards Latour's Philosophy of a Cautious Prometheus?

Stephanie Koerner

In considering the influential thought of Bruno Latour, I would like to explore some questions that are likely to interest contributors to this volume. I met Latour almost a decade ago in Vienna. I remember this fairly clearly because his book *Pandora's Hope* (1999) was in press. He was discussing with my brother, Joseph Leo Koerner, plans for the exhibition and catalogue (edited volume) *Iconoclash. Beyond the Image Wars in Science, Religion and Art* (2002). In the introduction to the volume, Latour posed the remarkable question:

> What has happened that has made images (and by images we mean any sign, work of art, inscription, object, picture that acts a s a mediation to access something else) the focus of so much passion? [Under what circumstances have] destroying them, erasing them, defacing them…been taken as the ultimate touchstone to prove the validity of ones faith, of one's science, of one's acumen, of one's artistic creativity? To the point where being an iconoclast seems the highest virtue, the highest piety, in intellectual circles? (Latour, *Iconoclash. Beyond the Image Wars in Science, Religion and Art* 2002: 14).

Iconoclasm, Latour (2002: 14) said, is when we feel that we can assume to know the motivations of acts of breaking images. *Iconoclash*, by contrast, is when we hesitate—are troubled by an action, and are unable to say whether the claim is that 'we must have images' or that 'we cannot have images' (cf. Galison 2002). Iconoclash is when we cannot say, without further inquiry, whether there is a 'crisis over representations' or a settlement, on the part of those in power, that the problem lies with the beliefs of the 'others' being excluded from negotiations. Since then I have come to work at the University of Manchester and have had opportunities to publish on the relevance for challenges nowadays facing anthropology and archaeology of themes of Latour's writings and to begin working on an edited volume based on lectures for an introductory core course in art history and visual culture studies. Throughout I have recurrently thought about possible implications of concerns that his work shares with Stephen Toulmin, especially in his *Cosmopolis: The Hidden Agenda of Modernity* (1990), including:

- the centrality amongst timbers of modern dualist forms of reasoning and practice of deterministic conceptions of nature,
- the relevance for appreciating the long history of alternative pedagogical and political ideals of shifting foci away from 'crises over representation' towards instantiations of settlements grounded in such presuppositions as those outlined above,
- 'ways ahead' against the grain of such supposed settlements.

All this is largely to say something about the ways in which the volume is structured, and to take a moment to consider some continuous themes in Latour's writings that have relevance for the contributions in this volume. They also compare with concerns that have motivated Toulmin's writings for some time. But in the present context, I think the theme of visualisation might be especially interesting to stress. In their introduction to a recent handbook on landscape archaeology, the editors note that:

> ...during the 1970s and 1980s landscape archaeology ceased to be simply a unit of analysis over and above the 'site' and became instead an object of investigation in its right. As a specialised term within the archaeological discipline, the word has witnessed a recent efflorescence, and with this a privileged if somewhat uneasy use. This is because what archaeologists have understood to be 'landscape archaeology' has shifted, so that today it does not mean exactly what it used to even 20 years ago...Indeed even within the works of individual archaeologists, the term may shift its connotations according to context (David and Thomas 2008: 27-28).

This does not mean that there are no patterns. Bruno David and Julian Thomas stress differences between today's approaches and both (a) culture historians' conceptions of 'sites' and 'artifacts' during the first half of the 20th century (e.g., Childe 1929), and (b) 'New' and 'processual' approaches to relationships between the archaeological record and dynamic processes in the past (e.g., Binford 1972; Clarke 1968, 1972). They also suggest that variability may relate to tensions around landscape's relationship to such dichotomies as those, which Latour has been challenging now for decades. Thomas notes that one problem with conceptions of landscape, which set 'visual perception' against an 'external world of extension' is that they are 'distinctively modern Western' and 'anachronistic when applied to the distant past.' But that is not all. There is a 'tension between the distanced visual landscape theorised in Western systems of knowledge and the landscapes that we experience' (Thomas 2008: 302).

Already in the 1980s, Latour's work was concerned about revisiting the roles of linear perspective in the histories of art and science. It seems to me that he was especially influenced by Svetlana Alpers brilliant book: *The Art of Describing. Dutch Art in the Seventeenth Century* (1983), and how her approach compares with approaches in science studies to practices of 'representing, intervening and persuading'(Latour and Woolgar 1979; Hacking 1983). These sorts of practices were at one time ignored by historians of art and of science. In the paper—'Visualisation and Cognition: Drawing Things Together' (Latour 1986: 1-2)—Latour argued that in order to arrive 'at parsimonious explanations' of some things that can be said about 'modern scientific culture…it is best not to appeal to supposed universal traits,' which are 'simply too grandiose, not to say hagiographic in most cases and plainly racist in more than a few others.' Instead we should try to:

> steer a course that avoids relativism and, by positing a few simple, empirically verifiable causes can account for the enormous differences in effects that everyone knows are real [that is, not 'mere' social constructs and/or somehow not real]. We need to keep the scale of effects but seek more mundane explanations than, for instance, a great divide of human consciousness (Latour 1984: 2).

This also brings us to the work of Peter Galison (1987, 2003), who has so long stressed the importance of—'positing a few simple causes' that might be explored through comparison of possible concrete instantiations for exploring such broad developments as changes that have taken place in the scales of scientific instruments and practices, their normative roles and associated implications for conceptions of 'objectivity' (Galison and Daston 2007). Latour's 1986 paper suggested that focusing on historically contingent connections of this sort could enable us to appreciate that the 'rationalisation that took place with the so-called scientific revolution was not of the mind, of the eye, of philosophy, but of the sight.' Or put another way, instead of trying to explain the values that became invested in the realism of Renaissance art and rationality of science in terms of highly generalised images of such things as a 'disembodied spectator' and a reality 'out there'—such as that which concerns Thomas (2008) in the work mentioned earlier, we might try to explore the complex socially and materially embedded proliferation of devices (linear perspective, telescopes, microscopes, log books, maps, globes, printed instructions, and so on) that made it possible to 'mobilise the supposed immutables of reality.'

Specifically, these assemblages of instruments, technical manuals, and specialised practices made it possible for artists and scientists to invent objects, with properties of being mobile, presentable and combinable in extremely novel ways by virtue of their participating in what W.M. Ivens (1953, 1973) called 'optical consistency.'

Maybe, then as now—at stake with many debates about visualisation are such questions as those Latour posed in 1986. For instance:

- What was so important about the images and inscriptions that map makers, engineers, landscape painters, and other practitioners were producing in the 17th and 18th centuries?
- What is so important about the further 'rationalisation of sight' today, for instance, through computer technologies that—to use Galison's (2002) terms—'scatter images into data and gather data into images'.

- What is so important about technologies that make possible other examples you mentioned already in 1986, including genetic chromatography, geological stratigraphy, calculations of responses to huge numbers of public questionnaires, etc.?

> 'You doubt of what I say? I'll show you.' And without moving more than a few inches, I unfold before your eyes figures, diagrams, plates, texts, silhouettes and then and there present things that are far away and with which connections… between supposed immutable are mobilized (Latour 1986: 3, 14).

In order to understand how 'rationalised sight' became invested with normative authority, we need to investigate how it was possible for a few scientists and or engineers to support claims, for instance, that they could master not only enormous machines, land drainage schemes and governance institutions, but also things that do not yet even exist—how they 'mobilised immutables'.

Latour's work indicates that the histories of quite innumerable such properties. For instance, in his discussion of industrial drawing, he stressed the importance of principles of linear perspective to the ways in which mechanics, economics, marketing and other fields (I would add archaeological drawing of all sorts)—present things as all 'right here' by variously 'flattening' them onto the same surface: 'the accumulation of drawings in an optically consistent space is, once again, the same 'universal exchanger' that allows work to be planned, dispatched, realised, and responsibility to be allocated (Latour 1986: 25). Relating especially directly to one of the themes that runs through this volume, you suggested that instruments and practice structured around 'optical consistency' have played crucial roles amongst devises that 'deny of the coevalness' of the 'West and the Rest' (Fabian 1993). Much attention has been devoted to text, but Latour's point draws attention to the importance of the devises and practices of pictorial representations for visualising cultural differences as space-time distances. A conversation with John Urry described in the volume touches upon this theme and its implications for a question that runs through many of your works, namely: 'how the few may dominate the many' (Latour 1986: 26, also, for instance, 1999).

Foucault has also been of major importance for fresh approaches. Foucault's (1970, 1972) analyses of the objects, practices, and social institutions that formed the authority of the 'panopticum' marked a turning point in approaches to this question and thus to the ways in which 'disciplinarity' can be conceptualised.

In ways that also facilitated awareness of the tenacity of whole systems of dichotomies, including those of the real versus the historically contingent. Latour and Lorraine Daston (2000) greatly stress that what some call 'mere social constructions' are as real as bricks! This relates to the assumptions shared by what some call 'influentially opposed paradigms'—including assumptions of nature-culture, moderns-others dichotomies. This brings us to a number of arguments in *We Have Never Been Modern* (Latour 1993). For instance, many problems summarised under the expression, the 'Great Divide', might be addressed in new ways if we discovered such things as those in the much cited passage:

> All human collectives are similar in that they simultaneously construct humans, divinities, and non-human ['actants']. None of them lives in a world of signs or symbols arbitrarily imposed onto an external Nature known to us alone…If there is one thing we do, it is surely that we construct both our human collectivities and the nonhumans that surround them. In constituting their collectivities, some societies mobilize ancestors, lions, fixed stars and the coagulated blood of sacrifice; in ours, we mobilize genetics, zoology, cosmology, and hæmatology (Latour 1993:106)…The fact that one society needs ancestors and fixed stars, while another one, more eccentric, needs genes and quasars, relates to the dimensions of the collective to be held together. The relation of modern scientific knowledge and power does not differ in that by dividing Nature from Society it has at last escaped the influences of the latter, but in that it has demanded increased numbers of nature-culture hybrids to recompose its social networks and extend their scale (Latour 1993:9).

These themes also relate to Latour's involvement in the projects that produced the exhibition and edited volumes, *Iconoclash* and *Making this Public* (Latour and Weibel (eds) 2002, 2005). One problem with many 'critiques of meta-narratives' has been the idea that the iconic figures of the so-called Scientific Revolution envisaged the motivations of their 'new systems' (to use Galileo's [1564-1642] term, 1968) as arising out of the 'best of all possible worlds' (to use Leibniz' [1642-1716] terms, 1989). This has resulted in tendencies to ignore the circumstances under which political power and moral authority became anchored to new cosmologies—the 'Modern Cosmopolis' described by Toulmin (1990) and/or 'cosmopolitics' that you (for instance Latour 2004) and Stengers (1997) have discussed. It also seems to obscure the extent to which such authority has depended on these cosmologies' contradictions.

Laurent Olivier and I organised a panel for the 2004 meeting of the European Association of Archaeologists around problems posed for the field by what John Urry calls 'timeless' time. We took as points of departure themes of Walter Benjamin's works, including his argument that 'state of emergency' was the norm for contemporary times. Olivier and I explored the implications of noting that in some works he seems to have envisaged the 'state of emergency' as a consequence not means to legitimise ideological claims to necessity. A very different picture is suggested by Toulmin's *Cosmopolis* (1990) and several contributions to the *Iconoclash* (Latour and Weibel 2002) volume—one which may throw new light on the importance to the authority of the 'Modern Cosmopolis'—and the eclipsing of alternative 'agendas'—of claims to necessity under the conditions of 'state of emergency' of contemporary times, of claims about these representing the 'worst of possible worlds'. This then may reveal the relevance of shifting attention from 'crises over representations' to supposed 'settlements'—circumstances of 'iconoclash'.

This is a new way of addressing the themes of this volume, and it relates to issues Latour talked about in the paper, 'A Cautious Prometheus: A few steps toward a philosophy of design' (2008).

At the very moment when the scale of what has to be remade has become infinitely larger (no political revolutionary committed to challenging capitalist modes of production has ever considered re-designing the earth's climate), what it means to make something is also being deeply modified (2008: 4).

The question that Latour raised to designers is:

> Where are the visualisation tools that allow the contradictory and controversial nature of matters of concern to be represented? A common mistake (a very post-modern one) is to believe that this goal will have been reached once the 'linear', 'objectified', and 'reified' view has been scattered through multiple view points and heterogeneous make shift assemblages. However breaking down the tyranny of the modernist point of view will lead nowhere since we never have been modern. Critique, deconstruction and iconoclasm, once again, will simply not do the job of finding alternative design. What are needed are tools that capture what have always been the hidden practices of modernist innovations: objects have always been projects; matters of fact have always been matters of concern (Latour 2008: 13).

References

Alpers, S. 1983 *The Art of Describing. Dutch Art in the Seventeenth Century*. Chicago: University of Chicago, 26-72 and 222-28.

Benjamin, W. 1994 [1940] Theses on the Philosophy of History, in H. Arendt (ed.), *Illuminations: works of Benjamin*, 245-55. London: Fontana Press.

Binford, L.P. 1972 *An Archaeological Perspective*. New York: Seminar Press.

Bintliff, J. (ed.) 2004 *A Companion to Archaeology.* Oxford: Blackwell.

Childe, V.G. 1929 *The Danube in Prehistory*. Oxford: Oxford University Press.

Clarke, D. 1968 Analytic *Archaeology*. London: Methuen.

—— 1973. Archaeology: the loss of innocence. *Antiquity* 47: 6-18.

Daston, L. (ed.) 2000 *Biographies of Scientific Objects*. Chicago: University Chicago Press.

Daston, L. and Galison, P. 2007 *Objectivity*. London: Zone Books.

David, B. and Thomas, J. 2008 Landscape Archaeology: Introduction, in B. David and J. Thomas (eds) *A Handbook of Landscape Archaeology*, 27-43. Walnut Creek: New Left Coast Press.

Foucault, M. 1970 *The Order of Things*. London: Tavistock.

—— 1972 *The Archaeology of Knowledge*. London: Tavistock.

Galison, P. 1987 *How Experiments End*. Chicago: University of Chicago Press.

—— 2002 Images scatter into data. Data gather into images, in B. Latour and P. Weibel (eds) *Iconoclash. Beyond the image wars in science, religion and art*, 300-323. London: MIT Press.

Galison, P.L. 2007 *Einstein's Clocks and Poincaré's Maps: empires of time*. London: Sceptre.

Hacking, I. 1983 *Representing and Intervening: Introductory topics in the philosophy of science*. Cambridge: Cambridge University Press.

Ivens, W.M. 1953 *Prints and Visual Communications*. Cambridge, MA: Harvard University Press.

Latour, B. 1986 Visualization and Cognition: drawing things together. *Knowledge and Society* 6: 1-40.

—— 1993 *We Have Never Been Modern*. Cambridge, MA: Harvard University Press.

—— 1999 *Pandora's Hope: essays on the reality of science studies*. Cambridge, MA.: Harvard University Press.

—— 2004 Whose Cosmos? Which Cosmopolitics? *Common Knowledge* 10(3): 450-62.

—— 2008 A Cautious Prometheus: A few steps toward a philosophy of design (with special attention to Peter Sloterdijk). Keynote lecture for the 3 September 2008 meeting of the Design History Society ('networks of Design'), Cornwall, UK.

Latour, B. and Weibel (eds) 2002 *Iconoclash. Beyond the image wars in science, religion and art*. London: MIT Press.

—— (eds) 2005 *Making Things Public. Atmospheres of democracy*. Cambridge: MIT Press.

Latour, B. and Woolgar, S. 1979 *Laboratory Life: the social construction of scientific facts*. London: Sage.

—— 1982 The Cycle of Credibility, in B. Barnes and D. Edge (eds) *Science in Context. Readings in the sociology of science*, 35-43. Milton Kenes: The Open University Press.

—— 1973 *On the Rationalization of Sight*. New York: Plenem Press.

Thomas, J. 2008 Archaeology, Landscape and Dwelling, in B. David and J. Thomas (eds) *A Handbook of Landscape Archaeology*, 300-306. Walnut Creek: New Left Coast Press.

Toulmin, S. 1990 *Cosmopolis. The hidden agenda of modernity*. Chicago: University of Chicago.

Urry, J. 2000 *Sociology beyond Societies: mobilities for the twenty-first century*. London: Routledge.

Epilogue
Putting Heritage Issues into Motion

Aspects of Ian Russell and Stephanie Koerner's Conversation, September 2009

Ian Russell (IR): We are nearly done with this project. Thank you so much for letting me take part in bringing together all of the wonderful contributions to the conference sessions you have organised over the last ten years. It would be great to have had been able to document some of the conversations we've had over the last few years organising the book

Stephanie Koerner (SK): Yes, but if we started that at this point it might make another volume, and you said we wanted to close this one off with something 'up beat'.

IR: A number of chapters in the final section do this. Many of them build upon presentations from the conference where we first discussed doing this volume. We were at the meeting of the European Association of Archaeologists (EAA) in Krakow, Poland, in September 2006. With John Carman you organised a session entitled 'The Unquiet Past: Cultural Heritage and the Uncertainties of 'Risk Society'' and Tim Darvill was the discussant.

From your abstract, the session proposed to interrogate current models of 'globalisation' and 'multiculturalism' that can sometimes obscure the complexities involved in our engagements with 'cultural heritage'. In a world that can feel chaotic, out of control and full of threats to sustainable lifeways, cultural heritage is often perceived as something we can 'hold onto' as a permanent, fixed point in an environment of ever more rapid and radical change. Thus, the ways in which we make sense of the past play a significant role in how we cope with the world today and construct and envisage possible futures (Carman and Koerner 2006).

Several participants at that EAA session responded to texts that you had distributed. I think these included excerpts from Toulmin's (1990) *Cosmopolis*, papers of Ulrich Beck, Bruno Latour and Viveiros de Casto in the 2004 edition of C*ommon Knowledge*, and excerpts from Renato Rosaldo's *Putting Culture into Motion* (1989).

SK: What a fantastic memory!

IR: I really enjoy how you make efforts to engage archaeological and anthropological writers with current and pressing theoretical issues in the social sciences. It is a good way to inspire new thinking and encourage debates about changes in the roles of archaeology—especially in areas of specialisation emerging at intersections of academic institutions and wider public affairs—political, economic, ecological, ethical, etc.

SK: Yes, for example, starting with Rosaldo's writings (the chapters 'Putting Culture into Motion' and 'Imperial Nostalgia' in particular), Rosaldo started off by stressing that, in everyday life we are continuously aware of the indeterminacy of our activities and relationships. We are conscious and occasionally highly alert to the contingency and fragility of our means to anchor our activities and commitments to one another, as well as to some forms of what you, following Vico (1948 [1944]) might call 'public grounds of truth'. 'The wise,' Rosaldo (1989: 92) said, 'guide themselves as often by waiting to see how events unfold as by plans and prediction…[W]e learn by doing, and make things up as we go along.' He goes on to say that, 'unfortunately, even those who have most influentially challenged universalising generalisations have envisaged culture as determined by some form of necessity—noting that Clifford Geertz (1973: 44) once argued that (1) culture is best seen not as complexes of concrete behaviour patterns…but as a set of control mechanisms—plans, recipes, rules, instructions (what computer programmers call 'programmes') for governing behaviour…), and (2) man is precisely the animal most desperately dependent upon such extra-genetic, outside-the skin control mechanisms, such cultural programmes'. This would mean that without 'cultural programmes' human beings would be 'unworkable monstrosities with very few useful instincts, fewer recognisable sentiments, and no intellect: mental basket cases' (Geertz 1973: 49). For Rosaldo, we need to ask:

> Must one agree that without cultural plans humans become grotesque creatures, disoriented beyond any capacity for desire, or feeling, or thought? Do our options really come down to the vexed choice between supporting cultural order or yielding to the chaos of brute idiocy? (Rosaldo 1989: 98).

IR: Indeed, several papers in the session were concerned with what alternatives to such 'vexed choices' might be and what effectiveness they might have in overcoming the 'crises over representation' that you and Toulmin have tried to illuminate.

SK: Yes…by focusing on settlements motivated by what Rosaldo (1989) calls 'imperial nostalgia' and/or what Latour (2004) might describe as a situation where 'maniacal destruction' is combined with 'equally maniacal conservation' for example.

IR: After returning to the 2006 session's papers and reading through your editorial introductions for this volume, I noted a passage that may compare quite directly with Toulmin's (1990) arguments for a 'return to practical reason' grounded in insights that are similar to those cited by Isabelle Stengers in the preface to the conversation I had with Alain Schnapp, namely:

> If there are more of us who regain the capacity to do our own sorting of the elements that belong to our time, we will rediscover the freedom of movement that modernism denied us a freedom that, in fact, we have never lost (Latour 2004: 462).

SK: I think that you have touched upon one of the volume's key 'upbeat themes'. In retrospect, I realised that Tim Darvill's wonderful final comments touched upon that insight. He went on to compare developments in areas of archaeological research on interstices of academic and economic and government sectors, which have been motivated by awareness of how much controversies over heritage turn on 'life quality issues' with developments in fields undergoing analogous changes (cf. Darvill 2007).

IR: Such developments compare interestingly with Toulmin's arguments concerning 'ways ahead'. For Toulmin (1990), as for Darvill (2007), we are not compelled to choose between vexed options of universalism and particularism. Archaeologists and anthropologists can provide and promote a diversity of knowledges—each of them developing their own relevance to and ways of addressing the contemporary world's needs and contemporary society's aspirations. To conclude with the words of Silvio Funtowicz and Jerry Ravetz' (1997), we can effect an inheritance 'of a world in which simplicity is a memory of a bygone age' (Funtowicz and Ravetz 1997).

References

Beck, U. 2004a The Truth of Others. A cosmopolitan approach. *Common Knowledge* 10(3): 430-61.

—— 2004b Neither Order nor Peace. *Common Knowledge* 11(1): 1-7.

Carman, J. and Koerner, S. 2006 The Unquiet Past: Cultural Heritage and the Uncertainties of 'Risk Society,' session organised for the meeting of the European Association of Archaeologists (EAA), Krakow, Poland, September 2006.

Darvill, T. 2007 Research Frameworks for World Heritage Sites and the Conceptualisation of Archaeological Knowledge. *World Archaeology* 39(3): 436-57.

Funtowicz, S.O. and Ravetz, J.R. 1997 The Poetry of Thermodynamics. Energy, entropy/exergy and quality. *Futures* 29(9): 791-810.

Latour, B. 2004 Whose Cosmos, Which Cosmopolis? *Common Knowledge* 10(3): 450-62.

Rosaldo, R. 1989 *Culture and Truth. The remaking of social analysis*. London: Routledge.

Toulmin, S. 1990 *Cosmopolis. The hidden agenda of modernity*. Chicago: University of Chicago Press.

Vico, G. 1948 [1744] *The New Science of the Common Nature of the Nations of Giambattista Vico,* translated by T.G. Bergin and M.H. Fisch. London: Cornell University Press.

Viveiros de Casto, E. 2004 Exchanging Perspectives: the transformation of objects into subjects in amerinadian Ontologies. *Common Knowledge* 10(3): 463-84.

Index